OT 57
Operator Theory: Advances and Applications
Vol. 57

Editor:
I. Gohberg
Tel Aviv University
Ramat Aviv, Israel

Editorial Office:
School of Mathematical Sciences
Tel Aviv University
Ramat Aviv, Israel

Editorial Board:

A. Atzmon (Tel Aviv)
J. A. Ball (Blacksburg)
L. de Branges (West Lafayette)
K. Clancey (Athens, USA)
L. A. Coburn (Buffalo)
R. G. Douglas (Stony Brook)
H. Dym (Rehovot)
A. Dynin (Columbus)
P. A. Fillmore (Halifax)
C. Foias (Bloomington)
P. A. Fuhrmann (Beer Sheva)
S. Goldberg (College Park)
B. Gramsch (Mainz)
J. A. Helton (La Jolla)

M. A. Kaashoek (Amsterdam)
T. Kailath (Stanford)
H. G. Kaper (Argonne)
S. T. Kuroda (Tokyo)
P. Lancaster (Calgary)
L. E. Lerer (Haifa)
E. Meister (Darmstadt)
B. Mityagin (Columbus)
J. D. Pincus (Stony Brook)
M. Rosenblum (Charlottesville)
J. Rovnyak (Charlottesville)
D. E. Sarason (Berkeley)
H. Widom (Santa Cruz)
D. Xia (Nashville)

Honorary and Advisory
Editorial Board:

P. R. Halmos (Santa Clara)
T. Kato (Berkeley)
P. D. Lax (New York)

M. S. Livsic (Beer Sheva)
R. Phillips (Stanford)
B. Sz.-Nagy (Szeged)

Springer Basel AG

Operator Calculus and Spectral Theory

**Symposium on Operator Calculus
and Spectral Theory
Lambrecht (Germany)
December 1991**

Edited by

**M. Demuth
B. Gramsch
B.-W. Schulze**

Springer Basel AG

Editors' addresses:

Michael Demuth
Max-Planck-Arbeitsgruppe
FB Mathematik
Universität Potsdam
Am Neuen Palais 10
D–O–1571 Potsdam, Germany

Bernhard Gramsch
FB Mathematik
Universität Mainz
D–6500 Mainz, Germany

Bert-Wolfgang Schulze
AG des Max-Planck-Institutes für Mathematik
»Partielle Differentialgleichungen und Komplexe Analysis«
Mohrenstr. 39
D–O–1197 Berlin, Germany

A CIP catalogue record for this book is available from the Library of Congress, Washington D.C., USA

Deutsche Bibliothek Cataloging-in-Publication Data

Operator calculus and spectral theory / Symposium on Operator Calculus and Spectral Theory, Lambrecht (Germany), December 1991. Ed. by M. Demuth ... –
Basel ; Boston ; Berlin : Birkhäuser, 1992
  (Operator theory ; Vol.57)
    ISBN 978-3-0348-9703-7   ISBN 978-3-0348-8623-9 (eBook)
    DOI 10.1007/978-3-0348-8623-9
NE: Demuth, Michael [Hrsg.]; Symposium on Operator Calculus
    and Spectral Theory <1991, Lambrecht>; GT

© 1992 Springer Basel AG
Originally published by Birkhäuser Verlag Basel in 1992
Softcover reprint of the hardcover 1st edition 1992
        Printed on acid-free paper, directly from the authors' camera-ready manuscripts
ISBN 978-3-0348-9703-7

Table of Contents

Preface

This volume contains the proceedings of the international conference on " Operator Calculus and Spectral Theory" in Lambrecht, Germany, December 9. - 14., 1991, sponsored by the Deutsche Forschungsgemeinschaft, and by the Karl-Weierstrass-Institute of Mathematics, Berlin.

The idea was to bring together specialists from different areas of modern analysis, geometry and mathematical physics, in a similar spirit as in the earlier series of conferences of the Karl-Weierstrass- Institute (Ludwigsfelde, 1976; Reinhardsbrunn, 1985; Holzhau, 1988; Breitenbrunn,1990).

Berlin, Mainz M. Demuth B. Gramsch B.-W. Schulze

List of talks given in Lambrecht, 8. - 14. December 91

Name	Title
S. Albeverio	Some recent developments in Dirichlet forms and associated processes
U. Bunke	On the spectral flow of Dirac operators with negative definite functions as symbols - applications in probability theory and mathematical physics
M. Combescure	Recurrent versus diffusive behaviour for time-dependent quantum Hamiltonian
E.B. Davies	Analysis on graphs and noncommutative geometry
M. Demuth	On stochastic spectral analysis and asymptotics for Schrödinger operators
P. Duclos	A global approach to the location of quantum resonances
Y.V. Egorov	On negative spectrum of elliptic operators
V. Enß	Non-threshold states of long-range N-body problems
B. Gramsch	-algebras and microlocal analysis
P.B. Gilkey	On the index of geometrical operators for Riemannian manifolds with boundary
B. Helffer	Spectral problems in statistical mechanics
R. Hempel	Second order pertubations of elliptic operators with periodic principal part
T. Ichinose	On the Weyl quantized relativistic Hamiltonian
V.Y. Ivrii	Scott conjecture for molecules
N. Jacob	Pseudodifferential operators with negative definite functions as symbols - applications in probability theory and mathematical physics
A. Jensen	Mapping properties of the wave operators in scattering theory

Name	Title
W. Kaballo	Decomposition of metromorphic Fredholm resolvents
W. Kirsch	One-dimensional Schrödinger operators with high potential barriers
P. Komech	The boundary value problems for elliptic operators in regions with corners
V.A. Kondratev	On the bounds of the first eigenvalue of elliptic problems
M. Lapidus	The spectrum of fractal drums, the Riemann hypothesis, analysis on fractals and the Weyl-Berry conjecture
M. Lorenz	Elliptic operators in domains with edges
A. Martinez	Resonances for diatomic molecules in the Born-Oppenheimer approximation
O.A. Oleinik	Asymptotic behaviour of solutions of nonlinear elliptic equations in unbounded domains
B.A. Plamenevskij	Index formula for pseudodifferential operators with discontinuous symbols; K-theoretic, homological interpretations
D. Robert	Time delay for long-range pertubations of the Laplace operator
J. Roßmann	Stable asymptotics for solution to the Dirichlet problem for second order differential equations
E. Schrohe	Fredholm criteria and functional calculus for boundary value problems on noncompact manifolds
B.-W. Schulze	The symbolic structures of elliptic operators in manifolds with corners
B. Shatalov, B. Sternin	Maslov's operational method
T. Sturm	Schrödinger semigroups with singular potentials

Name	Title
A. Teta	N-particle systems with zero-range interactions
A. Unterberger	Operator calculus and spectral theory: the relativistic oscillator
J. van Casteren	On differences of generalized Schrödinger semigroups: a trace class and a Hilbert-Schmidt property
M. van den Berg	Capacity, measure and spectrum of Dirichlet Laplacian
D. Vassiliev	Fourier integral operators with complex phase
D.R. Yafaev	Radiation conditions and scattering theory for N-particle quantum systems

Operator Theory:
Advances and Applications, Vol. 57
© 1992 Birkhäuser Verlag Basel

Heat equation asymptotics of a generalized Ahlfors Laplacian on a manifold with boundary

Thomas P. Branson

Peter B. Gilkey

Bent Ørsted

Antoni Pierzchalski

Abstract: Let M be a smooth Riemannian manifold with boundary and let $P = ad_0\delta_0 + b\delta_1 d_1 - \epsilon\rho$ be the generalized Ahlfors Laplacian on $C^\infty T^*M$ where a and b are positive constants and where $\epsilon\rho$ is an arbitrary constant multiple of the Ricci tensor. We use functorial methods to compute the heat equation asymptotics $a_n(P, \mathcal{B})$ for $n \le 3$ with absolute or relative boundary conditions $\mathcal{B}$. MOS classification number 58G25 primary 53A30 secondary. Key words: heat equation asymptotics, nonminimal leading symbol, Ahlfors Laplacian.

1. Introduction

In recent years, there has been a renewed interest in the study of the differential geometry of Riemannian manifolds with boundary and of corresponding elliptic boundary value problems. In this paper, we address a special class of such problems, originally motivated by quasiconformal geometry and pertaining to the study of operators whose leading symbol is not (as for the Laplacian on forms) scalar. The spectral geometry in this situation presents new questions, and we show how one may combine previous techniques of calculation in order to find the first few terms in the asymptotic expansion for the trace of the fundamental solution of the heat equation.

Let M be a compact Riemannian manifold of dimension $m \geq 2$ with smooth boundary ∂M. Let TM and T^*M be the tangent and cotangent bundles of M. Let ρ be the Ricci tensor. Let a and b be positive constants and let ϵ be an arbitrary constant. Let

$$d_k : C^\infty \Lambda^k M \to C^\infty \Lambda^{k+1} M \tag{1.1}$$

be exterior differentiation and let

$$\delta_k : C^\infty \Lambda^{k+1} M \to C^\infty \Lambda^k M \tag{1.2}$$

be the dual, interior multiplication. We define the generalized Ahlfors Laplacian

$$P = P(a, b, \epsilon) = a d_0 \delta_0 + b \delta_1 d_1 - \epsilon \rho \text{ on } C^\infty T^*M. \tag{1.3}$$

We define P_0 by setting $\epsilon = 0$; $P = P_0 - \epsilon \rho$. We impose relative or absolute boundary conditions $\mathcal{B}$ to define the domain of $P_\mathcal{B}$; see §3 for details. The operator $P_\mathcal{B}$ is elliptic and self-adjoint; the fundamental solution of the heat equation $e^{-tP_\mathcal{B}}$ is infinitely smoothing on $L^2 T^*M$ and is of trace class. As $t \to 0^+$, there is an asymptotic expansion of the form:

$$\text{Tr}_{L^2}(e^{-tP_\mathcal{B}}) \sim \sum_{n=0}^{\infty} a_n(P, \mathcal{B}) t^{(n-m)/2}. \tag{1.4}$$

The heat equation asymptotics $a_n(P, \mathcal{B})$ are locally computable and are a fundamental tool in spectral geometry. If $\partial M = \emptyset$ there is no boundary condition and we denote the corresponding invariants by $a_n(P)$; $a_n(P)$ vanishes if n is odd.

If $a = b$, then the leading symbol of P is scalar and the relevant heat equation asymptotics are well understood; we refer to Branson and Gilkey (1990) for details. If $a \neq b$, the situation is much more complicated. The first two invariants $a_0(P)$ and $a_2(P)$ for manifolds without boundary were calculated using functorial methods in Branson, Fulling, and Gilkey (1991); we refer to Gusynin, Gorbar, and Romankov (1991) for another approach. In this paper, we will extend these results to manifolds with boundary to compute $a_n(P, \mathcal{B})$ for $n \leq 3$ with relative $\mathcal{B} = \mathcal{B}^r$ and absolute $\mathcal{B} = \mathcal{B}^a$ boundary conditions; these results are contained in Theorems 4.2 and 4.3.

Theorems 4.2 and 4.3 generalize to operators on $C^\infty \Lambda^k M$ which have the form

$$P = a d_{k-1} \delta_{k-1} + b \delta_k d_k - \epsilon \rho \tag{1.5}$$

where a and b are positive constants and where $\epsilon \rho$ is an arbitrary constant multiple of the Ricci tensor; such operators are said to be "nonminimal". We shall omit details in the interests of brevity as the formulas become somewhat messy for forms of higher degree.

The classical Ahlfors operator S and the Ahlfors Laplacian

$$L = S^* S = (1 - \tfrac{1}{m}) d_0 \delta_0 + \tfrac{1}{2} \delta_1 d_1 - \rho \text{ on } C^\infty T^*M. \tag{1.6}$$

play a fundamental role in the study of quasiconformal geometry; see for example Ahlfors (1974, 1976). The invariants $a_0(L, \mathcal{B})$ and $a_1(L, \mathcal{B})$ were computed by Ørsted and Pierzchalski (1990) using entirely different methods for three boundary conditions called (E), (N), and (D) in that paper. The boundary conditions (N) differ from absolute boundary conditions by a 0th order term involving the action of the second fundamental form L. Since a_0 and a_1 are not affected by the presence of such a term, Theorem 5.16 (N) of [ØP] follows from Theorem 4.2 of this paper; similarly Theorem 5.16 (E) of [ØP] follows from Theorem 4.3 of this paper. Theorem 5.16 (D) of [ØP] has no counterpart in this paper.

In §2, we discuss manifolds without boundary to compute the interior integrands; our analysis follows that of Branson, Fulling, and Gilkey (1991). In §3, we define relative and absolute boundary conditions. We conclude in §4 by extending the results of §2 to manifolds with boundary.

2. Heat equation asymptotics for manifolds without boundary

We suppose the boundary of M empty in §2. We review some results from Branson, Fulling, and Gilkey (1991). We introduce the following notational conventions. Let Roman indices i, j, ... range from 1 through m and index a local orthonormal frame e_i for the tangent bundle TM. We use the metric to identify TM with T^*M. Let R_{ijkl} be the components of the Riemann curvature tensor of the Levi-Civita connection ∇. Let

$$\rho_{ij} = R_{ikkj} \text{ and } \tau = \rho_{ii} \tag{2.1}$$

be the Ricci tensor and scalar curvature; we adopt the Einstein convention and sum over repeated indices. If σ is a scalar invariant of the metric, we integrate with respect to the Riemannian measure $d\text{vol}$ to define

$$\sigma[M] = \int_M \sigma d\text{vol}. \tag{2.2}$$

Thus, for example, $\text{vol}(M) = 1[M]$.

We first suppose $a = b = 1$ and $\epsilon = 0$. Let

$$\Delta_k = d_{k-1}\delta_{k-1} + \delta_k d_k \tag{2.3}$$

be the Laplacian on $C^\infty \Lambda^k M$. Patodi (1970, see Prop 2.1) calculated the invariants $a_n(\Delta_k)$ for $n = 0$, 2, 4 and arbitrary k. We specialize his results to the cases $k = 0$ and $k = 1$ to see:

Lemma 2.1 (Patodi): *If the boundary of M is empty, then:*

(a) $a_0(\Delta_0) = (4\pi)^{-m/2} 1[M]$.

(b) $a_0(\Delta_1) = (4\pi)^{-m/2} m[M]$.

(c) $a_2(\Delta_0) = \frac{1}{6}(4\pi)^{-m/2} \tau[M]$.

(d) $a_2(\Delta_1) = \frac{1}{6}(4\pi)^{-m/2}(m - 6)\tau[M]$.

The following technical result is at the heart of our analysis.

Lemma 2.2 (Branson-Fulling-Gilkey): *Let $P_0 = ad_0\delta_0 + b\delta_1 d_1$. If the boundary of M is empty, then:*

$$a_n(P_0) = a^{(n-m)/2}a_n(\Delta_0) + b^{(n-m)/2}a_n(\Delta_1) - b^{(n-m)/2}a_n(\Delta_0).$$

Proof: We take a spectral resolution of Δ_k. Let

$$E(\Delta_k, \lambda) = \{\omega \in C^\infty \Lambda^k M : \Delta_k \omega = \lambda\omega\} \tag{2.4}$$

be the eigenspaces of the Laplacian. The $E(\Delta_k, \lambda)$ are finite dimensional and there is an orthogonal direct sum decomposition:

$$L^2 \Lambda^k M = \text{kernel}(\Delta_k) \oplus \{\oplus_{\lambda>0} E(\Delta_k, \lambda)\}. \tag{2.5}$$

If $\omega \in E(\Delta_k, \lambda)$ and if $\lambda > 0$, let

$$\begin{aligned}
\pi(d_{k-1}\delta_{k-1}, \lambda)\omega &= \lambda^{-1} d_{k-1}\delta_{k-1}\omega, \\
\pi(\delta_k d_k, \lambda)\omega &= \lambda^{-1} \delta_k d_k \omega.
\end{aligned} \tag{2.6}$$

These are complementary projections whose range gives a splitting

$$E(\Delta_k, \lambda) = E(d_{k-1}\delta_{k-1}, \lambda) \oplus E(\delta_k d_k, \lambda). \tag{2.7}$$

This defines the Hodge decomposition:

$$\begin{aligned}
L^2 \Lambda^k M &= \text{kernel}(\Delta_k) \oplus \{\oplus_{\lambda>0} E(d_{k-1}\delta_{k-1}, \lambda)\} \oplus \{\oplus_{\lambda>0} E(\delta_k d_k, \lambda)\} \\
&= \text{kernel}(\Delta_k) \oplus \text{image}(d_{k-1}) \oplus \text{image}(\delta_k).
\end{aligned} \tag{2.8}$$

Denote the kth Betti number by β_k and define:

$$\begin{aligned}
f(d_{k-1}\delta_{k-1}, t) &= \text{Tr}_{L^2}(e^{-t\Delta_k}|_{\text{image}(d_{k-1})}) \\
&= \Sigma_{\lambda>0} e^{-t\lambda} \dim E(d_{k-1}\delta_{k-1}, \lambda) \\
f(\delta_k d_k, t) &= \text{Tr}_{L^2}(e^{-t\Delta_k}|_{\text{image}(\delta_k)}) \\
&= \Sigma_{\lambda>0} e^{-t\lambda} \dim E(\delta_k d_k, \lambda) \\
f(\Delta_k, t) &= \text{Tr}_{L^2}(e^{-t\Delta_k}) \\
&= \beta_k + f(d_{k-1}\delta_{k-1}, t) + f(\delta_k d_k, t).
\end{aligned} \tag{2.9}$$

As $\lambda \neq 0$, d_{k-1} is an isomorphism from $E(\delta_{k-1}d_{k-1}, \lambda)$ to $E(d_{k-1}\delta_{k-1}, \lambda)$. Consequently $f(\delta_{k-1}d_{k-1}, t) = f(d_{k-1}\delta_{k-1}, t)$. Therefore:

$$\begin{aligned}
f(d_0\delta_0, t) &= f(\delta_0 d_0, t) = f(\Delta_0, t) - \beta_0 \\
f(\delta_1 d_1, t) &= f(\Delta_1, t) - \beta_1 - f(d_0\delta_0, t) \\
&= f(\Delta_1, t) - \beta_1 - f(\Delta_0, t) + \beta_0.
\end{aligned} \tag{2.10}$$

Let $P_0 = ad_0\delta_0 + b\delta_1 d_1$. We compute and equate coefficients of t in the asymptotic expansions to complete the proof:

$$\begin{aligned}
f(P_0, t) &= \text{Tr}_{L^2}(e^{-tP_0}) = \beta_1 + f(d_0\delta_0, at) + f(\delta_1 d_1, bt) \\
&= \beta_1 + f(\Delta_0, at) - \beta_0 + f(\Delta_1, bt) - \beta_1 - f(\Delta_0, bt) + \beta_0. \quad \blacksquare
\end{aligned} \tag{2.11}$$

We use Lemmas 2.1 and 2.2 to compute the heat equation asymptotics for manifolds without boundary:

Theorem 2.3 (Branson-Fulling-Gilkey): *Let $P = ad_0\delta_0 + b\delta_1 d_1 - \epsilon\rho$. If the boundary of M is empty, then:*

(a) $a_0(P) = (4\pi)^{-m/2}(a^{-m/2} + (m-1)b^{-m/2})[M]$.

(b) $a_2(P) = \frac{1}{6}(4\pi)^{-m/2}\big\{a^{(2-m)/2} + (m-7)b^{(2-m)/2}$
$\qquad\qquad +6\epsilon m^{-1}\{a^{-m/2} + (m-1)b^{-m/2}\}\big\}\tau[M]$.

Proof: The basic existence theorem for the asymptotic expansion of the heat equation trace gives functorial properties of the heat invariants; see Gilkey (1984, Lemma 1.7.5) for details. Since P is natural, there exist local invariants $\tilde{a}_n(P)(x)$ which are homogeneous of degree n in the covariant derivatives of the curvature tensor so that

$$a_n(P) = \tilde{a}_n(P)[M]. \tag{2.12}$$

We use invariance theory to see that there exist universal constants

$$\alpha_i = \alpha_i(a, b, m) \text{ and } \beta_i = \beta_i(a, b, m) \tag{2.13}$$

so that:

$$a_0(P) = \alpha_0[M] \text{ and } a_2(P) = (\alpha_2 + \epsilon\beta_2)\tau[M]. \tag{2.14}$$

To evaluate these constants, we may take $M = S^m$ with a standard constant curvature metric; we choose the normalization so $\rho = g$ and $\tau = m$. Let $P_0 = ad_0\delta_0 + b\delta_1 d_1$ so $P = P_0 - \epsilon I$. Then:

$$\text{Tr}_{L^2}(e^{-tP}) = \text{Tr}_{L^2}(e^{-t(P_0 - \epsilon I)}) = e^{t\epsilon}\,\text{Tr}_{L^2}(e^{-tP_0}). \tag{2.15}$$

We equate terms in the asymptotic expansion to see

$$\begin{aligned}
a_0(P) &= a_0(P_0), \text{ and} \\
a_2(P) &= a_2(P_0) + \epsilon a_0(P_0).
\end{aligned} \tag{2.16}$$

We use Lemma 2.2 to express $a_n(P_0)$ in terms of the quantities $a_n(\Delta_0)$ and $a_n(\Delta_1)$ given in Lemma 2.1. We complete the proof by computing:

$$\begin{aligned}
\epsilon a_0(P_0) &= \epsilon(4\pi)^{-m/2}\{a^{-m/2} + (m-1)b^{-m/2}\}1[S^m] \\
&= \epsilon\frac{1}{m}(4\pi)^{-m/2}\{a^{-m/2+}(m-1)b^{-m/2}\}\tau[S^m]. \quad \blacksquare
\end{aligned} \tag{2.17}$$

3. Boundary conditions

If the boundary of M is not empty, we must impose suitable boundary conditions. Let

$$i^* : \Lambda M \to \Lambda \partial M. \tag{3.1}$$

be the pullback via the inclusion $i : \partial M \to M$. If $\omega \in C^\infty \Lambda M$, decompose

$$\omega|_{\partial M} = \omega^T + \omega^N \text{ where } \omega^T = i^*\omega \in C^\infty \Lambda \partial M. \tag{3.2}$$

Let π^T and π^N be the corresponding projection operators. Let

$$\mathcal{B}^a\omega = \pi^N\omega \oplus \pi^N d\omega. \tag{3.3}$$

Let $\star$ be the Hodge operator and let $\mathcal{B}^r\omega = \mathcal{B}^a \star \omega$. We say ω satisfies **absolute boundary conditions** if $\mathcal{B}^a\omega = 0$; ω satisfies **relative boundary conditions** if $\mathcal{B}^r\omega = 0$.

If $P = ad_{k-1}\delta_{k-1} + b\delta_k d_k - \epsilon\rho$ on $C^\infty\Lambda^k M$ and $\mathcal{B} \in \{\mathcal{B}^a, \mathcal{B}^r\}$, let

$$\text{domain}(P_{\mathcal{B}}) = \{\omega \in C^\infty\Lambda^k M : \mathcal{B}\omega = 0\}; \tag{3.4}$$

$P_{\mathcal{B}}$ is a self-adjoint elliptic operator. We use the Hodge decomposition theorem to identify

$$\begin{aligned}
\text{kernel}(\Delta_{k,\mathcal{B}^a}) &= H^k(M), \\
\text{kernel}(\Delta_{k,\mathcal{B}^r}) &= H^k(M, \partial M)
\end{aligned} \tag{3.5}$$

so these boundary conditions are natural ones to consider.

Lemma 3.1: *Let $\mathcal{B} \in \{\mathcal{B}^a, \mathcal{B}^r\}$, let $\omega \in C^\infty\Lambda M$ satisfy $\Delta\omega = \lambda\omega$, and let $\mathcal{B}\omega = 0$. Then $\mathcal{B}d\omega = 0$ and $\mathcal{B}\delta\omega = 0$.*

Proof: We recall that

$$\star\star = \pm 1, \quad \star\pi^N = \pm\pi^T\star, \quad \star\pi^T = \pm\pi^N\star, \text{ and } \delta = \pm \star d \star. \tag{3.6}$$

Suppose first $\mathcal{B} = \mathcal{B}^a$. If $\mathcal{B}^a\omega = 0$, then $\pi^N d\omega = 0$. Since $d^2 = 0$, $\pi^N dd\omega = 0$. This shows $\mathcal{B}^a d\omega = 0$. We use (3.6) to see:

$$\pi^N\delta\omega = \pm\pi^N \star d \star \omega = \pm \star \pi^T d \star \omega. \tag{3.7}$$

Let $d^T = \pi^T d\pi^T$ be the tangential exterior derivative; $\pi^T d = d^T\pi^T$. Consequently,

$$\pi^N\delta\omega = \pm \star d^T\pi^T \star \omega = \pm \star d^T \star \pi^N\omega. \tag{3.8}$$

Since $\pi^N\omega$ vanishes on ∂M, $\star\pi^N\omega$ vanishes on ∂M and hence $d^T \star \pi^N\omega$ vanishes on ∂M. This shows $\pi^N\delta\omega = 0$. Since $\pi^N d\omega$ also vanishes on ∂M, the same argument shows $\pi^N\delta d\omega = 0$. Since $\Delta\omega = \lambda\omega$, $d\delta\omega = \lambda\omega - \delta d\omega$ so $\pi^N d\delta\omega = 0$. This completes the proof if $\mathcal{B} = \mathcal{B}^a$; the case $\mathcal{B} = \mathcal{B}^r$ follows by duality. ∎

We generalize Lemma 2.2 to this setting:

Lemma 3.2: *Let* $\mathcal{B} \in \{\mathcal{B}^a, \mathcal{B}^r\}$ *and let* $P = ad_0\delta_0 + b\delta_1 d_1$. *Then*

$$a_n(P, \mathcal{B}) = a^{(n-m)/2} a_n(\Delta_0, \mathcal{B}) + b^{(n-m)/2} a_n(\Delta_1, \mathcal{B})$$
$$- b^{(n-m)/2} a_n(\Delta_0, \mathcal{B}).$$

Proof: As in the proof of Lemma 2.2, we take a spectral resolution of Δ_k. Let

$$E(\Delta_k, \lambda) = \{\omega \in C^\infty \Lambda^k M : \Delta_k \omega = \lambda \omega \text{ and } \mathcal{B}\omega = 0\} \qquad (3.9)$$

be the eigenspaces of $\Delta_{k,\mathcal{B}}$. The $E(\Delta_k, \lambda)$ are finite dimensional and there is an orthogonal direct sum decomposition:

$$L^2 \Lambda^k M = \text{kernel}(\Delta_{k,\mathcal{B}}) \oplus \{\oplus_{\lambda > 0} E(\Delta_{k,\mathcal{B}}, \lambda)\}. \qquad (3.10)$$

As before, if $\omega \in E(\Delta_{k,\mathcal{B}}, \lambda)$ and if $\lambda > 0$, let

$$\begin{aligned}
\pi(d_{k-1}\delta_{k-1}, \lambda)\omega &= \lambda^{-1} d_{k-1}\delta_{k-1}\omega, \\
\pi(\delta_k d_k, \lambda)\omega &= \lambda^{-1} \delta_k d_k\omega.
\end{aligned} \qquad (3.11)$$

We use Lemma 3.1 to see the boundary conditions are preserved so these are complementary projections on $E(\Delta_{k,\mathcal{B}}, \lambda)$. The remainder of the proof is the same as that given for Lemma 2.2. ∎

4. Heat equation asymptotics for manifolds with boundary

Let $P = ad_0\delta_0 + b\delta_1 d_1 - \epsilon\rho$ and let $\mathcal{B} \in \{\mathcal{B}^a, \mathcal{B}^r\}$. We will use the argument of §2 and the results of §3 to compute the heat equation asymptotics $a_n(P, \mathcal{B})$ in this section for $n \leq 3$.

Let Roman letters a, b, ... range from 1 through $m - 1$ and index a local orthonormal frame for the tangent space of the boundary. Let L_{ab} be the second fundamental form. Define constants:

$$\begin{aligned}
c(m, k) &= \binom{m}{k} = \frac{m!}{k!(m-k)!}, \\
c_0(m, k) &= c(m, k) - 6c(m - 2, k - 1), \\
d_0(m, k) &= c(m - 1, k) - c(m - 1, k - 1), \\
d_1(m, k) &= 16d_0(m, k) - 96d_0(m - 2, k - 1), \\
d_2(m, k) &= 8d_0(m, k) - 192d_0(m - 2, k - 1), \\
d_3(m, k) &= 3c(m, k) + 10d_0(m, k) - 96d_0(m - 2, k - 1), \\
d_4(m, k) &= 6c(m, k) - 4d_0(m, k) + 96d_0(m - 2, k - 1).
\end{aligned} \qquad (4.1)$$

Set $c(m, k) = c_0(m, k) = d_\nu(m, k) = 0$ for $k < 0$ or $k > m$. We refer to Blažić, Bokan, and Gilkey (1992) for the proof of the following result which generalizes Theorem 2.1 to manifolds with boundary:

Lemma 4.1 (Blažić-Bokan-Gilkey):

(a) $a_0(\Delta_k, \mathcal{B}^a) = (4\pi)^{-m/2} c(m,k)[M]$.

(b) $a_1(\Delta_k, \mathcal{B}^a) = \frac{1}{4}(4\pi)^{-(m-1)/2} d_0(m,k)[\partial M]$.

(c) $a_2(\Delta_k, \mathcal{B}^a) = \frac{1}{6}(4\pi)^{-m/2} c_0(m,k)\{\tau[M] + 2L_{aa}[\partial M]\}$.

(d) $a_3(\Delta_k, \mathcal{B}^a) = \frac{1}{384}(4\pi)^{-(m-1)/2}\{d_1(m,k)\tau + d_2(m,k)R_{amam}$
$\qquad\qquad + d_3(m,k)L_{aa}L_{bb} + d_4(m,k)L_{ab}L_{ab}\}[\partial M]$.

Remark: We use the Hodge $\star$ operator to see $a_n(\Delta_k, \mathcal{B}^r) = a_n(\Delta_{m-k}, \mathcal{B}^a)$.

We use Lemma 4.1 to prove the main results of this paper:

Theorem 4.2: *Let* $P = a d_0 \delta_0 + b \delta_1 d_1 - \epsilon\rho$ *and let* $\mathcal{B} = \mathcal{B}^a$.

(a) $a_0(P, \mathcal{B}^a) = (4\pi)^{-m/2}(a^{-m/2} + (m-1)b^{-m/2})[M]$.

(b) $a_1(P, \mathcal{B}^a) = \frac{1}{4}(4\pi)^{-(m-1)/2}(a^{(1-m)/2} + (m-3)b^{(1-m)/2})[\partial M]$.

(c) $a_2(P, \mathcal{B}^a) = \frac{1}{6}(4\pi)^{-m/2}\big\{\{a^{(2-m)/2} + (m-7)b^{(2-m)/2}$
$\qquad\qquad + 6\epsilon m^{-1}(a^{-m/2} + (m-1)b^{-m/2})\}\tau[M]$
$\qquad\qquad + 2\{a^{(2-m)/2} + (m-7)b^{(2-m)/2}\}L_{aa}[\partial M]\big\}$.

(d) $a_3(P, \mathcal{B}^a) = \frac{1}{384}(4\pi)^{-(m-1)/2}\big\{a^{(3-m)/2}(16\tau + 8R_{amam} + 13L_{aa}L_{bb} + 2L_{ab}L_{ab})$
$\qquad\qquad + b^{(3-m)/2}\{(16m - 144)\tau + (8m - 216)R_{amam}$
$\qquad\qquad + (13m - 129)L_{aa}L_{bb} + (2m + 102)L_{ab}L_{ab}\}$
$\qquad\qquad + \epsilon(\beta_4\tau + \beta_5 R_{amam})\big\}[\partial M]$.

(e) $\beta_4 = 96(m-1)^{-1}(a^{(1-m)/2} + (m-2)b^{(1-m)/2})$.

(f) $\beta_5 = 96(m-1)^{-1}\{a^{(1-m)/2} + (2m-3)b^{(1-m)/2}\}$.

Theorem 4.3: *Let* $P = a d_0 \delta_0 + b \delta_1 d_1 - \epsilon\rho$ *and let* $\mathcal{B} = \mathcal{B}^r$.

(a) $a_0(P, \mathcal{B}^r) = (4\pi)^{-m/2}(a^{-m/2} + (m-1)b^{-m/2})[M]$.

(b) $a_1(P, \mathcal{B}^r) = -\frac{1}{4}(4\pi)^{-(m-1)/2}(a^{(1-m)/2} + (m-3)b^{(1-m)/2})[\partial M]$.

(c) $a_2(P, \mathcal{B}^r) = \frac{1}{6}(4\pi)^{-m/2}\big\{\{a^{(2-m)/2} + (m-7)b^{(2-m)/2}$
$\qquad\qquad + 6\epsilon m^{-1}(a^{-m/2} + (m-1)b^{-m/2})\}\tau[M]$
$\qquad\qquad + 2\{a^{(2-m)/2} + (m-7)b^{(2-m)/2}\}L_{aa}[\partial M]\big\}$.

(d) $a_3(P, \mathcal{B}^r) = \frac{1}{384}(4\pi)^{-(m-1)/2}\big\{a^{(3-m)/2}(-16\tau - 8R_{amam} - 7L_{aa}L_{bb} + 10L_{ab}L_{ab})$
$\qquad\qquad + b^{(3-m)/2}\{(144 - 16m)\tau + (216 - 8m)R_{amam}$
$\qquad\qquad + (123 - 7m)L_{aa}L_{bb} + (10m - 114)L_{ab}L_{ab}\}$
$\qquad\qquad - \epsilon(\beta_4\tau + \beta_5 R_{amam})\big\}[\partial M]$ where the β_i are as above.

Proof: Let $\mathcal{B} = \mathcal{B}^a$; the calculation for relative boundary conditions is similar and is therefore omitted. The invariance theory of Gilkey (1984) shows there exist universal constants $\alpha_\nu = \alpha_\nu(a, b)$ and $\beta_\nu = \beta_\nu(a, b)$ so:

$$
\begin{aligned}
a_0(P, \mathcal{B}^a) &= (4\pi)^{-m/2} \alpha_0 [M] \\
a_1(P, \mathcal{B}^a) &= \tfrac{1}{4} (4\pi)^{(1-m)/2} \alpha_1 [\partial M] \\
a_2(P, \mathcal{B}^a) &= \tfrac{1}{6} (4\pi)^{-m/2} (\alpha_2 + \epsilon \beta_2) \tau [M] + \alpha_3 L_{aa} [\partial M] \\
a_3(P, \mathcal{B}^a) &= \tfrac{1}{384} (4\pi)^{(1-m)/2} ((\alpha_4 + \epsilon \beta_4) \tau + (\alpha_5 + \epsilon \beta_5) R_{amam} \\
&\quad + \alpha_6 L_{aa} L_{bb} + \alpha_7 L_{ab} L_{ab}) [\partial M].
\end{aligned}
\tag{4.2}
$$

In particular, only the coefficients of τ and R_{amam} depend on ϵ. We use Theorem 2.3 to compute α_0, α_2, and β_2. We combine Lemmas 3.2 and 4.1 to compute α_1, α_3, α_4, α_5, α_6, and α_7. This leaves only the coefficients β_4 and β_5 to be determined. We adopt absolute boundary conditions henceforth and suppress them from the notation.

We use product formulas to determine β_4. Give $S = S^{m-1}$ the constant curvature metric with $\rho^S = g^S$. Give $M = S \times [0, \pi]$ the product metric; ρ is projection on T^*S. We adopt the notation of Lemma 3.2 and let $\pi(\Delta_k, 0)$ be projection on the kernel of Δ_k. Define:

$$
\begin{aligned}
f(\rho, d_{k-1} \delta_{k-1}, t) &= \Sigma_{\lambda > 0} e^{-t\lambda} \ \mathrm{Tr}(\pi(d_{k-1} \delta_{k-1}, \lambda) \rho) \\
f(\rho, \delta_k d_k, t) &= \Sigma_{\lambda > 0} e^{-t\lambda} \ \mathrm{Tr}(\pi(\delta_k d_k, \lambda) \rho) \\
f(\rho, \Delta_k, t) &= \mathrm{Tr}(\pi(\Delta_k, 0) \rho) + f(\rho, d_{k-1} \delta_{k-1}, t) \\
&\quad + f(\rho, \delta_k d_k, t) \\
f(\rho, P_0, t) &= \mathrm{Tr}_{L^2}(\rho e^{-t P_0}).
\end{aligned}
\tag{4.3}
$$

Standard analytic techniques show there are asymptotic series as $t \to 0^+$:

$$
\begin{aligned}
f(\rho, \Delta_1, t) &\sim \Sigma_n a_n(\rho, \Delta_1) t^{(n-m)/2} \\
f(\rho, P_0, t) &\sim \Sigma_n a_n(\rho, P_0) t^{(n-m)/2}.
\end{aligned}
\tag{4.4}
$$

The constant β_4 is determined by the following Lemma:

Lemma 4.4: *On $M = S^{m-1} \times [0, \pi]$ with the normalizations as above,*

(a) *There exists asymptotic series as $t \to 0^+$ of the form*

$$
f(\rho, d_0 \delta_0, t) \sim \Sigma_n a_n(\rho, d_0 \delta_0) t^{(n-m)/2}.
$$

(b) $a_1(\rho, d_0 \delta_0) = \tfrac{1}{4} (4\pi)^{(1-m)/2} [\partial M]$.

(c) $a_1(\rho, P_0) = \tfrac{1}{384} (4\pi)^{-(m-1)/2} \beta_4 (m - 1) [\partial M]$.

(d) $a_1(\rho, P_0) = b^{(1-m)/2} a_1(\rho, \Delta_1) + (a^{(1-m)/2} - b^{(1-m)/2}) a_1(\rho, d_0 \delta_0)$.

(e) $a_1(\rho, \Delta_1) = \tfrac{1}{4} (4\pi)^{(1-m)/2} (m - 1) [\partial M]$.

Proof: We take a spectral resolution of the Laplacian Δ_0^S on $L^2(S)$:

$$\{\phi_\nu, \lambda_\nu\}_{\nu=0}^\infty \ \text{ for } 0 = \lambda_0 < \lambda_1 \le \cdots. \tag{4.5}$$

Similarly, we take a spectral resolution of $\Delta_0^{[0,\pi]}$ on $L^2[0,\pi]$ with absolute (i.e. Neumann) boundary conditions:

$$\{\psi_\mu, \sigma_\mu\} \ \text{ for } 0 = \sigma_0 < \sigma_1 < \cdots. \tag{4.6}$$

We combine these two resolutions to construct a spectral resolution of Δ_0^M with absolute boundary conditions:

$$\{\phi_\nu \psi_\mu, \lambda_\nu + \sigma_\mu\}_{\nu,\mu}. \tag{4.7}$$

We differentiate and normalize the resulting eigenforms to see

$$\{(\psi_\mu d\phi_\nu + \phi_\nu d\psi_\mu)/\sqrt{\lambda_\nu + \sigma_\mu}, \lambda_\nu + \sigma_\mu\}_{\nu+\mu>0} \tag{4.8}$$

is a spectral resolutions of $d_0 \delta_0$ on range(d_0) with absolute boundary conditions. Since

$$\rho(\psi_\mu d\phi_\nu + \phi_\nu d\psi_\mu) = \psi_\mu d\phi_\nu, \tag{4.9}$$

and since $\psi_\mu d\phi_\nu$ is pointwise orthogonal to $\phi_\nu d\psi_\mu$, we conclude that:

$$\begin{aligned}
f(\rho, d_0\delta_0, t) &= \Sigma_{\nu+\mu>0} e^{-t\lambda_\nu} e^{-t\sigma_\mu} (\psi_\mu d\phi_\nu, \psi_\mu d\phi_\nu)_{L^2}/(\lambda_\nu + \sigma_\mu) \\
&= \Sigma_{\nu+\mu>0} e^{-t\lambda_\nu} e^{-t\sigma_\mu} \lambda_\nu/(\lambda_\nu + \sigma_\mu).
\end{aligned} \tag{4.10}$$

We differentiate with respect to t

$$\partial_t f(\rho, d_0\delta_0, t) = -\Sigma_\nu \lambda_\nu e^{-t\lambda_\nu} \Sigma_\mu e^{-t\sigma_\mu} = \partial_t\{\Sigma_\nu e^{-t\lambda_\nu}\}\Sigma_\mu e^{-t\sigma_\mu}. \tag{4.11}$$

We differentiate the asymptotic series for $f(\Delta_0^S, t)$ to see:

$$\partial_t\{\Sigma_\nu e^{-t\lambda_\nu}\} \sim \tfrac{1}{2}\Sigma_n(n-m)a_n(\Delta_0^S)t^{(n-m-2)/2}; \tag{4.12}$$

this can be justified using results of Gilkey and Smith (1983). Since $f(\Delta_0^{[0,\pi]}, t)$ has an asymptotic expansion, $\partial_t f(\rho, d_0\delta_0, t)$ has an asymptotic expansion. We integrate and add a suitable constant term to prove (a); there is no t^{-1} term and hence no $\log(t)$ term. We further compute:

$$\begin{aligned}
\tfrac{1}{2}\Sigma_n(n-m)a_n(\rho, d_0\delta_0)t^{(n-m-2)/2} &\sim \partial_t f(\rho, d_0\delta_0, t) \\
&= \partial_t\{f(\Delta_0^S, t)\} \cdot f(\Delta_0^{[0,\pi]}, t) \\
&\sim \tfrac{1}{2}\Sigma_{i,j}(i+1-m)a_i(\Delta_0^S)a_j(\Delta_0^{[0,\pi]})t^{(i+j-m-2)/2}.
\end{aligned} \tag{4.13}$$

Since S is without boundary, $a_i(\Delta_0^S)$ vanishes for i odd. We compare the coefficients of $t^{-(m+1)/2}$; only the terms with $n=1$, $i=0$, and $j=1$ contribute. We complete the proof of (b) by checking:

$$a_1(\rho, d_0, \delta_0) = a_0(\Delta_0^S)a_1(\Delta_0^{[0,\pi]}) = \tfrac{1}{4}(4\pi)^{(1-m)/2}[\partial M]. \tag{4.14}$$

To prove (c), we take the variation of the asymptotic series to see:

$$\begin{aligned}
\Sigma_n \tfrac{d}{d\epsilon}\big|_{\epsilon=0} a_n(P)t^{(n-m)/2} &\sim \tfrac{d}{d\epsilon}\big|_{\epsilon=0} \mathrm{Tr}_{L^2}(e^{-tP_0+t\epsilon\rho}) \\
&= t\,\mathrm{Tr}_{L^2}(\rho e^{-tP_0}) \sim t\Sigma_n a_n(\rho, P_0)t^{(n-m)/2}.
\end{aligned} \tag{4.15}$$

We equate terms in the asymptotic expansions to see

$$\tfrac{d}{d\epsilon}\big|_{\epsilon=0} a_3(P) = a_1(\rho, P_0). \tag{4.16}$$

Since $\tau = m - 1$ and $R_{amam} = 0$, (c) follows from (4.2).

We see from the definition that:

$$\begin{aligned}
f(\rho, P_0, t) &= \mathrm{Tr}(\rho\pi(\Delta_k, 0)) + f_1(\rho, d_0\delta_0, at) + f_1(\rho, \delta_1 d_1, bt) \\
&= f(\rho, \Delta_1, bt) + f(\rho, d_0\delta_0, at) - f_1(\rho, d_0\delta_0, bt).
\end{aligned} \tag{4.17}$$

We equate terms in the asymptotic expansions to prove (d).

We take $a = b = 1$ and use (4.16) to see

$$\tfrac{d}{d\epsilon}\big|_{\epsilon=0} a_3(\Delta_1 - \epsilon\rho) = a_1(\rho, \Delta_1). \tag{4.18}$$

We use Branson and Gilkey (1990, Theorem 7.2) to compute:

$$\tfrac{d}{d\epsilon}\big|_{\epsilon=0} a_3(\Delta_1 - \epsilon\rho) = \tfrac{1}{4}(4\pi)^{(1-m)/2}\,\mathrm{Tr}(\psi\rho)[\partial M] \tag{4.19}$$

where $\psi = +1$ on $T^*(S)$ and $\psi = -1$ on $T^*([0,\pi])$; (e) follows since

$$\mathrm{Tr}(\psi\rho) = m - 1. \quad\blacksquare \tag{4.20}$$

We take M to be the upper hemisphere in the sphere S^m to determine β_5. We normalize the metric so $\rho = g$. We argue as in the proof of Theorem 2.3 to see

$$a_3(P, \mathcal{B}^a) = a_3(P_0, \mathcal{B}^a) + \epsilon a_1(P_0, \mathcal{B}^a). \tag{4.21}$$

This shows $m\beta_4 - \beta_5 = 96(a^{(1-m)/2} + (m-3)b^{(1-m)/2})$. Consequently:

$$\begin{aligned}
\beta_5 &= m\beta_4 - 96(a^{(1-m)/2} + (m-3)b^{(1-m)/2}) \\
&= 96(m-1)^{-1}\{(m-(m-1))a^{(1-m)/2} \\
&\quad + (m(m-2) - (m-1)(m-3))b^{(1-m)/2}\} \\
&= 96(m-1)^{-1}\{a^{(1-m)/2} + (2m-3)b^{(1-m)/2}\}. \quad\blacksquare
\end{aligned} \tag{4.22}$$

Remark: The most general natural second order elliptic operator on $C^\infty T^*M$ with positive definite leading symbol takes the form:

$$\mathcal{P} = ad_0\delta_0 + b\delta_1 d_1 - \epsilon\rho - \epsilon_1\tau \tag{4.23}$$

where a and b are positive constants, where $\epsilon\rho$ is an arbitrary multiple of the Ricci tensor, and where $\epsilon_1\tau$ is an arbitrary multiple of the scalar curvature. Let

$$P = ad_0\delta_0 + b\delta_1 d_1 - \epsilon\rho \text{ so } \mathcal{P} = P - \epsilon_1\tau. \tag{4.24}$$

The argument used to prove Theorem 2.3 may be used to extend Theorems 4.2 and 4.3 to this setting and to show that if $\mathcal{B} \in \{\mathcal{B}^a, \mathcal{B}^r\}$ then:

$$
\begin{aligned}
a_n(\mathcal{P},\mathcal{B}) &= a_n(P,\mathcal{B}) \text{ for } n = 0,\ 1, \\
a_2(\mathcal{P},\mathcal{B}) &= a_2(P,\mathcal{B}) + \epsilon_1(4\pi)^{-m/2}(a^{-m/2} + (m-1)b^{-m/2})\tau[M], \\
a_3(\mathcal{P},\mathcal{B}^a) &= a_3(P,\mathcal{B}) + \epsilon_1\tfrac{1}{4}(4\pi)^{(1-m)/2}(a^{(1-m)/2} + (m-3)b^{(1-m)/2})\tau[\partial M], \\
a_3(\mathcal{P},\mathcal{B}^r) &= a_3(P,\mathcal{B}) - \epsilon_1\tfrac{1}{4}(4\pi)^{(1-m)/2}(a^{(1-m)/2} + (m-3)b^{(1-m)/2})\tau[\partial M].
\end{aligned}
\tag{4.25}
$$

Acknowledgements

Research of P. Gilkey partially supported by the N. S. F.

References

L. Ahlfors (1974), *Conditions for quasiconformal deformations in several variables* in: **Contributions to analysis, a collection of papers dedicated to L. Bers**, Academic Press, New York; (1976), *Quasiconformal deformations and mappings in* $\mathbf{R}^n$, J. Analyse Math. **30**, 74-97.

N. Blažić, N. Bokan, and P. Gilkey (1992), *Spectral geometry of the form valued Laplacian for manifolds with boundary*, (to appear Indian Journal of Mathematics).

T. Branson and P. Gilkey (1990), *The asymptotics of the Laplacian on a manifold with boundary*, Comm. in PDE **15**, 245-272.

T. Branson, S. Fulling, and P. Gilkey (1991), *Heat equation asymptotics of nonminimal operators on differential forms*, J. Math. Phys. **32**, 2089-2091.

P. Gilkey (1984), **Invariance theory, the heat equation, and the Atiyah Singer index theorem**, Publish or Perish Press.

P. Gilkey and L. Smith (1983), *The eta invariant for a Class of Elliptic Boundary Value Problems*, Comm. on Pure and Appl. Math. XXXVI, 85-131.

V. P. Gusynin, E. V. Gorbar, and V. V. Romankov (1991), *Heat kernel expansion for nonminimal differential operators and manifolds with torsion*, Nuclear Physics B362, 449-471.

B. Ørsted and A. Pierzchalski (1990), *The Ahlfors Laplacian on a Riemannian manifold with boundary*, (preprint).

V. K. Patodi (1970), *Curvature and the eigenforms of the Laplace operator*, J. Indian Math. Soc **34**, 269-285.

Authors' address:

Thomas P. Branson
Department of Mathematics
University of Iowa
Iowa City IA 52242 USA
branson@math.uiowa.edu

Bent Ørsted
Institut for Matematik og Datalogi
Odense Universitet
5230 Odense M Danmark
orsted@imada.ou.dk

Peter B. Gilkey
Department of Mathematics
University of Oregon
Eugene OR 97403 USA
gilkey@bright.math.uoregon.edu

Antoni Pierzchalski
Department of Mathematics
University of Lódź
90238 Lódź Poland

Operator Theory:
Advances and Applications, Vol. 57
© 1992 Birkhäuser Verlag Basel

RECURRENT VERSUS DIFFUSIVE QUANTUM BEHAVIOR FOR TIME DEPENDENT HAMILTONIANS

Monique Combescure

I want to report on some recent results in a subject where, up to now, only a few rigorous results are known, although it is the field of intense investigations. These results have been obtained by myself or other people, but I want to review over many of them in order to make a perspective, and to present to the audience the main routes that have already been explored and the numerous problems that are left open along or around these routes.

The aim of these approaches is the study of the long time behavior for the classical dynamics and for the corresponding quantum dynamics, when the hamiltonian has some time dependence. These approaches have some experimental background, namely observations made on some experiments.

One of these experiments is the so-called Bayfield and Koch experiment, which is the microwave ionization of hydrogen or Rydberg atoms, when the frequency of the field is well below the ionization frequency (see Bayfield et al. 1989). The second experiment is the escape of an electron initially bound to a dielectric-surface, when some electric alternating-current field is applied perpendicular to the surface (see Yücel et al. 191).

The main observations which emerge from these experiments are :

- an irregular diffusive behavior of the electron in phase space.

- and suitable decorrelations in time between some observables taken at times 0 and t.

and the question is :

Can these observations be the manifestation of a "chaotic long time behavior" at the *quantum* level ?

When studying these systems on a theoretical point of view, it appears that most of the time-dependent (non quadratic) hamiltonian systems manifest *classical chaos* . The classical chaos has a precise mathematical definition in terms of positive Lyapunov exponents, which immediately implies a very sensitive dependence of the classical orbits on the initial conditions.

Now the question is : what happens at the quantum level ? The physical intuition is that, somehow, the quantum dynamics *smoothes out* the classical instabilities associated with the sensitive dependence on initial conditions. This smoothing effect can perhaps be understood in terms of the summing over classical paths which is implicit in the quantum dynamics.

Typically, the time-dependence of the experimental situations described above is a *periodic* one. This periodicity in time generally produces a chaotic behavior of the classical dynamics, whereas the corresponding quantum dynamics appears to be generically *more stable and predictive* (at least away from possible resonances). The quantum long time behavior for these systems is best understood in terms of the spectral properties of the Floquet operator U(T, 0), namely the unitary quantum evolution operator between times 0 and T(T being the period).

Another type of time dependence is the purely random one. In this case, the "quantum interference effects" which are expected to enforce stability and predictability in the time-periodic case, are destroyed. This produces typically a "diffusive" behavior in phase space at the quantum level.

Halfway between these two extreme cases are the cases of more or less disordered time dependences such as

1) the quasi periodic time dependence

2) the case of a "deterministic disorder" induced by deterministic aperiodic sequences like the Thue-Morse on Rudin-Shapiro sequences.

3) the case where the time dependence is implemented by some stochastic process t $\rightarrow \theta(t)$, like, for example, a Markov process.

The aim of this talk is to present rigorous results for the quantum long time behavior, more precisely to show what is the *quantum* response, as regards the long time behavior, to the more or less disordered time dependences 1) to 3) of the Hamiltonian; I shall :

- first recall more precise quantum stability results for the time-periodic case

- indicate an extension of the Floquet analysis of the periodic to the quasi-periodic case 1); this extension has been proposed recently by Jauslin et al. (1991)

- show the relevance of a suitably defined "quantum autocorrelation function" in cases 1) and 2).

- present a rigorous result obtained recently by Cheremshantsev (1991)in the case 3).

1) Time-periodic case

The "prototype" of models with time periodic dependence is the so-called kicked-rotator model (see Chirikov 1979), whose Hamiltonian is

$$\mathcal{H}(q, p, t) = p^2 + \lambda \cos q \sum_{n=-\infty}^{+\infty} \delta(t - nT)$$

where $q \in \mathbb{T}$ (one dimensional torus) for the classical dynamics, whereas the Hilbert space of quantum states is $L^2(\mathbb{T})$ with $p = -i \frac{\partial}{\partial q}$. Classically, the dynamics becomes completely chaotic for λT above some critical value. The numerical approach of the quantum dynamics shows the so-called "quantum suppression of the classical chaos", also known as "quantum dynamical localization" effect (see Fishman et al. 1982) .However the mathematical knowledge of this system is rather poor, up to now. The spectral analysis of the Floquet operator $U(T, 0)$ leads to distinguish three cases :

- the *resonant case* $(T/4\pi \in \mathbb{Q})$ where the spectrum of $U(T, 0)$ is known to be purely absolutely continuous for any λ (see Chirikov 1979)

- the *non-resonant case* $(T/4\pi$ obeying some diophantine condition) where the nature of the spectrum is not known, although we expect it to be pure-point

- the *"almost-resonant case "* $(T/4\pi$ suitably well approximated by rational numbers) where the spectrum of $U(T, 0)$ is known to be generically continuous , presumably singularly continuous (see Casati et al. 1984).

For a "smoothed" version of the kicked rotator called by him the "pulsed rotator", Bellissard (1985) has established the pure-point nature of the spectrum of $U(T, 0)$, in the non-resonant case, and for the coupling λ being small enough. His idea was to perform a perturbative approach using a super-convergent scheme "à la Kolmogorov Arnold Moser" to treat the small divisors arising in this problem.

In the same vein, I have treated the case of time-periodic perturbations of the harmonic oscillator (see Combescure 1987) :

$$H(t) = \frac{p^2 + \omega^2 x^2}{2} + \lambda V(t),$$

$V(t)$ belonging to some class of time-periodic bounded operators in $L^2(\mathbb{R})$. For this problem, which arises from the physical stability problem of the quantum motion inside radio-frequency ions' traps, I could establish the pure point nature of the spectrum of the Floquet operator provided :

- the time-frequency $\frac{2\pi}{T}$ is outside some "resonant set" of frequencies

- the coupling λ is small enough.

A question which is left open is the following : what happens if λ increases ? Will, in particular, the spectral nature of the Floquet operator change above some critical coupling λ ?

Now let us come back to the kicked rotator model but assume that the local cosq potential is replaced by some rank one projection $\varphi > < \varphi$, φ being an element of the Hilbert space. This simplified model appears to be exactly solvable, as regards its spectral properties, allowing a complete non perturbative treatment, and exhibiting the appearance of a singular continuous component in the spectrum of the Floquet operator. More generally, if

$$H(t) = H_0 + \lambda \, \varphi > < \varphi \sum_{n = -\infty}^{+\infty} \delta \, (t - nT)$$

in some Hilbert space $\mathcal{H}$, where H_0 has a purely discrete spectrum, then if $(a_m)_{m \in \mathbb{Z}}$ are the coordinates of the state φ on the basis of the eigenstates of H_0, the following result holds (see Combescure 1990) :

- if $\sum_{m=\infty}^{+\infty} |a_m| < \infty$ and φ is cyclic for H_0, then the spectrum of the Floquet operator is pure point for almost every λ (even large !)

- if $\sum_{-\infty}^{+\infty} |a_m| = \infty$, the spectrum of the Floquet operator has a singular continuous component for any λ, which leads to a diffusive increase of the energy.

In fact, the time-periodic perturbations of the harmonic oscillator appears to be a border case of more general quantum stability results. The intuition behind this fact is that the equal spacing of the levels of the unperturbed hamiltonian leads to *strong resonance phenomena* . Namely, it has been established by Howland (1989) that if $H(t) = H_0 + V(t)$ in some Hilbert space $\mathcal{H}$ where H_0 has pure point spectrum with simple eigenvalues $0 < e_1 < e_2 < ... e_n < ...$ satisfying

$$e_{n+1} - e_n \geq Cn^\alpha \quad \text{for some } \alpha > 0$$

and when the bounded operator $V(t)$ has suitable regularity properties in t, depending on α , then this resonance phenomena disappear, and the Floquet operator has pure point spectrum for almost every λ (even large !).

This last result confirms the physical intuition of the "quantum dynamical localization" for most time-periodic systems.

2. The quasi-periodic time-dependence

The simplest example is the "quasi-periodically driven kicked rotator", with Hamiltonian

$$H(q, p, t) = p^2 + \lambda \cos q \cos \omega t \sum_{n=-\infty}^{+\infty} \delta(t - \frac{2\pi}{\omega'} n) \tag{1}$$

where ω and ω' are incommensurate, i.e. $\omega/\omega' \notin \mathbb{Q}$. A complete mathematical treatment of this system has not been obtained yet, but some *numerical evidence* is provided of some ergodicity in the quantum long time behavior (see Milonni et al. 1987). It was shown, in particular that the "quantum autocorrelation function"

$$C_\psi(t) = \lim_{T\to\infty} T^{-1} \int_0^T ds < \psi, U(t + s,s)\psi > \tag{2}$$

(where $U(t, s)$ is the unitary quantum evolution operator associated with (1)) decays, numerically, when $|t|$ becomes large.

Another simple quasi-periodically driven quantum system has been studied by Luck et al. (1988)

$$H(t) = E \sigma_z + \lambda \sigma_x \sum_{n=-\infty}^{+\infty} \gamma_n \delta(t - nT) \tag{3}$$

where σ_x and σ_z are the usual 2×2 Pauli matrices, and $\gamma_n \in \{0,1\}$ is the nth element of the quasi-periodic Fibonacci sequence, produced by the infinite application of the following substitution rule

$$1 \to 10$$
$$0 \to 1.$$

The authors have provided numerical evidence that the quantum evolution induced by (3) is somehow intermediate between quasi-periodic and random.

The first general mathematical approach of quasi periodically driven quantum systems has been proposed recently by Jauslin et al.(1991). Using ideas borrowed from Yajima, and Howland, they propose to define a generalized "quasi-energy operator" on an enlarged Hilbert space, where an auxiliary degree of freedom is associated to each fundamental frequency of the driving. More precisely, let

$$H(q, p, t) = H_0(q, p) + V(q, \theta(t)) \quad \text{on Hilbert space } \mathcal{R} \tag{4}$$

where $\theta(t) = g_t\, \theta$ is some quasi periodic flow on some space Ω, with invariant measure μ. Typically,

$$\theta_j(t) = \theta_j + \omega_j\, t \quad , \quad (j = 1, ...n)$$

$\omega_1, ...\omega_n$ being n incommensurate frequencies. The enlarged Hilbert space is then

$$\kappa = \mathcal{R} \otimes L^2\, (\Omega, d\mu)$$

and if $U(t, s; \theta)$ is the quantum unitary evolution induced by (4), we can define the following map in κ :

$$(W(t)\Psi)(\theta) = T_{-t}\, U(t, 0; \theta)\, \Psi(\theta)$$

where $(T_{-t}\, \Psi)(\theta) = \Psi(g_{-t}\, \theta)$
The $W(t)$ so defined is a strongly family of unitary operators in κ, so that by Stone theorem

$$W(t) = e^{-iKt}$$

K being self-adjoint in κ. K is called the "quasi-energy operator". We recall that in the periodic case $\theta(t) = \theta + \omega t$ on the one dimensional torus, the spectrum of K in κ is in one to one correspondance with the spectrum of the Floquet operator $U(\frac{2\pi}{\omega}, 0; 0)$ in $\mathcal{R}$. Then the main result of Jauslin et al. (1991) is as follows :

Denote by κ_{ac} (resp. κ_{sc} or κ_{pp}) the absolutely continuous (resp. singular continuous or pure point) subspaces of K in κ :

Theorem 1   (see Jauslin et al. (1991)

(i)   if $\psi \otimes 1 \in \kappa_{a.c}$ then $C_\psi(t) \underset{t \to 0}{\to} 0$ (a.e θ)

(ii)   if $\psi \otimes 1 \in \kappa_{s.c}$ then $C_\psi(t) \to 0$ in Cesaro mean
$$\text{when } t \to \infty \text{ (a.e } \theta)$$

(iii) if $\psi \otimes 1 \in \kappa_{pp}$ then $C_\psi(t)$ is almost periodic (a.e θ) where $C_\psi(t)$ is defined by Eq.(2).

This type of general results is an extension from the time-periodic to the quasi-periodic case of the spectral approach of the quantum stability problem; it also shows that a suitable decay in time of the "quantum autocorrelation function" $C_\psi(t)$ can be viewed as a signature of some quantum long time delocalization.

3. Case of "deterministic disorder" in time

We come back to the aperiodic Hamiltonian (3) for the quantum two level system, where the sequence γ_n modulating the kicks is no longer the quasiperiodic Fibonnacci sequence, but the so-called Thue-Morse sequence. This Thue-Morse sequence is obtained by infinite iteration of the following substitution rule

$$1 \rightarrow 10$$
$$0 \rightarrow 10$$

starting from $\gamma_0 = 1$.

The remarkable feature of this sequence is that

$$\sum_{n=0}^{2^N-1} (2\gamma_n - 1)\, e^{i\omega n} = (1 - e^{i\omega})(1 - e^{2i\omega})...(1 - e^{2^{N-1}i\omega})$$

so that

$$\left| \sum_{n=0}^{2^N-1} (2\gamma_n - 1)\, e^{i\omega n} \right|^2 = 2^N (1 - \cos \omega)...(1 - \cos 2^{N-1} \omega)$$

is such that the weak-star limit of $\left| \sum_{n=0}^{2^N-1} (2\gamma_n - 1)\, e^{i\omega n} \right|^2$ when $N \rightarrow \infty$ is the so-called Riesz measure known to be singular continuous with respect to the Lebesgue measure. It is nothing but the "correlation measure" of the Thue-Morse sequence.

Remark 1 We can have more *disordered* deterministic substitution sequences, like for example, the "Rudin-Shapiro sequence" whose correlation measure is absolutely continuous with respect to Lebesgue measure (and even equal to it !).

Now the question is :

how does the deterministic disorder of the sequence manifest itself in the quantum long time behavior ?

And the answer, for the hamiltonian (3) with Thue-Morse sequence γ_n is as follows :

Theorem 2 (see Combescure 1991)

For a non-trivial (but thin) set $\mathcal{E}$ of parameter (E, λ), and any quantum state ψ, the quantum autocorrelation function $C\psi(t)$ is the Fourier transform of a well defined positive measure called the "quantum autocorrelation measure", which splits into a pure point part, and a singular continuous part directly connected with the correlation measure of the substitution sequence itself.

Remark 2 The set $\mathcal{E}$ is expected to be everywhere dense but of measure 0 in the whole domain $[0, 2\pi]^2$ of parameters (E, λ), although I have not been able to prove this point. Its description is very reminiscent of the transfer matrix problem for aperiodic Thue-Morse crystals, namely for the discretized one-dimensional Schrödinger equation :

$$\psi_{n+1} + \psi_{n-1} + \gamma_n\psi_n = E\,\psi_n$$

where $\gamma_n \in \{0, 1\}$ is the deterministic value of the potential at site $n \in \mathbb{Z}$. The study of products of 2×2 *transfer* matrices in this problem is replaced by the study of products of 2×2 *unitary* matrices in our quantum 2 level problem.

The previous situation was a purely quantum 2-level system. However, the above approach can be extended to some quantum systems with infinite number of levels, namely the following simple example

$$H(q, p, t) = \frac{p^2 + \omega^2 q^2}{2} + \lambda q \sum_{-\infty}^{+\infty} \gamma_n\, \delta(t - n) \tag{5}$$

where γ_n is again the nth term of a deterministic substitution sequence (see Combescure 1992). We take it to be either the Thue-Morse sequence (which has purely singular continuous correlation measure) or the Rudin-Shapiro sequence (with Lebesgue correlation measure).

Due to the quadratic characters of (5), the classical and quantum dynamics can be treated on an equal footing. Furthermore, we know an everywhere dense set of "resonant frequencies", namely those frequencies ω of unperturbed oscillator such that the base-2 expansion of $\omega/2\pi$ is finite. It is not hard to show that on the classical as well as on the quantum level, diffusion in phase space occurs via the growth in N of

$$\sum_{n=0}^{N} \gamma_n \, e^{i\omega n}$$

This expression being simply connected with the correlation measure of the sequence γ_n. More precisely, we define the quantum autocorrelation function $C_\psi(n)$ as

$$C_\psi(n) \ = \ \lim_{N\to\infty} \ N^{-1} \sum_{p=1}^{N} \ <\psi, U(p + n, p)\,\psi>$$

where ψ is any quantum state in $L^2(\mathbb{R})$, and $U(t, s)$ the unitary evolution operator generated by (5). Then $C_\psi(n)$ is the Fourier transform of a positive measure

$$C_\psi(n) \ = \ \int_0^1 e^{2i\pi xn}\, d\sigma_\psi(x)$$

(which would be simply the spectral measure of the Floquet operator in the state ψ for the periodic case $\gamma_n \equiv 1$), and we have :

Theorem 3 (see Combescure 1992)

Assume $\exists\, n \in \mathbb{N} : \omega 2^n = 0 \pmod{2\pi}$. Then $\forall\, r \in \mathbb{N}$

$$U(r\, 2^{n+1}, 0)\,\psi = \psi \qquad \text{(any } \psi) \tag{6}$$

and

$$\frac{d\sigma_\psi}{d\lambda}(x) = \text{pure point measure} \qquad\qquad + \text{ singular continuous measure}$$
$$\text{(located at } x = 0,\ 2^{-n-1},\ 2^{-n},\ 3.2^{-n-1},...) \qquad \text{(dilation of Riesz measure)}$$
$$\text{due to recurrent property (6)}$$

(this is true without any condition on the size λ of the kicks).

Question : what happens *inbetween* these resonant frequencies ? We can show that, in fact, the energy growth w.r.t. time depends *very sensitively* and *very weirdly* on ω. Define the energy growth as

$$\Delta E(t, \omega) = \ <U(t, 0)\psi, H_\omega\, U(t, 0)\psi> - \ <\psi, H_\omega\psi>$$

for any $\psi \in \mathscr{D}(H_\omega)$, $H_\omega = \dfrac{p^2 + \omega^2 q^2}{2}$ and look at the following sequence of times $t_n = 2^n$. Then we prove in Combescure (1992) that :

$$\frac{\text{Log } \Delta E(t_n, \omega)}{\log t_n} \to \alpha > 1 \quad \text{as } n \to \infty, \text{ for } \nu \text{ a.e.} \omega \ (\nu \text{ Riesz measure})$$

$$\to 0 \qquad \text{as } n \to \infty \text{ for a.e.} \omega \text{ (w.r. to Lebesgue measure)}$$

However $\forall$ I interval $\subset [0, 1]$

$$\int_I \frac{d\omega}{2\pi} \Delta E(t_n, \omega) \underset{n \to \infty}{\sim} \nu(I) \, t_n \quad (\nu(I) : \text{Riesz measure of I})$$

Therefore the "linear growth" of energy is not true pointwise (due to strong recurrence effects for ω dyadic) but only in the mean. This *highly delicate* behavior in ω is related to the very fine multi-scale structure of the Riesz measure. Therefore we think that such aperiodic substitution sequences provide us with a good scenario of what could be thought of as a "stochastic quantum long-time behavior".

We can compare with the purely random case, namely the case where γ_n are i.i.d. random variables (e.g. a Bernouilli process). In that case, the escape to infinity in phase space occurs via a "random-walk-like process", and the return probability at time t wildly oscillates when t becomes large, almost surely.

4. Stochastic time dependence via Markov process.

I now come to a case recently studied by Cheremshantsev (1991), namely that of a Hamiltonian of the following form :

$$H(q, p, t) = p^2 + V(q) \, \omega(t) \quad \text{in } L^2(\mathbb{R}) \tag{7}$$

where $V(q) \sim |q|^{-\alpha}$ as $|q| \to \infty$ for $0 < \alpha < 1$, and $t \to \omega(t) \in \{-1, 1\}$ is the path of some Markov process, μ being the probability measure on the space of paths. If $U(t, s; \omega)$ is the unitary evolution operator generated by (7), and

$$\psi(t, \omega) = U(t, 0; \omega)\psi \qquad (\psi \in L^2(\mathbb{R}))$$

the following result holds :

Theorem 4 (see Cheremshantsev 1991)

$\psi(t, \omega)$ is μ-almost surely *delocalized* as t becomes large in the following sense :

(i) $\displaystyle\lim_{T\to\infty} T^{-1} \int_0^T dt \; \|F(|x| < t^\beta) \psi(t, \omega)\|^2 = 0.$ μ.a. surely

$\qquad\qquad\qquad$ (some $\beta > 0$ related to α)

(ii) $< x^2(t) > \overset{\text{def}}{=} < \psi(t, \omega), x^2\psi(t, \omega) >$ (suitable ψ)

is such that

$$\liminf_{T\to\infty} T^{-1} \int_1^T dt \; t^{-\gamma} < x^2(t) > \; > 0 \qquad\qquad \mu\text{.a. surely}$$

$\qquad\qquad$ (some $\gamma > 0$ related to α).

5. Conclusion

We have seen that, for time-dependent hamiltonians, the quantum dynamics can manifest a "stochastic long time-behavior" comparable to some classically chaotic motion. Here, the quantum autocorrelation function (or measure) appears to be a good indicator of stochasticity and /or diffusion in phase space. The quantum observables at time t may exhibit very delicate *sensitive dependence* on some parameters (compare with the sensitive dependence in classical chaos which holds with respect to *initial data* !) .We think that this highly delicate dependence can impose severe limitations to the validity of numerical investigations.

Finally, this area of research leaves open a number of interesting mathematical problems.

REFERENCES

Bayfield J.E. et al (1989), Phys. Rev. Lett., **63**, 364.

Bellissard J. (1985), *Trends and Developments in the Eighties* , S. Albeverio and Ph. Blanchard eds., World Scientific, Singapore.

Casati G. and Guarneri I. (1984), Commun. Math. Phys. **95**, 121.

Cheremshantsev S. (1991), Delocalization for slowly decaying potentials randomly depending on time, Bochum preprint.

Chirikov B.V. (1979), Phys. Rep. **52**, 263.

Combescure M. (1987), Ann. Inst. Henri Poincaré, **47**, 63.

Combescure M. (1990), J. Stat. Phys. **59**, 679.

Combescure M. (1991), J. Stat. Phys. **62**, 779.

Combescure M. (1992) to appear in Ann. Inst. Henri Poincaré.

Fishman S. et al (1982), Phys. Rev. Lett. **49**, 509.

Howland J. (1989), Ann. Inst. Henri Poincaré, **49**, 325 .

Jauslin H. and Lebowitz J.(1991), Spectral and stability aspects of quantum chaos, Chaos (1991) **1**, 114.

Luck J.M. et al (1988), J. Stat. Phys. **53**, 551.

Milonni P.W. et al (1987), Phys. Rev. A **35**, 1714.

Yücel S. and Andrei E.Y. (1991), Phys. Rev. B **43**, 12 029.

Laboratoire de Physique Théorique et Hautes Energies*, Université de Paris-Sud, Bâtiment 211, 91405 Orsay cedex FRANCE.

*Laboratoire associé au Centre National de la Recherche Scientifique.

Operator Theory:
Advances and Applications, Vol. 57
© 1992 Birkhäuser Verlag Basel

Perturbations of spectral measures
for Feller operators

Michael Demuth

Abstract. In stochastic spectral analysis of selfadjoint Feller operators quantitative error estimates are given for the corresponding spectral measures. Both regular and singular perturbations are considered.

1. The background idea of stochastic spectral analysis

Many spectral properties of selfadjoint operators can be studied indirectly by the spectral behaviour of their semigroups. For Schödinger operators of the form $H_0 + V$, H_0 selfadjoint realization of $-\Delta$, V a Kato–class potential, in $L^2(\mathbf{R}^n)$ a summary of these results can be found by Simon (1982) or by Demuth (1991). The spectral consequences are mainly based on the explicit representation of the integral kernels for the semigroups or resolvents. That means for instance the resolvent $(H_0 + V + z1)^{-1}$ is an integral operator for appropriate regular values, the kernel of which is given by

$$\left[(H_0 + V + z1)^{-1}\right](x,y) = \int_0^\infty d\lambda \ e^{-\lambda z} \ E_x^{y,\lambda} \left\{ e^{-\int_0^\lambda V(\omega(s))ds} \right\} , \tag{1}$$

$x, y \in \mathbf{R}^n$. $E_x^{y,\lambda}\{.\}$ denotes the conditional Wiener measure. $\omega(.)$ are the trajectories of the Wiener process.

For Kato–class potentials the kernel of $e^{-t(H_0+V)}$ can be estimated by the free kernel (where $V \equiv 0$) i.e. by the Wiener density function

$$p_W(t, x, y) = (4\pi t)^{-n/2} \ e^{-|x-y|^2/4t} . \tag{2}$$

Therefore the proofs of spectral properties of $H_0 + V$ using the Feynman–Kac–representation in (1) are mainly based on several characteristic features of the Wiener density function p_W. This density function has many "nice" properties. For instance it is symmetric with respect to any coordinate $x_i \in \mathbf{R}$, $i = 1, \ldots, n$. It is decreasing if any $|x_i| \to \infty$, even with the same rate of convergence. It is uniformly bounded by $|x - y|^{-n}$.

But for many spectral theoretical aspects it is not necessary to have or to use the whole variety of these p_W-properties. Therefore the first main task in *Stochastic Spectral Analysis* was to select a set of sufficient (and almost necessary) assumptions on a transition density function p, which are rich enough to provide spectral results and which are poor enough to include an interesting class of operators $K_0 + V$, where K_0 is *not* canonical to the Laplacian.

That was done by Demuth, van Casteren (1989 and 1992). We have established the following basic assumptions on stochastic spectral analysis, shortly denoted as BASSA.

BASSA:

1. *Existence*
   Let $(E, \mathcal{E})$ be a second countable locally compact Hausdorff space E with the Borel field $\mathcal{E}$. Assume a continuous function

   $$p \quad \text{mapping} \quad (0, \infty) \times E \times E \to [0, \infty)$$

   with the properties

   $$\int_A p(t, x, y) dy \le 1 \text{ for all } t > 0, x \in E, A \subset E ,$$

   and

   $$\int_E p(t, x, u) p(s, u, y) du = p(s + t, x, y) .$$

2. *Continuity*
   Let $C_\infty(E)$ be the set of continuous functions vanishing at infinity. For any $f \in C_\infty$ and any $x \in E$ we assume

   $$\lim_{t \to 0} \int_E f(y) p(t, x, y) dy = f(x) .$$

3. *Symmetry*
   For all $t > 0$, $x, y \in E$ we assume

   $$p(t, x, y) = p(t, y, x) .$$

4. *Feller property*
   For any $f \in C_\infty(E)$ we assume

   $$x \to \int_E f(y) p(t, x, y) dy \in C_\infty(E) .$$

$\square$

2. Free Feller operators

Definition:
Assuming BASSA the function p corresponds to a semigroup. Its generator is denoted with K_0, i.e.

$$(e^{-tK_0}f)(x) = \int_E f(y)p(t,x,y)dy \;.$$

Because e^{-tK_0} satisfies the Feller property K_0 is called the free Feller operator. (This corresponds to the name "free Schrödinger operator" for the Laplacian).   □

Remarks concerning the assumptions.
The density function is one central link between operator theory and stochastic analysis. The existence and the Feller property ensure that the underlying process $(\mathbf{R}_+; \Omega, \mathcal{F}, P_F; \omega(.))$ is a strong Markov process with the Feller property. Together with the continuity assumption it implies that the process has rigth continuous path with left–hand limits. The symmetry condition is equivalent to the selfadjointness of K_0.   □

Examples:
The most crucial condition in BASSA is the Feller property. This property has its own interest and is studied seperately in the literature. Let me mention here only two examples.

Davies (1991) studied locally finite Riemannian manifolds where K_0 is formally given by

$$K_0 f = -\frac{1}{\sigma^2}\,\nabla\left(\sigma^2\,\nabla f\right)\;.$$

Here $\sigma = \sigma(x)$ are strictly positive measurable functions on E such that $\sigma \in L^\infty_{loc}(E)$ and $\sigma^{-1} \in L^\infty_{loc}(E)$. K_0 is defined correctly via closable Dirichlet forms. e^{-tK_0} is then a positivity preserving strongly continuous semigroup on L^p, $1 \le p < \infty$. It is an integral operator with a kernel p_D. The Feller property is proved if

$$\lim_{|x|\to\infty} p_D(t,x,y) = 0$$

for all $y \in E$, $t > 0$. This is shown by pointwise estimates for $p_D(t,x,y)$. It is remarkable that the conditions for σ are very general, in particular it is not necessary to have any differentiability of σ.

The next example is given by Jacob (1992) in the present proceedings. In a series of articles he considered pseudo–differential operators defined as extension of an operator $a(x,D)$:

$$(a(x,D)u)(x) := (2\pi)^{-n/2} \int_{\mathbf{R}^n} e^{ix\xi}a(x,\xi)\widehat{u}(\xi)d\xi$$

$u \in C_0^\infty(\mathbf{R}^n)$ for special classes of $a(x,D)$, in particular for $a(x,\xi)$ of the form

$$a(x,\xi) = \sum_{j=1}^n b_j(x)|\xi_j|^{2r}, \qquad 0 < r \le 1$$

with $b_j \in C^\infty(\mathbf{R}^n)$, b_j independent of x_j. Then the corresonding extension generates a Feller semigroup. Further examples by Jacob include the relativistic Hamiltonian, where $a(x, \xi) = (|\xi|^2 + m^2)^{1/2} - m$. $\square$

3. Feller operators

Definition: *Kato–Feller–potentials*
Assume a density function satisfying BASSA. Let V be a real–valued function on E, $V = V_+ - V_-$. V is called a Kato–Feller–potential if

$$\lim_{\tau \to 0} \sup_x \int_0^\tau ds \int_E p(s, x, y)[V_-(y) + \chi_B(y)V_+(y)]dy = 0 \; , \tag{3}$$

where B is a compact subset of E. $\square$

These Kato–Feller–class is optimal for

$$\lim_{\lambda \to \infty} \|(K_0 + \lambda)^{-1}V_-\|_{\infty,\infty} = 0 \; , \tag{4}$$

which determines the relative form bound of V_- with respect to K_0. Then the right–hand side of the generalized Feynman–Kac–formula

$$E_x \left\{ e^{-\int_0^t V(\omega(s))ds} f(\omega(t)) \right\} \tag{5}$$

(here E_x is the exspectation with respect to the Feller measure P_F) yields a strongly continuous semigroup on L^2, its generator is the selfadjoint operator $K_0 \dot{+} V$. $e^{-t(K_0 + V)}$ is again an integral operator. The kernel can be estimated by

$$(e^{-t(K_0 \dot{+} V)})(x, y) \leq c \, e^{ct} \, p^{1/2}(t, x, y) \sup_{x,y \in E} p^{1/2}(t, x, y) \tag{6}$$

(see van Casteren (1989)). If $p(t, x, y)$ is uniformly bounded in x and y the last estimate implies the Feller property for the semigroup $e^{-t(K_0 + V)}$, too.

Definition: *Feller operator*
The generator of a Feller semigroup is denoted as Feller operator. $\square$

Therefore $K_0 \dot{+} V$ is a Feller operator with a regular perturbation. This denotation corresponds to the name for generators of Schrödinger semigroups.

Singular perturbations can also be included. Let Γ be a subset of E, $|\Gamma| > 0$. With S we denote the penetration time

$$S := \inf\{\tau > 0 : \int_0^\tau 1_\Gamma(\omega(s))ds > 0\} \; . \tag{7}$$

Then we define the absorption semigroup

$$E_x \left\{ e^{-\int_0^t V(\omega(s))ds} f(\omega(t)), \; S > t \right\}$$
$$=: \; (U(t)f)(x) \; . \tag{8}$$

Let $\Sigma = E \setminus \Gamma$, then $U(t) \uparrow L^2(\Sigma)$ is again a Feller semigroup, its generator is denoted with $(K_0 \dot{+} V)_\Sigma$. It is a selfadjoint operator in $L^2(\Sigma)$.

4. Spectral measures

The spectral measure plays a fundamental role in characterizing the different parts of the spectrum for selfadjoint operators. For the selfadjoint Feller operators $K_0 \dot{+} V$, $K_0 \dot{+} W$, considered here, we denote the spectral measures with $E_{K_0 \dot{+} V}(.)$, $E_{K_0 \dot{+} W}(.)$, respectively. Instead of considering the potential dependence of matrix elements of these spectral measures in weighted L^2–spaces, we study the operator norm of sandwiched spectral measures.

Because we have assumed Kato–Feller–potentials a natural norm is the Kato–Feller–norm, which is defined by

$$\|V\|_{KF} := \sup_{x \in E} \int_0^1 ds \int_E p(s, x, y)|V(y)|dy .\tag{9}$$

The objective of the following theorem is to control explicitly the changes of the spectral measure in terms of $\|V - W\|_{KF}$.

Theorem 1: Assume BASSA and two Kato–Feller–potentials V and W. Let φ be a multiplication operator with a nonvanishing continuous function $\varphi : E \to \mathbf{R}_+$ with $\varphi^{-1} \leq 1$. Let $\Delta = (\alpha_1, \alpha_2)$ be an interval on $\mathbf{R}_+$ such that α_1 and α_2 are no eigenvalues of $K_0 \dot{+} V$ or $K_0 \dot{+} W$, respectly. Assume that

$$\sup_{\substack{\lambda \in \Delta \\ \varepsilon \to (0,1]}} \|\varphi^{-1}(K_0 - \lambda \pm i\varepsilon)^{-1}\varphi^{-1}\| =: d_\Delta < \infty .\tag{10}$$

Moreover, suppose a pointwise estimate

$$\int_E p(t, x, y)\varphi^4(y)dy \leq c_0(1 + t^m)\varphi^4(x)\tag{11}$$

with some $m \in \mathbf{N}$.

Let us denote positive constants c_V and A_V by

$$E_x \left\{ e^{-\int_0^\lambda V(\omega(s))ds} \right\} \leq c_V \, e^{A_V \lambda} .\tag{12}$$

Take the Kato–Feller–norms $\|V\varphi^2\|_{KF}$, $\|W\varphi^2\|_{KF}$ small enough, i.e. take for instance

$$\|V\varphi^2\|_{KF} < \frac{1}{12} \frac{1}{b_V} \frac{1}{c_0^{1/2}} \frac{1}{c_{4V}^{1/4}} \frac{1}{1 + 3d_\Delta}\tag{13}$$

with $b_V > \max\{m, \alpha_2, 2A_{4V}\}$. ($\|W\varphi^2\|_{KF}$ correspondingly small).
Then the difference of the spectral measures can be estimated by

$$\|\varphi^{-1}(E_{K_0 \dot{+} V}(\Delta) - E_{K_0 \dot{+} W}(\Delta))\varphi^{-1}\| \leq c(V, W, \Delta)\|(V - W)\varphi^2\|_{KF} .\tag{14}$$

The constant $c(V, W, \Delta)$ depending on V, W, Δ and on the geometry of E can be estimated explicitly. If the condition in (13) is satisfied one has

$$\begin{aligned}
c(V, W, \Delta) \;\leq\; & 2|\Delta|\, c_0^{1/2}\, c_{4V}^{1/4}\, c_{4W}^{1/4}\, a_0^{-1} \\
& (1 + 3a_0\gamma_V^{-1}(d_\Delta + 1))(1 + 3a_0\gamma_W^{-1}(d_\Delta + 1)) ,
\end{aligned} \tag{15}$$

with $a_0 > \max\{m, \alpha_2, 2A_{4V}, 2A_{4W}\}$ and with

$$\gamma_V = 1 - 12 b_V c_0^{1/2} c_{4V}^{1/4}(1 + 3d_\Delta)\|V\varphi^2\|_{KF} .$$

$\square$

Remark: Note that V and W are not assumed to be bounded. The condition on $\|W\varphi^2\|_{KF}$, corresponding to (13), could be neglected. It would follow from the condition for $\|V\varphi^2\|_{KF}$ and an analogous result for $(K_0+V-\lambda\pm i0)^{-1}-(K_0+W-\lambda\pm i0)^{-1}$. $\square$

Proof of Theorem 1: The spectral measure of a selfadjoint operator H on a bounded open interval $\Delta = (\alpha_1, \alpha_2)$ where neither α_1 or α_2 is an eigenvalue of H is given by

$$E_H(\Delta) = s - \lim_{\varepsilon\to 0} (2\pi i)^{-1} \int_{\alpha_1}^{\alpha_2} [(H - \lambda - i\varepsilon)^{-1} - (H - \lambda + i\varepsilon)^{-1}]d\lambda . \tag{16}$$

We set $R_V(\lambda \pm i\varepsilon) := (K_0 + V - \lambda \mp i\varepsilon)^{-1}$. Then

$$\begin{aligned}
& \|\varphi^{-1}(E_{K_0+V}(\Delta) - E_{K_0+W}(\Delta))\varphi^{-1}\| \\
\leq\; & \lim_{\varepsilon\to 0} (2\pi)^{-1} \int_{\alpha_1}^{\alpha_2} d\lambda\, \{\|\varphi^{-1}[R_V(\lambda + i\varepsilon) - R_W(\lambda + i\varepsilon)]\varphi^{-1}\| \\
& + \|\varphi^{-1}[R_V(\lambda - i\varepsilon) - R_W(\lambda - i\varepsilon)]\varphi^{-1}\|\} .
\end{aligned} \tag{17}$$

The first term in (17) is estimated by:

$$\begin{aligned}
& \|\varphi^{-1}[R_V(\lambda + i\varepsilon) - R_W(\lambda + i\varepsilon)]\varphi^{-1}\| \\
\leq\; & (1 + |\lambda + i\varepsilon - a|\, \|\varphi^{-1}R_V(\lambda + i\varepsilon)\varphi^{-1}\|) \\
& (1 + |\lambda + i\varepsilon - a|\, \|\varphi^{-1}R_W(\lambda + i\varepsilon)\varphi^{-1}\|) \\
& \|\varphi[R_V(-a) - R_W(-a)]\varphi\| ,
\end{aligned} \tag{18}$$

where a is any regular value for K_0+V and K_0+W. The rest of the proof is splitted into several lemmata. The objective is to estimate the terms in (18) uniformly in λ and ε.

Lemma 2: Take the assumptions of Theorem 1. Then

$$\|\varphi[R_V(-a) - R_W(-a)]\varphi\| \leq 4\, c_0^{1/2}\, c_{4V}^{1/4}\, c_{4W}^{1/4} \cdot \|(V - W)\varphi^2\|_{KF} . \tag{19}$$

if $a > \max\{m, 2A_{4V}, 2A_{4W}\}$. $\square$

Proof of Lemma 1: Demuth, van Casteren (1991) have shown that

$$\begin{aligned}
\|\varphi\, [R_V(-a) - R_W(-a)]\, \varphi\| \;\leq\; & \|R_0(-a)\, |V - W|\, R_{2V}(-a)\, \varphi^2\|_\infty^{1/2} \\
& \cdot \|R_0(-a)\, |V - W|\, R_{2W}(-a)\, \varphi^2\|_\infty^{1/2} .
\end{aligned}$$

The first factor squared is smaller than

$$\sup_x \int_0^\infty d\lambda \ e^{-a\lambda} E_x \left\{ |V(\omega(\lambda)) - W(\omega)(\lambda))| \ [R_{2V}(-a)\varphi^2](\omega(\lambda)) \right\} .$$

But

$$(R_{2V}(-a)\varphi^2)(x) \le \left[\int_0^\infty d\lambda \ e^{-a\lambda} E_x \left\{ e^{-4\int_0^\lambda V(\omega(s))ds} \right\} \right]^{1/2}$$

$$\left[\int_0^\infty d\lambda \ e^{-a\lambda} E_x \{\varphi^4(\omega(\lambda))\} \right]^{1/2}$$

$$\le \ c_{4V}^{1/2} \ c_0^{1/2} (a - A_{4V})^{-1/2} \left(\frac{1}{a} + \frac{m!}{a^{m+1}} \right)^{1/2} \varphi^2(x)$$

$$\le \ 2 \ c_0^{1/2} \ c_{4V}^{1/2} \ a^{-1} \ \varphi^2(x) ,$$

if $a > m$, $a > 2A_{4V}$.
Therefore

$$\|\varphi(R_V(-a) - R_W(-a))\varphi\| \le 2 \ c_0^{1/2} \ c_{4V}^{1/4} \ c_{4W}^{1/4} \ a^{-1} \|R_0(-a)|V - W|\varphi^2\|_\infty$$

$$\le \ 2 \ c_0^{1/2} \ c_{4V}^{1/4} \ c_{4W}^{1/4} \ a^{-1} \sum_{k=0}^\infty e^{-ak} \ \|(V - W)\varphi^2\|_{KF}$$

which proves Lemma 2.   q.e.d.

Corollary 3: Setting $W \equiv 0$ Lemma 2 provides

$$\|\varphi[R_V(-a) - R_0(-a)]\varphi\| \le 4 \ c_0^{1/2} \ c_{4V}^{1/4} \ a^{-1} \ \|V\varphi^2\|_{KF} \qquad (20)$$

if $a > \max\{m, 2A_{4V}\}$.   □

In the next lemma we estimate the perturbed sandwiched resolvent near the real axis.
It is a consequence of Lemma 2.

Lemma 4: Take the assumptions of Theorem 1. Then

$$\|\varphi^{-1} R_V(\lambda + i\varepsilon)\varphi^{-1}\| \le \gamma_V^{-1}(d_\Delta + \frac{1}{3}) , \qquad (21)$$

with

$$\gamma_V := 1 - 12 \ b_V \ c_0^{1/2} \ c_{4V}^{1/4}(1 + 3d_\Delta)\|V\varphi^2\|_{KF}$$
$$b_V > \max\{m, \alpha_2, 2A_{4V}\} ,$$

where $\Delta = (\alpha_1, \alpha_2)$ and d_Δ is given in (10). γ_V is greater than zero because of the
assumption in (13).   □

Proof of Lemma 4: Using again (18) and Corollary 3 we obtain

$$\|\varphi^{-1} R_V(\lambda + i\varepsilon)\varphi^{-1}\| \ \le \ d_\Delta + (1 + 3b_V d_\Delta) \ 4 \ c_0^{1/2} \ c_{4V}^{1/4} \ \|V\varphi^2\|_{KF} \ b_V^{-1}$$
$$\cdot \ (1 + 3b_V \|\varphi^{-1} R_V(\lambda + i\varepsilon)\varphi^{-1}\|) .$$

$\|V\varphi^2\|_{KF}$ is chosen small enough. Then (21) follows obviously.   q.e.d.

Rest of the proof of Theorem 1:
From the Lemmata 2 and 4 follows

$$\|\varphi^{-1}[R_V(\lambda + i\varepsilon) - R_W(\lambda + i\varepsilon)]\varphi^{-1}\|$$
$$\leq \ (1 + |\lambda + i\varepsilon - a_0|\gamma_V^{-1}(d_\Delta + \frac{1}{3}))$$
$$(1 + |\lambda + i\varepsilon - a_0|\gamma_W^{-1}(d_\Delta + \frac{1}{3}))$$
$$4 \ c_0^{1/2} \ c_{4V}^{1/4} \ c_{4W}^{1/4} \ a_0^{-1}\|(V - W)\varphi^2\|_{KF}$$

with $a_0 > \max\{m, \alpha_2, 2A_{4V}, 2A_{4W}\}$, which implies (14) with the constant in (15).
q.e.d.

The next and last objective in this article is to analyse perturbations of the spectral measures for infinitely high potentials. As mentioned in (8)

$$E_x \left\{ e^{-\int_0^t V(\omega(s))ds} f(\omega(t)), S > t \right\}$$

establishes a strongly continuous semigroup in $L^2(\Sigma)$, $\Sigma = E \setminus \Gamma$, the generator of which is $(K_0\dot{+}V)_\Sigma$. For the singularity region Γ we assume that the regular points of Γ and the regular points of the interior of Γ form the same set.

On the other hand $e^{-t(K_0\dot{+}V)_\Sigma}$ is the limit of a family of semigroups $e^{-t(K_0+V+\beta U)}$ as $\beta \to \infty$. Here U is an additional positive potential with support Γ. $(K_0\dot{+}V)_\Sigma$ and $K_0\dot{+}V + \beta U$ are selfadjoint operators in $L^2(\Sigma)$ and $L^2(E)$, respectively. Their spectral measures are denoted with $E_\Sigma(.)$ and $E_\beta(.)$, respectively. Because these are operators in different Hilbert spaces we introduce an embedding operator by $(Jf)(x) := \chi_\Gamma(x)f(x)$.

Theorem 5: Assume BASSA and a Kato–Feller–potential V. Let φ be a multiplication operator with a continuous function satisfying $|\varphi^{-1}| \leq 1$,

$$\int_E p(t, x, y)\varphi^8(y)dy \leq c_0 \left(1 + t^m\right) \varphi^8(x) , \tag{22}$$

$m \in \mathbf{N}$, and for arbitrary large R let

$$\sup_{|x| \geq R} \varphi^2(x)\left[E_x\{S < \lambda\}\right]^{1/2} < \varepsilon \tag{23}$$

where ε is chosen arbitrarily small.
Let $\Delta = (\alpha_1, \alpha_2)$ and assume

$$\sup_{\substack{\lambda \in \Delta \\ \varepsilon \in (0,1]}} \|\varphi^{-1}J^*((K_0)_\Sigma - \lambda \pm i\varepsilon)^{-1}J\varphi^{-1}\| =: d_{\Delta,\Sigma} < \infty . \tag{24}$$

For the Kato–Feller–potential we assume, according to Theorem 1 (see (13)),

$$\|V\varphi^2\|_{KF} < \frac{1}{12 \ c_0^{1/2} \ c_{4V}^{1/4}} \ \frac{1}{b_V} \ \frac{1}{1 + 3d_{\Delta,\Sigma}} \tag{25}$$

with $b_V > \max\{m, \alpha_2, 2A_{4V}\}$. And we set

$$\gamma_{V,\Sigma} := 1 - 12\, b_V\, c_0^{1/2}\, c_{4V}^{1/4}\, (1 + 3d_{\Delta,\Sigma})\|V\varphi^2\|_{KF}\ . \tag{26}$$

Then we have the following assertions:
a) If we denote

$$\rho(\beta) := \int_0^\infty d\lambda\ e^{-\lambda} \sup_x \varphi^2(x) \left[E_x\left\{ e^{-\beta \int_0^\lambda U(\omega(s))ds}, S < \lambda \right\} \right]^{1/2}, \tag{27}$$

then $\rho(\beta)$ tends to zero as $\beta \to \infty$.
b) The difference of the spectral measures can be estimated quantitatively if β is sufficiently large:

$$\begin{aligned}
&\|\varphi^{-1}[E_\beta(\Delta) - J^* E_\Sigma(\Delta)J]\varphi^{-1}\| \\
\leq\ & \pi^{-1}|\Delta|\, \gamma_{V,\Sigma}^{-2}\, [1 + 3a_0(d_{\Delta,\Sigma} + 2)]^2 \\
& \|\varphi[(K_0\dot+V + \beta U + a_0)^{-1} - J^*((K_0\dot+V)_\Sigma + a_0)^{-1}J]\varphi\|
\end{aligned} \tag{28}$$

$$\begin{aligned}
\leq\ & (2\pi)^{-1}|\Delta|\, \gamma_{V,\Sigma}^{-2}\, [1 + 3a_0(d_{\Delta,\Sigma} + 2)]^2 \\
& (c_{2V}^{1/2} + c_0^{1/4}\, c_{4V}^{1/4}\, (1 + m^m) \cdot \rho(\beta)
\end{aligned} \tag{29}$$

with $a_0 > \max\{4, \alpha_2, A_{2V}, A_{4V}\}$. $\qquad\qquad\square$

Remark: The estimation of $\rho(\beta)$ is a difficult problem. One first quantitative estimate of $\rho(\beta)$ in the case $K_0 = -\Delta$ is given Demuth, Jeske, Kirsch (1992). The rate of convergence depends on the size of the boundary $\delta\Gamma$. $\qquad\qquad\square$

Proof of Theorem 5: As above we set

$$R_{V,\Sigma}(-a) := ((K_0\dot+V)_\Sigma + a)^{-1}\ .$$

For $a > 1$, $a > \alpha_2$ one has to estimate the product

$$[1 + 3a\,\|\varphi^{-1}J^* R_{V,\Sigma}(\lambda \pm i\varepsilon)J\varphi^{-1}\|] \tag{30}$$
$$\cdot\ [1 + 3a\,\|\varphi^{-1} R_{V+\beta U}(\lambda \pm i\varepsilon)\varphi^{-1}\|] \tag{31}$$
$$\cdot\ \|\varphi(R_{V+\beta U}(-a) - J^* R_{V,\Sigma}(-a)J)\varphi\| \tag{32}$$

uniformly in λ and ε.

The factor in (30) corresponds to Lemma 4 with $d_{\Delta,\Sigma}$ indead of d_Δ. The only point is that

$$\|\varphi J^*(R_{V,\Sigma}(-b) - R_{0,\Sigma}(-b))J\varphi\| \leq \|\varphi(R_V(-b) - R_0(-b))\varphi\|$$

if $b > 1$ and $b > \alpha_2$.

The second factor (31) can be estimated using the fact that $\|\varphi^{-1} R_{V+\beta U}(\lambda \pm i\varepsilon)\varphi^{-1}\|$ converges to $\|\varphi^{-1}J^* R_{V,\Sigma}(\lambda \pm i\varepsilon)J\varphi^{-1}\|$ as $\beta \to \infty$. This convergence will be considered in Lemma 6. Hence for sufficiently large β

$$\|\varphi^{-1} R_{V+\beta U}(\lambda \pm i\varepsilon)\varphi^{-1}\| \leq 1 + \|\varphi^{-1}J^* R_{V,\Sigma}(\lambda \pm i\varepsilon)J\varphi^{-1}\|\ .$$

The main intersting factor is (32). It will be considered separately.

Lemma 6: Under the assumptions of Theorem 5 it holds

$$\|\varphi\left(R_{V+\beta U}(-a_0) - J^* R_\Sigma(-a_0)J\right)\varphi\|$$
$$\leq \frac{1}{2}(c_{2V}^{1/2} + c_0^{1/4}\, c_{4V}^{1/4}\,(1+m^m))$$
$$\int_0^\infty d\lambda\, e^{-\lambda} \sup_x \varphi^2(x) \left[E_x\left\{e^{-\beta\int_0^\lambda U(\omega(s))ds}, S < \lambda\right\}\right]^{1/2}, \tag{33}$$

$a_0 > \max\{4, \alpha_2, A_{2V}, A_{4V}\}$, and it tends to zero as $\beta \to \infty$. $\qquad\square$

Proof: $\varphi(R_{V+\beta U}(-a) - J^* R_\Sigma(-a)J)\varphi$ is an integral operator with a symmetric kernel. Therefore its norm is smaller than

$$\sup_x \int_0^\infty d\lambda\, e^{-a\lambda}\varphi(x)E_x\left\{e^{-\int_0^\lambda V(\omega(s))ds}e^{-\beta\int_0^\lambda U(\omega(s))ds}\varphi(\omega(\lambda)), S < \lambda\right\}$$
$$\leq \frac{1}{2}\int_0^\infty d\lambda\, e^{-a\lambda} \sup_x \varphi^2(x)\left[E_x\left\{e^{-2\int_0^\lambda V(\omega(s))ds}\right\}\right]^{1/2}$$
$$\cdot\left[E_x\left\{e^{-2\beta\int_0^\lambda U(\omega(s))ds}, S < \lambda\right\}\right]^{1/2}$$
$$+ \frac{1}{2}\int_0^\infty d\lambda\, e^{-a\lambda} \sup_x \left[E_x\left\{e^{-4\int_0^\lambda V(\omega(s))ds}\right\}\right]^{1/4} \cdot \left[E_x\{\varphi^8(\omega(\lambda))\}\right]^{1/4}$$
$$\cdot\left[E_x\left\{e^{-2\beta\int_0^\lambda U(\omega(s))ds}, S < \lambda\right\}\right]^{1/2}.$$

For $|x| \geq R$ it is assumed that

$$\int_0^\infty d\lambda\, e^{-\lambda} \sup_{|x|\geq R} \varphi^2(x)[E_x\{S < \lambda\}]^{1/2}$$

is arbitrarily small independently of β. For $|x| \leq R$ notice that

$$\varphi^2(x)\left[E_x\left\{e^{-\beta\int_0^\lambda U(\omega(s))ds}, S < \lambda\right\}\right]^{1/2}$$

is monotoneously decreasing in β for any fixed x, continuous in x, and x is in a compact set. Hence the Theorem of Dini provides the convergence of the integral in (33). The constant factor in (33) follows in a similar way as for regular perturbations mentioned above. $\qquad$ q.e.d.

References

Davies, E.B. (1991), Heat kernel bounds, conservation of probability and the Feller property, Preprint, King's College, London.

Demuth, M. (1991), On topics in spectral and stochastic analysis for Schrödinger operators, in "Recent Developments in Quantum Mechanics".
Eds. A. Boutel de Monvel et. al., Kluwer, 223–242.

Demuth, M.; Jeske, F.; Kirsch, W. (1992), Quantitative estimates of the rate of large coupling convergence for Schrödinger operators (to be published).

Demuth, M.; van Casteren, J. (1989), On spectral theory for selfadjoint Feller generators, Rev. Math. Phys. $\underline{1}$, 325–414.

Demuth, M.; van Casteren, J. (1991), Perturbations for generalized Schrödinger operators in stochastic spectral analysis, Preprint Univ. Antwerp, No. 91–52.

Demuth, M.; van Casteren, J. (1992), On spectral theory for selfadjoint Feller generators II (to be published).

Simon, B. (1982), Schrödinger semigroups, Bull. Amer. Math. Soc. $\underline{7}$, 447–526.

van Casteren, J. (1989), A pointwise inequality for generalized Schrödinger semigroups, in "Partial Differential Operators", Teubner Texte zur Mathematik $\underline{112}$, 298–312.

Author's address:

Michael Demuth
Max–Planck–Arbeitsgruppe
FB Mathematik, Universität Potsdam
Am Neuen Palais 10
O–1571 Potsdam

Operator Theory:
Advances and Applications, Vol. 57
© 1992 Birkhäuser Verlag Basel

A GLOBAL APPROACH TO THE LOCATION
OF QUANTUM RESONANCES

Pierre Duclos

Abstract : We propose very elementary tools to locate subsets of the complex energy plane where quantum resonances can only take place. The methods uses positivity of some auxiliary operators.

1. Introduction

a. Quantum Resonances

Quantum resonances are usually defined as long living states of a system described by a Schrödinger equation :

$$-\frac{\hbar}{i}\frac{\partial}{\partial t}\psi = H\psi \quad , \quad H = -\hbar^2\Delta + V. \tag{1.1}$$

The hamiltonian H acts on $L^2(\Omega)$, where Ω in the configuration space of the system. Ω will be typically $\mathbf{R}^n$ or $\mathbf{R}_+ := (0,\infty)$. $\hbar$ is the Planck constant and V will be time independant.

There are various ways to define quantum resonances mathematically but we shall adopt the following widely accepted one. The potential V will be bounded and dilation analytic :

$\theta \to V(e^\theta x) =: V_\theta(x)$ is bounded analytic as an operator family on $L^2(\Omega)$, for all

θ in the strip $S_\alpha := \{\theta \in \mathbf{C}, \mid Im\theta \mid < \alpha\}$ for some strictly positive α. \hfill (1.2)

The following properties are well known (see Aguilar and Combes 1971 and Reed and Simon 1978) : the family of operators

$$H_\theta := -e^{-2\theta}\hbar^2\Delta + V_\theta \quad , \quad \theta \in S_\alpha \qquad (1.3)$$

is type A analytic (see Kato 1966 for this terminology) and

$$\forall\, \alpha > \beta_2 \geq \beta_1 > 0, \quad \sigma_d(H_{i\beta_1}) \subset \sigma_d(H_{i\beta_2}) \qquad (1.4)$$

where σ_d denotes the discrete spectrum. Then we have the :

definition 1. A *resonance* for H is a discrete eigenvalue of $H_{i\beta}$ for some β such that: $0 < \beta < \alpha$.

Instead of looking directly for the resonances we shall try to determine some *resonance free domains* for H, in short $RFD(H)$. Due to (1.4) we obviously have :

$$\bigcup_{0<\beta<\alpha} \rho(H_{i\beta}) \text{ is a } RFD(H), \qquad (1.5)$$

where ρ stands for resolvent set.

b. Motivations and questions

Let us start by briefly reviewing what is known on the location of resonances.

The shape resonances

If the potential has a compact well separated from an unbounded well by a barrier, the discrete eigenvalues of the compact well that one gets by making the barrier infinite, give rise to *shape resonances*. These resonances are typically very close to the real axis. We depict such a case in Figure 1 below. A candicate for such a potential can be for example : $V(x) := \frac{\frac{1}{2}+x^2}{1+x^3}$ on $\mathbf{R}_+$. The picture on the left shows typical positions of the shape resonances in the complex energy E plane.

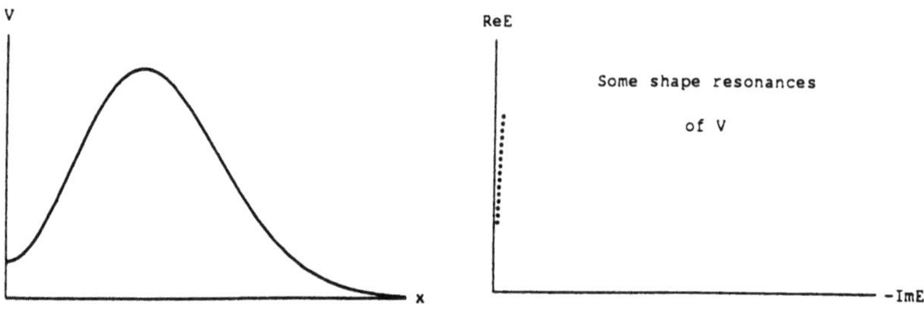

Figure 1.

Shape resonances are studied in Helffer and Sjöstrand (1986) and Combes et al. (1987).

The barrier top resonances

It has been proved by Briet et al. (1987 b) and Sjöstrand (1987) that if the potential exhibits a non degenerate maximum plus some regularity properties then the corresponding Schrödinger operator possesses resonances close to the maximum value of the potential. Actually Sjöstrand has shown that this phenomenon occurs also for non degenerate saddle points of the potential. But since we shall mostly consider here, one-dimensional situation we shall not elaborate on such a case. We depict this kind of resonances in Figure 2 below. The potential can be for example $V(x) := \frac{1}{1+x^2}$ on $\mathbb{R}_+$.

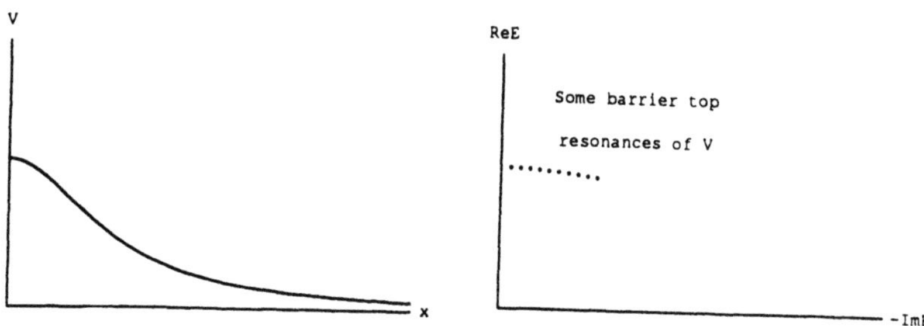

Figure 2.

Obstacle scattering resonances

To simplify, suppose you are considering the Laplace equation in $\mathbb{R}^3$ outside a spherical obstacle of radius 1, on the boundary of which you have imposed Dirichlet boundary condition ; then passing to the spherical coordinates the equation reads

$$\left(-\Delta_r + \frac{\Lambda}{(1+r)^2}\right)\psi = E\,\psi \tag{1.6}$$

where r is the distance to the obstacle, Δ_r the radial laplacian and Λ the Laplace-Beltrami operator on S^2. Since Δ_r and Λ commute, we can make the analysis of (1.6) in each spectral subspace of Λ so that we are led to the one dimensional problems :

$$\left(-\lambda^{-2}\Delta_r + \frac{1}{(1+r)^2}\right)\varphi = e\varphi \text{ with}$$

$$\varphi \in L^2\left(\mathbb{R}_+^*\right) \quad, \quad \varphi(0) = 0 \quad, \quad e = \lambda^{-2}E \quad, \quad \lambda \in \sigma(\Lambda). \tag{1.7}$$

We shall see in section 3 that the Schrödinger operator $-\hbar^2\Delta + \frac{1}{(1+x)^2}$ on $L^2\left(\mathbb{R}_+^*\right)$ possesses resonances as depicted in Figure 3 below.

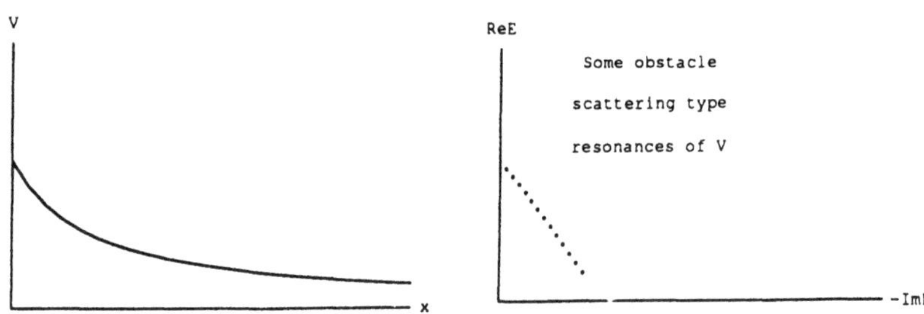

Figure 3.

Notice that the large λ limit is a semiclassical type limit.

Complex threshold and string of resonances

In Korsch (1984) one can see a numerical analysis of the one dimensional model

$$V(r) := \left(\frac{1}{2}\, r^2 - J \right) e^{-\lambda r^2} + J \quad , \quad J := 0.8 \quad , \quad \lambda := 0.1 \tag{1.8}$$

that we reproduce in Figure 4 below.

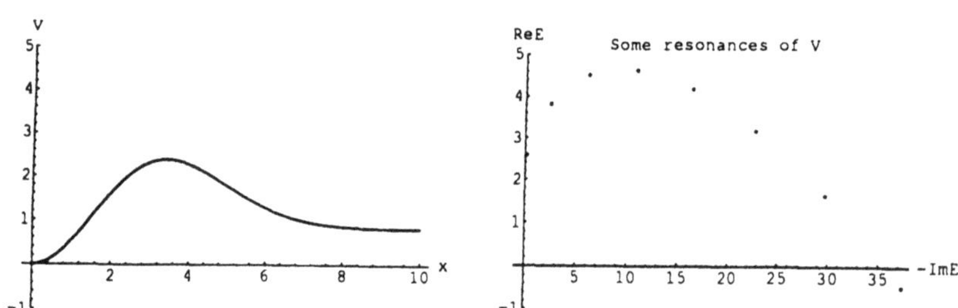

Figure 4.

There are several features of this picture that we want to discuss now. First of all we see that the resonances lies on a regular curve and accordingly one speaks of *string of resonances*. Then we see that the real part of the resonances are bounded above and one speaks of a *complex threshold*. More precisely we propose the

definition 2. The greatest real part of the resonances of H is called the compex threshold for H.

Finally we would like to speculate on the shape of the string of resonances. Model like those of Figures 1 and 4 have potentials with three different parts, the interior well responsible for the shape resonances, the middle part containing the barrier giving rise to barrier top resonances and the exterior tail which might generate obstacle scattering type resonances. Thus by matching the set of resonances of Figures 1, 2, 3 we obtain more or less a string of resonances as in Figure 4.

We are now ready to ask the following

questions 3.

Q1. When does a potential lead to a complex threshold ?

Q2. Does there exist a curve (for one dimensional model) which resonances spread on ?

Q3. Can one associate (in a sense to be defined) to the different parts of the string of resonances, regions of the configuration space ?

Q4. Does there exist a limiting curve for the string of resonances as $\hbar$ goes to zero ?

The purpose of this article is to answer these questions, at least very partially. Let us mention the semiclassical formula for the equation of the string of resonances given in the article of Korsch. Writing a complex Bohr-Sommerfeld quantization formula for resonances gives

$$\mathcal{N}(E) = n + \frac{1}{2}. \tag{1.9}$$

$\mathcal{N}(E)$ is called the *quantum number function* and has to be real :

$$Im \; \mathcal{N}(E) = 0 \tag{1.10}$$

which is the equation for the string.

c. Content of the rest of this article

In section 2, we find a sufficient condition (corollary 5) for H to possess a complex threshold and we propose methods to compute the boundary Γ of a large resonance free domain for H being a neighborhood of $e^{-2i\beta}\infty$ for $0 \le \beta < \alpha$.

In section 3, we show that a one dimensional generalised (i.e. not necessarily quadratic) barrier top of potential generates a string of resonance near the top. This string is in agreement with the boundary of the $RFD(H)$ found in the previous section, at least near the top and for some explicit models.

In section 4, we propose a method to extend the $RFD(H)$ in parts of the domain lying beyond the curve Γ found in section 2.

d. Related works

In Erdmann and Cycon (1983) a method close the one of section 2.a is initiated. Let us also mention a technique based on Birman-Schwinger type bounds in Siedentop (1983).

2. Upper bounds on resonances

Since we are dealing with bounded analytic potential, it is natural to try the regular pertubation methods to locate the resonances.

a. Perturbation methods

Let H be as defined in section 1 and let

$$\nu_\beta := \left\{ z \in \mathbb{C}, \ \|V_{i\beta}\| < dist(z, e^{-2i\beta}\mathbb{R}_+) \right\} \tag{2.1}$$

then we have the (ν_β^c denotes the complement of ν_β)

theorem 4. One has $\nu := \bigcup_{0 < \beta < \alpha} \nu_\beta$ is a $RFD(H)$ and for z in ν_β one has

$$\|(H_{i\beta} - z)^{-1}\| \leq 1/dist\left(z, \nu_\beta^c\right).$$

Proof. Of course : $z \in \rho(H_{i\beta}) \Leftarrow \|V_{i\beta}\| \, \|(H_{0i\beta} - z)^{-1}\| < 1$, where $H_{0\theta} := e^{-2\theta}H_0$. Since $\|(H_{0i\beta} - z)^{-1}\| = \|(H_0 - e^{2i\beta}z)^{-1}\| = dist(e^{2i\beta}z, \mathbb{R}_+)^{-1} = dist(z, e^{-2i\beta}\mathbb{R}_+)^{-1}$ one can easily conclude.

Since ν_β is the half plane delimited by the straight line :

$$D_\beta : cos \, 2\beta \, Imz + sin \, 2\beta \, Rez = \|V_{i\beta}\| \tag{2.2}$$

one has obviously the

Corollary 5. If V is bounded analytic on S_α, with α bigger than $\frac{\pi}{4}$, then H has a complex threshold less or equal to $\|V_{\frac{i\pi}{4}}\|$.

Since the $RFD(H)$ ν given by theorem 4, is a union of half plane we can look for the boundary Γ of ν by computing the envelop of the family of straight lines D_β. We shall consider a family of models to check our methods :

$$V(x, n) := (1 + x^n)^{-1} \text{ on } \mathbb{R}_+ \tag{2.3}$$

with dirichlet boundary condition at zero. Then an elementary calculation gives for the maximum of $\|V_{i\beta}\|$:

$$\|V_{i\beta}\| = \begin{cases} 1 & \text{if} \quad 0 \leq \beta \leq \frac{\pi}{2n} \\ (sin \, n\beta)^{-1} & \text{if} \quad \frac{\pi}{2n} \leq \beta < \frac{\pi}{n} \end{cases}. \tag{2.4}$$

Notice that V is bounded analytic on $S_{\frac{\pi}{n}}$.

From (2.4) it is straight forward to derive the parametric equations of the boundary Γ_n of the $RFD(H)$ given by theorem 4. We give in the figure 5 below, the graphs of Γ_n for $n = 1, 2, \ldots, 5$.

For the case $n = 2$, we can even easily get the cartesian equation of Γ_2 :
$x = 1 - \frac{y^2}{4}$.

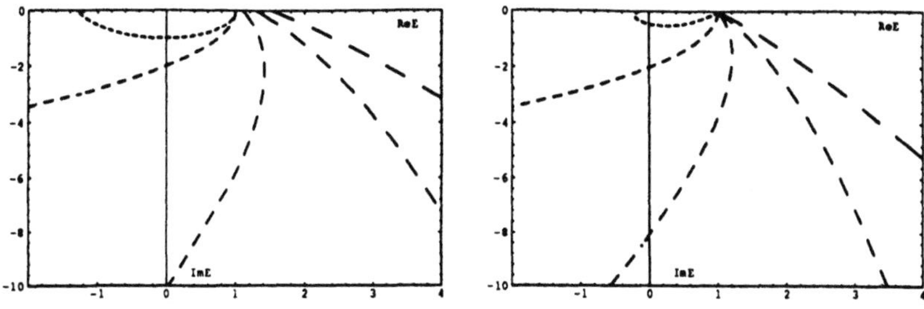

Figure 5.   Figure 6.

The bigger n is, the wider the dashes are.

But all these results can significantly be improved by the

b. Method of the imaginary part.

Obviously we have in the form sense

$$\mid H_{i\beta} - z \mid = \mid e^{2i\beta}(H_{i\beta} - z)\mid \geq Im\ e^{2i\beta}(z - H_{i\beta}) = Im\ e^{2i\beta}(z - V_{i\beta}),$$

so that

$$Im\ e^{2i\beta}z > Im\ e^{2i\beta}V_{i\beta} \Rightarrow z \in \rho(H_{i\beta}),$$

hence we have proven the

theorem 6. Let $\nu_\beta := \{z \in \mathbf{C},\quad Im\ e^{2i\beta}(z - V_{i\beta}) > 0\}$ then $\nu := \bigcup_{0 < \beta < \alpha} \nu_\beta$ is a $RFD(H)$. In addition for every z belonging to ν_β one has $\|(H_{i\beta} - z)^{-1}\| \leq 1/dist(z, \nu_\beta^c)$.

This theorem is taken from Briet et al. (1987 a).

Since $\|V_{i\beta}\| \geq Im\ e^{2i\beta}V_{i\beta}$, theorem 6 is a priori stronger than theorem 4. We can check that it is effectly stronger on our example (2.3). Again the envelop of the straight lines

$$D_\beta\ :\ Im\ e^{2i\beta}z = \sup_x Im\ e^{2i\beta}V_{i\beta} \tag{2.5}$$

can be analytically computed knowing that

$$\sup_x Im\ e^{2i\beta}V_{i\beta} = \begin{cases} sin\ 2\beta & if\quad 0 < \beta \leq \frac{\pi}{n+2} \\ \dfrac{cos\left(\frac{n-2}{2}\beta\right)}{sin\ n\beta} & if\quad \frac{\pi}{n+2} \leq \beta < \frac{\pi}{n} \end{cases} \tag{2.6}$$

We give the graphs of Γ_n, $n = 1, \ldots, 5$ in Figure 6 above.

Not only the $RFD(H)$ are bigger than the previous one (except for $n = 2$ for which they are equal) but we shall see in the next section, that they are optimal at least close to $z = 1$.

We finish this section by considering a

c. question on complex threshold.

One may ask

Q5. Does one have : V bounded $\Leftrightarrow$ H has a complex threshold ?

The answer is no, in both directions of the implication. On the one hand, the case of the square barrier is known to have resonances with arbitrarily large real part (see for example Nussenzveig 1972) and on the other hand $H := -\Delta + \frac{1}{x^2}$ has no resonances.

3. One dimensional barrier top resonances.

We extend the results of Briet et al. (1987 b) for more general barrier top resonances to show that the result of theorem 6 is optimal near the top of the potential. We assume that there exists $n \in \mathbf{N}^*$ such that :

$$V \text{ is a bounded real function defined on } \mathbf{R} \text{ and has}$$
$$\text{an absolute maximum } v_0 \text{ at } x = 0, \tag{3.1}$$

$$V_\theta(x) := V(e^\theta x), \ \theta \in \mathbf{R} \text{ admits an analytic extension in the strip}$$
$$S_{\alpha_M} \ , \ \alpha_M > 0, \text{ as a family of bounded operators}, \tag{3.2}$$

$$\exists \ \alpha_{NT} \ , \ 0 < \alpha_{NT} \leq \alpha_M \ , \ \forall \ \delta > 0 \ , \ \forall \ |x| \geq \delta$$
$$Im \ e^{2i\alpha_{NT}}(v_0 - V_{i\alpha_{NT}}) > 0, \tag{3.3}$$

$$\exists \ n \in \mathbf{N}^* \text{ and } v_n > 0 \text{ such that near } x = 0$$
$$V_\theta(x) = v_0 - e^{2n\theta}v_n|x|^n + 0(|x|^{n+1}) \text{ for any } \theta \text{ in } S_{\alpha_M}. \tag{3.4}$$

Assumptions (3.1,4) make precise the type of absolute maximum we are considering. Assumption (3.2) defines the analytic properties of V. Assumption (3.3) expresses that the potential V is non trapping at energy v_0 at least quantum mechanically.

It is straight forward to check that our model potentials (2.3) when considered on $\mathbf{R} : V(x) := (1 + |x|^n)^{-1}$, fulfil all these assumptions.

Then we have the

theorem 7. Let $H := -\hbar^2\Delta + V$ on $L^2(\mathbf{R})$ with V obeying (3.1-4). Then H has resonances which converge to the spectrum of $v_0 + e^{-i\frac{2\pi}{n+2}}\hbar^{\frac{2n}{n+2}}K_n$ as $\hbar$ goes to zero, where

$$K_n := -\Delta + v_n|x|^n. \tag{3.5}$$

Proof. It can be done along the same lines as in Briet et al. (1987 b) with some obvious modifications.

The importance of such a theorem is to show that in the semiclassical limit, H has a string of resonances which is tangent at v_0 to the straight line of equation :
$$\mathbf{R} \ni x \to v_0 + exp\left(-i\frac{2\pi}{n+2}\right)x.$$

We can now explain formula (2.6). In our example (2.3), the straight line D_β given by (2.5) sweeps the lower half plane by turning around its fixed point (1,0) as long as β is smaller than $\frac{\pi}{n+2}$. Just for this value, H has resonances in the direction given by $D_{\frac{\pi}{n+2}}$, that is with polar angle $\frac{-2\pi}{n+2}$. This is the reason why, D_β is forced to quit its fixed point and starts to generate the curve Γ_n.

Notice also that for $n = 1$, which corresponds to a cusp at the maximum, the direction of departure for the resonance is $-\frac{2\pi}{3}$. This justifies Figure 3 for obstacle scattering type resonances.

So we have obtained that the envelop Γ_n of the family of straight lines D_β given by (2.5-6) is actua.ly a natu.al boundary for the $RFD(H)$ at least in the vicinity of the top of our examples. It would be of definite value, if one can also prove that the domain (or at least part of the domain) lying on the other side of Γ_n is also free from resonances. This is the purpose of the next section.

4. Lower bounds on resonances.

The basic tool here, is a theorem already given in Briet et al. (1987 a). The proof is so simple that we reproduce it here : let $g := e^{2i\beta}(\chi^2 + i\widetilde{\chi}^2)$ where $\chi^2 + \widetilde{\chi}^2 = 1$, χ and $\widetilde{\chi}$ having bounded derivatives. Then in form sense one has

$$|g(H_{\imath\beta} - z)| \geq Re\ g(H_{\imath\beta} - z) \geq Re\ g(V_{i\beta} - z) - 2\hbar^2\chi^{'2}.$$

If one lets

$$\nu_\beta := \{z \in \mathbf{C}\ ,\ Re\ gz < Re\ gV_{i\beta} - 2\hbar^2\chi^{'2}\} \tag{4.1}$$

then one has proven the

theorem 8. For any $0 < \beta < \alpha$, ν_β is a $RFD(H)$ and for any z in ν_β one has $\|(H_{i\beta} - z)^{-1}\| \leq 1/dist(z,\nu_\beta^c)$.

We apply theorem 8 to our example (2.3). Assume for the moment that there exists x_0 in $\mathbf{R}_+$ so that

$$Re\ e^{2\imath\beta}V_{i\beta}(x_0) \leq Re\ e^{2i\beta}V_{i\beta}(x) \text{ if } x \leq x_0 + \varepsilon$$
$$Im\ e^{2i\beta}V_{i\beta}(x_0) \geq Im\ e^{2i\beta}V_{i\beta}(x) \text{ if } x \geq x_0 - \varepsilon \tag{4.2}$$

for some positive ε, then by choosing $\chi = 1$ on $(0, x_0 - \varepsilon)$ and $\widetilde{\chi} = 1$ on $(x_0 + \varepsilon, \infty)$, one obviously has

$$\nu_\beta^0(x_0) + O_\beta(\hbar^{2/3}) \text{ is a } RFD(H) \tag{4.3}$$

where

$$\nu_\beta^0(x_0) := \{z \in \mathbb{C} \ , \ Re \ e^{2i\beta}z < Re \ e^{2i\beta}V_{i\beta}(x_0), Im \ e^{2i\beta}z > Im \ e^{2i\beta}V_{i\beta}(x_0)\}. \tag{4.4}$$

There is a simple geometric interpretation of $\nu_\beta^0(x_0)$. This is the subset delimited by the straight lines passing by the point $V_{i\beta}(x_0)$ with polar angle -2β and $\frac{\pi}{2} - 2\beta$, see Figure 7 below.

Finally suppose that

$$Re \ e^{2i\beta}V_{i\beta} \text{ and } Im \ e^{2i\beta}V_{i\beta} \text{ are monotonically decaying} \tag{4.5}$$

then one has that $\displaystyle\bigcup_{x_0 \in \mathbf{R}_+} \nu_\beta^0(x_0) + O_\beta(\hbar^{2/3})$ is a $RFD(H)$. This shows that modulo an error of order $\hbar^{2/3}$ the subset delimited by the graph of $V_{i\beta}$ and the real axis is a $RFD(H)$ provided (4.5) is true. For $n = 2$ one can check easily that (4.5) is true for $\beta = \frac{\pi}{8}$. We show in Figure 8 below the domain where the resonances, for the case $n = 2$, can only be found.

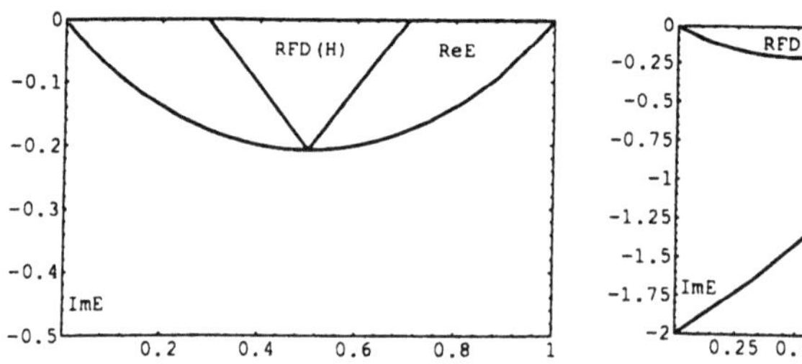
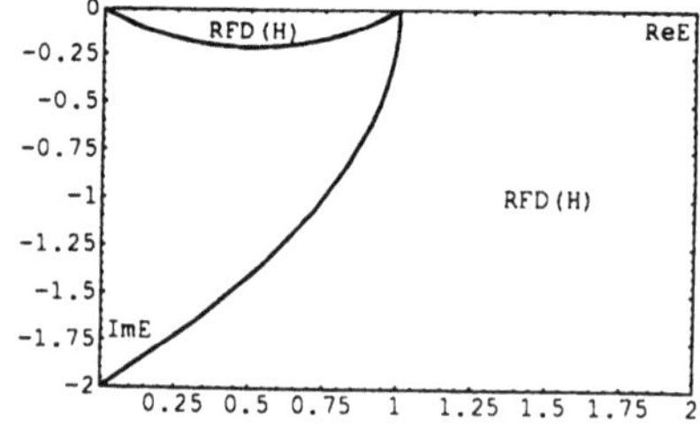

Figure 7.   Figure 8.

Acknowledgements. I would like to thank Ph. Briet, O. Cohendet, C. Essoh and F. Rigaudie with whom I had profitable discussions at various stages of this work. Let me thank also M. Demuth for giving me the opportunity to report on this work at the Lambrecht International Symposium on Operator Calculus and Spectral Theory. All pictures and some algebraic computations has been done with the help of Mathematica (1991).

References.

Aguilar J. and Combes J.M. (1971), A class of analytic pertubation for one-body Schrödinger Hamiltonians. Commun. Math. Phys. **22**, 269-279.

Briet et al. (1987 a), On the location of resonances for Schrödinger operators in the semiclassical limit I : Resonance free domains, Journ. Math. Anal. Appl. **125**, 90-99.

Briet et al. (1987 b), On the location of resonances for Schrödinger operators in the semiclassical limit II : Barrier top resonances, Commun. P.D.E. **12**, 201-227.

Combes et al. (1987), The Shape Resonance, Commun. Math. Phys. **110**, 215-236.

Erdmann C. and Cycon H.L., Absence of high energy resonances for many body Schrödinger operators, T.U. Berlin preprint **119**.

Helffer B. and Sjöstrand J. (1986), Résonances en limite semiclassique. Bulletin Soc. Math. France **114**.

Kato T. (1966), Perturbation Theory for Linear Operators. Berlin, Heidelberg, New-York, Springer.

Korsch H.J. (1984), Semiclassical theory of resonances, Proceeding of the Conference "Resonances-Models and Phenomena", Springer Lecture Notes in Physics **211**, 217-234.

Mathematica 2.0 (1991), A system for doing Mathematics by computer. Wolfram Research Inc.

Nussenzveig H.M. (1972), Causality and dispersion relations, Academic Press New-York and London.

Reed M. and Simon B. (1978), Methods of Modern Mathematical Physics IV, Analysis of operators. New-York, San Francisco, London, Academic Press.

Siedentop H.K.H. (1983), Bounds on resonance eigenvalue of Schrödinger operators-local Birman Schwinger bound, Phys. Lett. **99**, 65.

Sjöstrand (1987), Semiclassical resonances generated by non-degenerate critical points. Lecture Notes in Math. **1256**, 402-429.

Author's address :

Pierre Duclos
Centre de Physique Théorique   and   Phymat
CNRS Luminy   Université de Toulon et du Var
Case 907   BP 132
F-13288 Marseille Cedex 9   F-83957 La Garde Cedex

Operator Theory:
Advances and Applications, Vol. 57
© 1992 Birkhäuser Verlag Basel

On estimates for the eigen-values in some elliptic problems

Yuri Egorov, Vladimir Kondrat'ev

Abstract

The estimates of the eigen-values are obtained for the Dirichlet boundary problem for elliptic equations. The special case of the Sturm-Liouville problem is considered more in details. The second part contains some ge-neralizations of the results of Ch.Fefferman and R.Kerman-E.Sawyer.

1.Introduction

We consider at first the following eigen-value problem:

$$\frac{d^n y}{dx^n} + \lambda Q(x) y = 0,$$

$$y(a_i) = y'(a_i) = \ldots = y^{(r_i - 1)}(a_i) = 0;$$

$$0 = a_1 < a_2 < \ldots < a_s = 1, \quad r_1 + \ldots + r_s = n.$$

Such a problem is in general not self-adjoint. M. Krein has proved that if r_i are odd for $1 \leq i \leq s$ and $Q(x) \in C[0,1]$, $Q(x) \cdot (-1)^{r_s+1} \geq 0$, then the eigen-values of the problem (1) are real and

$$0 < \lambda_1(Q) < \lambda_2(Q) < \ldots < \lambda_k(Q) < \ldots$$

Our aim is to estimate the eigen-values, i.e. to find constants $c_k(\alpha), C_k(\alpha)$ independent of Q and such that

$$c_k(\alpha) \leq \lambda_k(Q) \leq C_k(\alpha), \quad k = 1, 2, \ldots,$$

assuming that

$$\int_0^1 |Q(x)|^\alpha dx = 1, \ \alpha \neq 0. \tag{1}$$

We state in below some rather elementary theorems for the Sturm-Liouville operator and some theorems for the operator $(-\Delta)^m + \lambda Q$. A part of our results has been published in Egorov, Kondrat'ev (1983, 1984, 1990a, 1990b). Similar problems were considered by J.Keller, A.Ramm, G.Talenti and others.

2.One-dimensional case

Consider the problem stated in above.

Theorem 1. *If $\alpha \geq 1$ then $\lambda_1 \geq C_\alpha$; if $\alpha < 1$ then λ_1 cannot be estimated from below.*
If $\alpha < 1/n$ then $\lambda_1 \leq C_\alpha$; if $\alpha \geq 1/n$, then λ_1 cannot be estimated from above.
If $\alpha \geq 1$ then the inequality

$$\lambda_k \geq C_\alpha k^{(n\alpha-1)/\alpha}, \quad k = 2, 3, \ldots$$

holds and the constant C_α is independent of Q, k; if $\alpha < 1$ then λ_k cannot be estimated from below.

Therefore, if $1/n \leq \alpha < 1$, then for any $L > 0$ there exists a function $Q(x)$ satisfying (1) such that $\lambda_1(Q) = L$.

Consider the Sturm-Liouville problem:

$$\frac{d^2y}{dx^2} + \lambda Q(x)y = 0, \quad y(0) = 0, \ y(1) = 0;$$

$$Q(x) \geq 0, \quad Q(x) \in C[0,1];$$

$$\int_0^1 Q(x)^\alpha dx = 1, \quad \alpha \neq 0.$$

Theorem 2. *If $\alpha > 1$ then*

$$\lambda_1(Q) \geq A_\alpha \equiv 4 \left| \frac{\alpha}{2\alpha - 1} \right|^{1/\alpha} \left(\int_0^1 \frac{dt}{\sqrt{|1 - t^{2\alpha/(\alpha-1)}|}} \right)^2.$$

If $\alpha = 1$ then

$$\lambda_1(Q) > 4 = \lim_{\alpha \to 1+0} A_\alpha.$$

In this case for any $\epsilon > 0$ there exists a function $Q(x)$, satisfying (1), such that $\lambda_1(Q) = 4 + \epsilon$. And if $\alpha < 1/2$ then $\lambda_1(Q) \leq A_\alpha$. Moreover,

$$\lambda_1(Q) = A_\alpha,$$

if

$$Q(x) = y^{2/(\alpha-1)},$$

where $y(x)$ is the solution of the following non-linear problem:

$$y'' + \lambda y^{(\alpha+1)/(\alpha-1)} = 0, \quad y(0) = 0, \ y(1) = 0, \tag{2}$$

$$\int_0^1 |y(x)|^{2\alpha/(\alpha-1)} dx = 1,$$

having the unique solution.

Proof. It is well-known that

$$\lambda_1(Q) = \inf_Y \frac{\int_0^1 y'^2 dx}{\int_0^1 Qy^2 dx},$$

where Y is the set of the functions $y \in C^1$ such that $y(0) = y(1) = 0$.

Using the Hölder inequality we get

$$\int_0^1 Q^\alpha dx = \int_0^1 Q^\alpha y^{2\alpha} \cdot y^{-2\alpha} dx \leq \left(\int_0^1 Qy^2 dx\right)^\alpha \cdot \left(\int_0^1 y^{2\alpha/(\alpha-1)} dx\right)^{1-\alpha}.$$

Hence

$$\lambda_1(Q) \geq \inf_Y \left(\int_0^1 y'^2 dx\right) / \left(\int_0^1 y^{2\beta} dx\right)^{1/\beta},$$

where $0 < \alpha < 1$ and $1/\alpha + 1/\beta = 1$.

It is easy to show that for $0 < \alpha < 1/2$

$$\inf_Y \frac{\int_0^1 y'^2 dx}{(\int_0^1 y^{2\beta} dx)^{1/\beta}} = A_\alpha.$$

The Euler equation for the last variational problem coincides with (2) and can be integrated. In the general case this method does not work, so the proof becomes more complicated.

We can prove that for $k = 1, 2, \ldots$

if $\alpha \geq 1$ then

$$\lambda_k \geq A_\alpha k^{(2\alpha-1)/\alpha};$$

if $0 < \alpha < 1/2$ then

$$\lambda_k \leq A_\alpha k^{1/\alpha};$$

if $\alpha < 0$ then

$$\lambda_k \leq A_\alpha k^{(2\alpha-1)/\alpha}.$$

These estimates are sharp. If $\alpha < 1$ then λ_2 (and other λ_k) cannot be estimated from below with a positive constant.

Such estimates are very useful in the qualitive theory of ordinary differential equations. For instance, from the Theorem 2 it follows that

If

$$\int_0^1 |Q(x)|^\alpha dx \leq A_\alpha, \ \alpha \geq 1,$$

then all solutions $y(x)$ of the equation

$$y'' + Q(x)y = 0 \tag{3}$$

are nonoscillating on $[0, 1]$ and the constant A_α is the best possible.

Another corollary is the following:

If $Q(x)$ is a 1-periodic function,

$$\int_0^1 Q(x)^\alpha dx \le A_\alpha, \quad Q(x) \ge 0, \ \alpha \ge 1,$$

then the solutions of (3) are stable in the Lyapunov sense.

Next we consider the problem

$$(p(x)y')' + \lambda y = 0, \quad y(0) = 0, \ y(1) = 0,$$

where

$$p(x) \ge 0, \ p(x) \in C[0,1], \ \int_0^1 p^\alpha dx = 1.$$

In this case we can prove that

If $\alpha > 1/2$ then $\lambda_1 \le C_\alpha$ and λ_1 cannot be estimated from below.
If $\alpha < -1$ then $\lambda_1 \ge C_\alpha$ and λ_1 cannot be estimated from above.
If $-1 \le \alpha \le 1/2$ then the eigen-value λ_1 may be equal to any positive real number.

We have succeeded to find the value C_α for $\alpha \ge 1$ only.

Consider the following Sturm-Liouville problem:

$$-y'' + p(x)y = \lambda y, \quad y(0) = 0, \ y(1) = 0,$$

$$p(x) \ge 0, \ p(x) \in C[0,1], \ \int_0^1 p^\alpha dx = 1.$$

Theorem 3. *If $\alpha \ge 1$ then $\pi^2 < \lambda_1 \le h_\alpha$. If $\alpha \le 1/2$ then $\lambda_1 \ge h_\alpha > \pi^2$. Here h_α is independent of p; it is the least eigen-value of the non-linear problem*

$$-y'' + y^{(\alpha+1)/(\alpha-1)} = h_\alpha y, \quad y(0) = 0, \ y(1) = 0,$$

$$\int_0^1 |y(x)|^{2\alpha/(\alpha-1)} dx = 1.$$

Remark that the spectrum of the last problem is discrete. This problem cannot be solved in the terms of elementary functions.

The same problem for the equation

$$-y'' - p(x)y = \lambda y,$$

where

$$p(x) \ge 0, \ p(x) \in C[0,1], \ \int_0^1 p^\alpha dx = 1,$$

has the following solution:

> *If $\alpha \geq 1$ then $\lambda_1 \geq H_\alpha$.*
> *If $\alpha < 1$ then λ_1 cannot be estimated from below.*
> *For all real α we have $\lambda_1 \leq H_\alpha$; if $\alpha < 1/3$ then $H_\alpha < 0$.*

Now we consider a more difficult problem:

$$(p(x)y')' + \lambda q(x)y = 0, \quad y(0) = 0, \ y(1) = 0,$$

where

$$p(x) \geq 0, \ p(x) \in C[0,1], \ \int_0^1 p^\alpha dx = 1.$$

$$q(x) \geq 0, \ q(x) \in C[0,1], \ \int_0^1 q^\beta dx = 1.$$

Theorem 4. *If $\alpha < -1$ and $\beta > 1$ then $\lambda_1 \geq B_{\alpha,\beta}$. For all other values of the parameters α, β it is impossible to estimate λ_1 from below.*

If $\alpha > 0$ and $1/\beta \leq 1/\alpha + 2$ or $-1/2 < \alpha < 0$ and $1/\beta \geq 1/\alpha + 2$, then $\lambda_1 \leq B_{\alpha,\beta}$. For all other values of the parameters α, β it is impossible to estimate λ_1 from above.

Many mathematicians (for example, A.S.Bratus', J.Keller and others) considered the problem: *to estimate λ_1 in the following boundary problem:*

$$(p(x)y'')'' - \lambda y = 0,$$

$$y(0) = 0, \ y'(0) = 0, \ y(1) = 0, \ y'(1) = 0,$$

where

$$p(x) \geq 0, \ p(x) \in C[0,1], \ \int_0^1 p^\alpha dx = 1.$$

This problem is of the great importance for the mechanics, especially for the values $\alpha = -1, -2, -3$. This problem remains open. We have not investigated it yet.

3.Many-dimensional case

Now we consider the multi-dimensional problem for the equation

$$(-\Delta)^m u = \lambda Q(x)u, \quad u(x) \in H_0^m(\Omega),$$

where Δ is the Laplacian, $Q(x) \geq 0$, $x \in R^n$, Ω is a bounded domain in R^n, $Q(x) \in C(\Omega)$. As in above we assume that

$$\int_\Omega Q(x)^\alpha dx = 1.$$

The same methods as in the section 2 allow to prove the following

Theorem 5.
1. If $n > 2m, \alpha > n/2m$ then $\lambda_1 \geq C_{\alpha,m,n}$.

2. If $n > 2m, \alpha > n/2m$ then λ_1 cannot be estimated from above.
3. If $n > 2m, \alpha \leq n/2m$ then λ_1 cannot be estimated from below.
4. If $n \leq 2m, \alpha \geq 1$ then $\lambda_1 \geq C_{\alpha,m,n}$.

If the boundary $\partial\Omega$ is smooth enough and $\alpha < 1/2m$ then $\lambda_1 \leq C_{\alpha,m,n}$. If $m = 1$ and $\alpha \geq n/2$ then $\lambda_1 \geq \Lambda$, where Λ is the least eigen-value in the following problem:

$$\Delta u + \Lambda u^{(\alpha+1)/(\alpha-1)} = 0$$

in the ball K, whose measure is equal to the measure of Ω, with the boundary conditions:

$$u = 0 \quad on \quad \partial K, \quad \int_K u^{2\alpha/(\alpha-1)} dx = 1.$$

Now let L be a symmetric elliptic differential operator of order $2m$ in the Sobolev space $H_0^s(\Omega)$, where Ω is an arbitrary domain in R^n,

$$(Lu,u) \geq a_0 \int \sum_{|\alpha|=m} |D^\alpha u|^2 dx, \quad u \in C_0^\infty(\Omega), \quad a_0 \geq 0.$$

Let $H = L - V$, where $V(x)$ is a locally summable function with real non-negative values. In the most important case of the Scrödinger operator, when $L = -\Delta$, we will write $H_0 = -\Delta - V(x)$.

In the Fefferman (1983) the following estimate for the first negative eigen-value of the operator H_0 was obtained:

$$c \cdot sup_{x,\delta}[Av_{B(x,\delta)}V - C\delta^{-2}] \leq -\lambda_1 \leq C \cdot sup_{x,\delta}[(Av_{B(x,\delta)}V^p)^{1/p} - c\delta^{-2}],$$

where $B(x,\delta)$ is the ball of radius δ with its center in the point x and

$$Av_{B(x,\delta)}f(x) = \delta^{-n} \int_{B(x,\delta)} f(x)dx,$$

$1 < p < \infty$, the constants c and C depend on n and p only.
This estimate was improved in Schechter(1989) to the form

$$sup_{x,\delta}[c \cdot Av_{B(x,\delta)}V - \delta^{-2}] \leq -\lambda_1 \leq sup_{x,\delta}[C \cdot (Av_{B(x,\delta)}V^p)^{1/p} - \delta^{-2}].$$

In Kerman-Sawyer(1986) the following estimate was obtained:

$$sup_Q(q^{-2} : F(V,Q) \geq C_1) \leq -\lambda_1 \leq sup_Q(q^{-2} : F(V,Q) \geq C_2),$$

where

$$F(V,Q) = \int_Q \int_Q V(x)V(y)|x - y|^{2-n} dx dy / \int_Q V(x)dx$$

and Q is a cube with the side length q. (All this concerns only the case $m = 1, L = -\Delta$.)
Now we prove the following rather elementary theorem.

Theorem 6. *Let $m \geq 1$, $n > 2m$ and λ_1 be the first negative eigen-value of the operator H. Then*

$$sup_Q \left(\delta^{-2m} : \int_Q V(x)dx \geq C_1 \delta^{n-2m} \right) \leq -\lambda_1 \leq$$

$$sup_Q \left(b_p \delta^{-2m} : \int_Q V(x)^p |x - x_0|^{2mp-n} dx \geq c_p \right),$$

where $p \geq n/2m$, Q is a cube with its center in the point x_0 and the side length δ.

Proof. It follows from the variational principle that

$$-\lambda_1 = sup_u \frac{\int V(x)|u(x)|^2 dx - (Lu, u)}{\int |u(x)|^2 dx}.$$

So we have $-\lambda_1 \geq C$, if C is a constant such that

$$\int V(x)|u_0(x)|^2 dx = (Lu_0, u_0) + C \int |u_0(x)|^2 dx$$

for some function $u_0 \in C_0^\infty(R^n)$.

Let Q' be the cube concentrated with Q with its side length 2δ, and $u_0(x)$ a function of $C_0^\infty(Q')$, equal to 1 in Q. Then

$$\int V(x)|u_0(x)|^2 dx \geq C_1 \delta^{n-2m}, \quad (Lu_0, u_0) \leq C_2 \delta^{n-2m}$$

and

$$\int |u_0(x)|^2 dx \geq C_3 \delta^n.$$

The constants C_2 and C_3 do not depend on δ, and we can take C_1 to be equal to $(C_2 + 1)C_3$. Therefore, we get

$$-\lambda_1 \geq (C_1 - C_2)/C_3 \delta^{-2m} = \delta^{-2m}.$$

Thus the left hand inequality in Theorem 6 has been proved.

On the other hand the variational principle gives the equality

$$-\lambda_1 = inf[\alpha > 0 : \int V(x)|u(x)|^2 dx \leq (Lu, u) + \alpha \int |u(x)|^2 dx].$$

Therefore $-\lambda_1 \leq C_4$, if C_4 is a constant such that

$$\int V(x)|u(x)|^2 dx \leq (Lu, u) + C_4 \int |u(x)|^2 dx, u \in C_0^\infty(R^n).$$

Let $q \geq n/2m$ and

$$\mu = sup_Q [\delta^{-2m} : \int_Q V(x)^p |x - x_0|^{2pm-n} dx \geq c_p],$$

where Q is a cube with side length δ. It is clear that in fact the sign $\geq$ in the definition of μ can be substituted by the sign of equality. Divide the space R^n on the cubes

Q_j with non-intersecting inner parts and the size $\mu^{-1/2m}$. From the definition of the number μ it follows that

$$\int_{Q_j} V(x)^p |x - x_j|^{2pm-n} dx \leq c_p,$$

where x_j is the center of the cube Q_j. We have

$$\int V(x)|u(x)|^2 dx = \sum_j \int_{Q_j} V(x)|u(x)|^2 dx \leq$$

$$\sum_j \left(\int_{Q_j} V(x)^p |x - x_j|^{2pm-n} dx\right)^{1/p} \left(\int_{Q_j} |u|^{2p'} |x - x_j|^{-n+p'(n-2m)} dx\right)^{1/p'} \leq$$

$$c_p^{1/p} \sum_j \left(\int_{Q_j} |u(x)|^{2p'} |x - x_j|^{-n+p'(n-2m)} dx\right)^{1/p'}.$$

Here $p' = p/(p-1)$. If Q_0 is the unit cube with its center in the origin, then there exists a constant a_p such that

$$\left(\int_{Q_0} |u(y)|^{2p'} |y|^{-n+p'(n-2m)} dy\right)^{1/p'} \leq a_p \int_{Q_0} \left(\sum_{|\alpha|=m} |D^\alpha u(y)|^2 + |u(y)|^2\right) dy.$$

After the substitution $x = x_j + \delta y$ we get the inequality

$$\left(\int_{Q_j} |u(x)|^{2p'} |x - x_j|^{-n+p'(n-2m)} dx\right)^{1/p'} \leq a_p \int_{Q_j} \left(\sum_{|\alpha|=m} |D^\alpha u(x)|^2 + \delta^{-2m} |u(x)|^2\right) dx,$$

valid for all $u \in H_m(Q_j)$. Therefore if $c_p^{1/p} = a_p^{-1} a_0$, then

$$\int V(x)|u(x)|^2 dx \leq (Lu, u) + a_p \delta^{-2m} \int |u(x)|^2 dx, u \in C_0^\infty(R^n).$$

Therefore, we have proved that $-\lambda_1 \leq a_p \delta^{-2m}$. The proof is complete.

The similar arguments imply the following generalization of our results in Egorov, Kondrat'ev (1990b). Let $V(x)$ be such a real potential in R^n that $V(x) \geq 0$ and

$$\int V(x)|x|^{2mp-n} dx < \infty, \ n > 2m, \ p \geq n/2m.$$

Let Q be such a cube that

$$\int_{R^n \setminus Q} V(x)|x|^{2mp-n} dx < a_p.$$

Using plane sections parallel to the coordinate planes and containing the center of the cube Q, we can divide Q on 2^n smaller cubes $Q_1, ..., Q_{2^n}$ and choose those of them, for which

$$\int_{Q_j} V(x)|x|^{2mp-n} dx \geq a_p.$$

Now we can divide these cubes in the same way, choose the smaller cubes with big values of integrals and stop the process only when the values of all integrals over obtained cubes are $\leq a_p$. Let K_α be the number of those obtained cubes whose side length is $\leq \alpha^{-1/2m}$.

Theorem 7. *Let $n > 2m$, $p \geq n/2m$, $\int_{R^n} V(x)dx < \infty$, N_α be the number of the eigen-values of the operator H, that are not greater than $-\alpha$. Then*

$$N_\alpha \leq C K_\alpha \quad and \quad \sum |\lambda_j| \leq C \sum q_j^{-2m},$$

where q_j is the length of the side of the cube Q_j from the chosen set of K_α cubes, for which $q_j \leq \alpha^{-1/2m}$, and both the sums are taken over such indices j, the constant C depends on n, m, p only.

To compare we state one of the results of Egorov, Kondrat'ev (1990b).

Theorem 8. *Let N_α be the number of the points in the spectrum of the operator L_α such that $\lambda < -\alpha$, $\alpha > 0$. Let $V_\alpha(x) = max(V(x) - \alpha, 0)$.*

If $n > 2m$, then for $p \geq n/2m$

$$N_\alpha \leq C_{n,m,p} \int V_\alpha(x)^p |x|^{2mp-n} dx.$$

If $n < 2m$ and n is odd, then for $p > 1$

$$N_\alpha \leq C_{n,m,p}\left(\int V_\alpha(x)^p |x|^{2mp-n} dx + 1\right).$$

If $n \leq 2m$ and n is even, then for $p > 1$

$$N_\alpha \leq C_{n,m,p}\left(\int V_\alpha(x)^p |x|^{2mp-n}(1 + |ln|x||)^{2p-1} dx + 1\right).$$

Applying the results of Fefferman (1983), Kerman, Sawyer (1986) it is possible to prove the following

Theorem 9. *Let $n > 2m$. There exist such positive constants c and C, depending on n only, that the following statements are valid:*

(A) Let $\alpha \geq 0$ and $Q_1, ..., Q_N$ be such a set of cubes with the side $\leq \alpha^{-1/2m}$, so their doubles do not intersect and $F(V, Q_j) \geq C$. Then the operator H has at least N eigen-values, not exceeding $-\alpha$.

(B) Inversely, let $\alpha \geq 0$ and H has at least CN eigen-values, which are $\leq -\alpha$. Then there are non-intersecting cubes $Q_1, ..., Q_N$ with its side lengths not exceeding $\alpha^{-1/2m}$, for which $F(V, Q_j) \geq c$, $j = 1, ..., N$.

Here

$$F(V,Q) = \int_Q \int_Q V(x)V(y)|x-y|^{2m-n}dxdy / \int_Q V(x)dx.$$

References

Egorov Yu.V., Kondrat'ev V.A. (1983) "*On an estimate of the first eigen-value for a self-adjoint elliptic operator*", Vestnik Mosk. un-ta, Mathem., Mechanics, 3, 46-52.
Egorov Yu.V., Kondrat'ev V.A. (1984) "*On estimates of the first eigen-value of the Sturm-Liouville problem*", Russian Math. Survey, 39 (2), 151-152.
Egorov Yu.V., Kondrat'ev V.A. (1990a) "*On an estimate for the first eigen- value of the Sturm-Liouville operator*", Vestnik Mosk. un-ta, Mathem., Mechanics, 6, 75-78.
Egorov Yu.V.,Kondrat'ev V.A. (1990b) "*On the negative spectrum of an elliptic operator*", Mathem. sbornik, 181(2), 147-166.
Fefferman Ch.(1983) " *The uncertainty principle*", Bulletin of the AMS, 9(2), 129-206.
Kerman R.,Sawyer E. (1986) "*The trace inequality and eigenvalue estimates for Schrödinger operator*", Annal. de l'Institut Fourier, 26(4), 207-228.
Schechter M. (1989) "*The spectrum of the Schrödinger operator*", Trans. of the AMS, 312(1), 115-128.

Authors' address:

Yu. V. Egorov, V. A. Kondrat'ev
Moscow State University
Department of Mathematics
Moscow V-234
119 899 Russia.

Operator Theory:
Advances and Applications, Vol. 57
© 1992 Birkhäuser Verlag Basel

Quantum Scattering with Long-Range Magnetic Fields

Volker Enss

Abstract. Loss and Thaller have shown that a quantum particle in a magnetic field with decay $F_{ij}(x) \sim |x|^{-1-\mu}$, $\mu > 1/2$ is an asymptotically complete short-range scattering system although the vector potential decays only like $A(x) \sim |x|^{-\mu}$. We include long-range electrostatic fields and simplify the proof of asymptotic completeness. The main tools are absorbing cells in phase space and a simpler intermediate time evolution.

1. Introduction

The distinction between short-range and long-range forces usually is marked by the need to modify the free time evolution for existence (and completeness) of wave operators. Scalar monotone potentials with decay like $|x|^{-1}$ at infinity are of long range while potentials with integrable decay w.r.t. the radius, e.g. $|x|^{(1+\varepsilon)}, \varepsilon > 0$, are of short range in any dimension. This boarderline is shifted towards slower decay if oscillations of the potentials yield a faster effective decay, see e.g. [1]. For magnetic fields the Schrödinger equation contains the unphysical vector potential which depends on the choice of gauge. Its decay rate does not have an invariant meaning. Moreover, the decay may be slow, apparently of long range, even for fast decay of the (physical) magnetic field: In a two-dimensional model a compactly supported magnetic field with nonzero flux cannot have a vector potential with isotropic decay rate faster than $|x|^{-1}$. M. Loss and B. Thaller have shown for Schrödinger-

and Diracoperators that the non-modified wave operators exist and are complete if the magnetic fields essentially decay faster than $|x|^{-3/2}$ and the vector potentials faster than $|x|^{-1/2}$ [4, 5]. In addition they showed analogies and differences between the corresponding classical and quantum motion. Recently, Nicoleau and Robert recovered their results for Schrödinger operators with smooth potentials within the framework of pseudodifferential operators using the Isozaki-Kitada modification [6]. We present here a version of the original proof which extends their results slightly. We include long-range electrostatic forces and assume less differentiability of the magnetic fields. The main point of this note, however, is a considerable simplification of the proof. We use an absorbing property of cells in phase space and a simpler intermediate time evolution.

In any space dimension ν the magnetic field $F_{ij}(x) = -F_{ji}(x)$ can be obtained from a vector potential $A_k(x)$ as $F_{ij}(x) = \partial_i A_j(x) - \partial_j A_i(x)$. Due to gauge invariance the vector potential is not unique and we are free to choose a suitable one. In our situation a transversal gauge which satisfies

$$x \cdot A(x) = 0 \tag{1.1}$$

is most convenient. For given field F such an A can be obtained as

$$A_i(x) = \sum_{k=1}^{\nu} \int_0^1 s\, ds\, F_{ki}(x\, s)\, x^k. \tag{1.2}$$

Some of our results and proofs require this gauge, others are gauge independent. The Hamiltonian of the system is

$$H = (1/2m)\,[\,p - A(x)\,]^2 + V(x). \tag{1.3}$$

For simplicity of presentation we assume that the potentials are bounded, then H is self-adjoint on $\mathcal{D}(H_0) = W^{2,2}(I\!R^\nu)$. The decay assumptions are:

$$|F_{ij}(x)| \leq \mathrm{const}\, (1 + |x|)^{-1-\mu}, \tag{1.4}$$

by (1.2) this implies for the vector potential

$$|A_k(x)| \leq \mathrm{const}\, (1 + |x|)^{-\mu}.$$

We require, in addition, integrable decay of *all* derivatives of A

$$|(\partial_j A_k)(x)| \leq \mathrm{const}\, (1 + |x|)^{-1-\epsilon}, \quad \varepsilon > 0. \tag{1.5}$$

This replaces condition (3.6) in [4]. The additional potentials should satisfy

$$V(x) = V^{\mathrm{short}}(x) + V^{\mathrm{long}}(x),$$

$$\| F(|x| > R)\, V^{\text{short}}(x) \| \in L^1([1, \infty), d\,R), \tag{1.6}$$

$$V^{\text{long}} \in C^1(\mathbb{R}^\nu), \quad |(\nabla V^{\text{long}})(x)| \le \text{const}\,(1 + |x|)^{-1-\gamma}. \tag{1.7}$$

Some results hold for any positive γ, μ, completeness for γ, $\mu > 1/2$. The treatment of V^{short} and of singular potentials is fairly standard, so we omit them in the proofs in this note altogether and concentrate on the slower decay.

The asymptotic evolution for any scattering state of this quantum system is well approximated by Dollard's modified free time evolution, the modified wave operators Ω^D are asymptotically complete. The modification uses only $V^{\text{long}}(x)$ and *not* the vector potentials. If $V(x)$ is of short range then the ordinary wave operators are asymptotically complete despite the slow decay of $A(x)$. We treat here only the limit $t \to +\infty$. Clearly, everything holds analogously for $t \to -\infty$. For a discussion of the corresponding classical system see Section 2 of [4].

2.　The Intermediate Time Evolution

The intermediate time evolution is generated by the free Hamiltonian and the tail part of the potentials. Let $\varphi \in C_0^\infty(\mathbb{R})$ satisfy

$$\varphi(z) = 1 \quad \text{for} \quad 0 \le z \le 1/2, \quad \varphi(z) = 0 \quad \text{for} \quad z \ge 1.$$

For some minimal speed $v_0 > 0$ (to be chosen in Section 3 depending on the scattering state Ψ) the tail part of the potentials is defined for $|t| \ge 1$ as

$$V_t(x) := V^{long}(x)\,[\, 1 - \varphi(|x|/v_0|t|)\,],$$

$$A_t(x) := A(x)\,[\, 1 - \varphi(|x|/v_0|t|)\,].$$

We have $V(x) = V_t(x)$ and $A(x) = A_t(x)$ outside a cone in space-time: $|x| \ge v_0|t|$. States with minimal speed above v_0 will be localized there for large times. The time dependent family of self-adjoint Hamiltonians with common domain $\mathcal{D}(H_0)$

$$H(t) = \frac{1}{2m}[\,p - A_t(x)\,]^2 + V_t(x)$$

$$= H_0 - \frac{1}{2m}(p \cdot A_t(x) + A_t(x) \cdot p) + \frac{1}{2m}(A_t(x))^2 + V_t(x)$$

generates the propagator $U(t, \tau)$

$$i\,\partial_t\,U(t,\tau) = H(t)\,U(t,\tau)\ ,\ \ U(\tau,\tau) = \mathbf{1},$$

which we will need for large $t \geq \tau \geq 1$. The existence of this intermediate time evolution operator follows from standard theorems because of the nice behaviour of the time derivative of $H(t)$. We will show in this section that one has good control of observables like position, velocity, and their correlation.

The cutoff in the potentials transfers the spatial decay into uniform decay depending on t:

$$\|V_t\| \leq c\,|t|^{-\gamma}\ ,\ \ \|\nabla V_t\| \leq c\,|t|^{-1-\gamma}, \tag{2.1}$$

$$\|A_t\| \leq c\,|t|^{-\mu},\ \ \|\partial_i(A_t)_j\| \leq c\,|t|^{-1-\epsilon}. \tag{2.2}$$

Proposition 2.1. *For any* $\beta \geq \max(1 - \gamma, 1 - \mu)$, $\Phi \in \mathcal{D}(x) \cap \mathcal{D}(p)$

$$\|\{x - t\,p/m\}\,U(t,\tau)\Phi\| \leq \mathrm{const}(\tau)\,(1 + t)^{\beta}. \tag{2.3}$$

Proof. Denote by $\mathbf{f}(t)$ the vector valued function

$$\mathbf{f}(t) := \sqrt{m}\{x - t\,(p - A_t(x))/m\}\,U(t,\tau)\,\Phi.$$

Since $\|A_t(x)\,t\| \leq \mathrm{const}\,(1+|t|)^{1-\mu}$ we have to show the bound (2.3) for $\|\mathbf{f}(t)\|$. One half of its square equals

$$\left(U(t,\tau)\,\Phi,\ \left\{\frac{m}{2}x^2 - \frac{t}{2}[x \cdot (p - A_t) + (p - A_t) \cdot x] + \frac{t^2}{2m}(p - A_t)^2\right\}U(t,\tau)\,\Phi\right).$$

$$\tag{2.4}$$

The time derivative of the first summand is

$$\frac{d}{dt}\left(U(t,\tau)\,\Phi,\ \frac{m}{2}\,x^2\,U(t,\tau)\,\Phi\right) \tag{2.5}$$

$$= \left(U(t,\tau)\,\Phi,\ i\left[H(t), \frac{m}{2}\,x^2\right]U(t,\tau)\,\Phi\right)$$

$$= \left(U(t,\tau)\,\Phi,\ \frac{1}{2}[(p - A_t) \cdot x + x \cdot (p - A_t)]\,U(t,\tau)\,\Phi\right)$$

$$= (U(t,\tau)\,\Phi,\ D_t\,U(t,\tau)\,\Phi).$$

With the shorthand

$$D_t = \left[(p - A_t(x)) \cdot x + x \cdot (p - A_t(x)) \right] / 2$$

we get for the derivative of the second term in (2.4)

$$-(d/dt)(U(t,\tau)\,\Phi,\, t\, D_t\, U(t,\tau)\,\Phi) \qquad\qquad (2.6)$$

$$= -(U(t,\tau)\,\Phi, (D_t + t(\partial_t D_t))\, U(t,\tau)\,\Phi)$$

$$-t\,(U(t,\tau)\,\Phi, [(1/m)(p - A_t)^2 - x \cdot (\nabla V_t)]\, U(t,\tau)\,\Phi).$$

Differentiating the final summand in (2.4) yields

$$(U(t,\tau)\,\Phi, [(t/m)(p - A_t)^2 + (t^2/2m)(\partial_t(p - A_t)^2)]\, U(t,\tau)\,\Phi)$$

$$-(U(t,\tau)\,\Phi, [(\nabla V_t) \cdot (p - A_t) + (p - A_t) \cdot (\nabla V_t)]\, U(t,\tau)\,\Phi).$$

Summing up we have

$$(d/dt)\,\|\mathbf{f}(t)\|^2/2$$

$$= \left((U(t,\tau)\,\Phi, \left[\frac{t}{2}\left(x - \frac{t}{m}(p - A_t) \right) \cdot ((\nabla V_t) + (\partial_t A_t)) + \leftrightarrow \right] U(t,\tau)\,\Phi \right).$$

Note that this calculation is independent of the gauge. In the transversal gauge some of the terms which cancel in general are zero individually, e.g. $2\,D_t = 2\,D = (x \cdot p + p \cdot x)$ does not depend on t explicitly. We have shown

$$(d/dt)\,\|\mathbf{f}(t)\|^2/2 = \|\mathbf{f}(t)\|\,(d/dt)\|\mathbf{f}(t)\|$$

$$= (\mathbf{f}(t) \cdot \mathbf{g}(t) + \mathbf{g}(t) \cdot \mathbf{f}(t))/2,$$

$$\mathbf{g}(t) := t\,[(\nabla V_t) + (\partial_t A_t)]\, U(t,\tau)\,\Phi.$$

With $(\partial_t A_t) = -A(x)\,\varphi'(|x|/t)\,(|x|/t^2) \sim t^{-1-\mu}$ and (1.7) we get

$$(d/dt)\|\mathbf{f}(t)\| \le \|\mathbf{g}(t)\| \le \text{ const } t^{\max(-\mu,-\gamma)}.$$

Integration yields (2.3). $\square$

3. Decomposition of States

We quote a well known result on asymptotic observables for scattering states which holds for any $\mu, \delta > 0$ independently of the chosen gauge.

Theorem 3.1. *For $\Psi \in \mathcal{H}^{cont}(H)$ there is a sequence of times $\tau_n \to \infty$ such that for any $g \in C_0^\infty(\mathbb{R})$, $f \in C_0^\infty(\mathbb{R}^\nu)$*

$$[g(H) - g(H_0)]\, e^{-iH\tau_n}\, \Psi \to 0, \tag{3.1}$$

$$[f(x/\tau_n) - f(p/m)]\, e^{-iH\tau_n}\, \Psi \to 0. \tag{3.2}$$

The proof proceeds essentially as in [2] with some obvious modifications due to the magnetic fields. For (3.2) the same calculation as in the proof of Proposition 2.1 with A and V instead of A_t and V_t in (2.5),(2.6) and $\tilde{D} = [(p - A) \cdot x + x \cdot (p - A)]/2$ yields

$$(m/2\,t^2)\left\{ \|\{x - t(p - A(x))/m\}\, e^{-iHt}\, \Psi\|^2 - \|x\,\Psi\|^2 \right\}$$

$$= -t^{-2} \int_0^t 2s\,ds\, \frac{d}{ds}\left(e^{-iHs}\, \Psi,\, \tilde{D}\, e^{-iHs}\, \Psi \right) + \left(e^{-iHt}\, \Psi,\, \frac{(p - A)^2}{2m}\, e^{-iHt}\, \Psi \right)$$

$$= -t^{-2} \int_0^t 2s\,ds\, \left(e^{-iHs}\, \Psi,\, [(p - A)^2/2m]\, e^{-iHs}\, \Psi \right)$$

$$+ \left(e^{-iHt}\, \Psi,\, [(p - A)^2/2m]\, e^{-iHt}\, \Psi \right)$$

$$+ t^{-2} \int_0^t s\,ds\, \left(e^{-iHs}\, \Psi,\, x \cdot (\nabla V)(x)\, e^{-iHs}\, \Psi \right).$$

The first equality follows by partial integration. Since $|x \cdot (\nabla V)(x)| \to 0$ as $|x| \to \infty$ it is relatively compact w.r.t. the Hamiltonian and thus the weighted time avarage vanishes as $|t| \to \infty$.

The weighted average of the bounded continuous function

$$h(s) = \left(e^{-iHs}\, \Psi,\, [(p - A)^2/2m]\, e^{-iHs}\, \Psi \right)$$

cannot stay away from $h(t)$ forever. Therefore, there is a sequence τ_n with

$$h(\tau_n) - \tau_n^{-2} \int_0^{\tau_n} 2s\,ds\, h(s) \to 0.$$

Clearly $(m/2t^2)\|x \exp(-iHt)\Psi\|^2$ vanishes for large times and we get

$$\|\{x/\tau_n - (p - A(x))/m\}\, e^{-iH\tau_n}\, \Psi\| \to 0. \tag{3.3}$$

Finally, relative compactness of $A(x)^2$ ensures decay of $\|A(x)\, e^{-iHt}\, \Psi\|^2$ in the time average and for a sequence of times which can be chosen to have the property (3.3) as well.

We consider a state Ψ from the dense subset of $\mathcal{H}^{cont}(H)$ for which there exists a $g \in C_0^\infty(I\!R)$, $\mathrm{supp}\, g \subset (0,\infty)$ with $g(H)\,\Psi = \Psi$, i.e. Ψ has bounded energy support away from the threshold value zero. The support of $g(H_0)$ characterizes a spherical shell in velocity space. Let $\{f_j\}$ be a finite smooth decomposition of the identity on this set where each f_j has support in a ball around some $v_j \neq 0$ with radius at most $v_0 := \min_j |v_j|/6$. A covering function $F_j(z) := \mathcal{F}(|z - v_j|^2/2)$ with $\mathcal{F}(\lambda) = 1$ for $\lambda \leq (2\,v_0)^2/2$, $\mathcal{F}(\lambda) = 0$ for $\lambda \geq (3\,v_0)^2/2$, $\mathcal{F} \in C_0^\infty(I\!R)$, $\mathcal{F}' \leq 0$ for $\lambda \geq 0$, is monotone nonincreasing in all directions away from v_j, has support outside a ball of radius $3\,v_0$ around the origin, and satisfies $F_j\, f_j = f_j$. With this construction

$$g(H_0) = \sum_j f_j(p/m)\, g(H_0) = \sum_j F_j(p/m)\, f_j(p/m)\, g(H_0)$$

where the sum is finite. With Theorem 3.1

$$e^{-iH\tau_n}\, \Psi = g(H)\, e^{-iH\tau_n} \approx \sum_j F_j(p/m)\, f_j(p/m)\, e^{-iH\tau_n}\, \Psi$$

$$\approx \sum_j F_j(x/\tau_n)\, f_j(p/m)\, e^{-iH\tau_n}\, \Psi$$

$$\approx \sum_j F_j(p/m)\, F_j(x/\tau_n)\, f_j(p/m)\, e^{-iH\tau_n}\, \Psi.$$

Next we show that each of the finitely many components is asymptotically stable and that uniformly for late times the correct time evolution is approximated well by the simpler intermediate evolution $U(t,\tau)$.

Proposition 3.2. *Let the potentials satisfy (1.4)-(1.7) with $\gamma, \mu > 0$. (a)*

$$\Psi_j := \lim_{t\to\infty} e^{iHt}\, F_j(x/t)\, f_j(p/m)\, e^{-iHt}\, \Psi \tag{3.4}$$

exists and (b) for $\tau \to \infty$

$$\sup_{t\geq\tau} \|(e^{-iH(t-\tau)} - U(t,\tau))\, F_j(x/\tau)\, f_j(p/m)\, e^{-iH\tau}\, \Psi\| \to 0. \tag{3.5}$$

Proof. (a) The proof relies on the absorbing property of F_j f_j, i.e. on the positivity of the total time derivative up to integrable corrections. It is essentially the same as in Section 2 of [3]. The additional terms due to the magnetic field are all integrable:

$$-(1/m)\, A(x)\, i[p,\, F_j(x/t)]\, f_j(p/m) \sim t^{-1-\mu}$$

because the gradient w.r.t. x of $F_j(x/t)$ decays in norm like t^{-1} and $A(x)$ is bounded by $c\,t^{-\mu}$ on the support of (the gradient of) $F_j(x/t)$. Similarly

$$(1/2m)\, F_j(x/t)\{(p - A)\cdot i[f_j(p/m),\, A] + \leftrightarrow\} \sim t^{-1-\epsilon}$$

because $\|\,[f_j(p/m),\, A(x)]\,\| \le \mathrm{const}(f_j)\, \max_{k,\ell}\|\partial_k\, A_\ell(x)\|$ decays integrably by (1.5) on the support of $F_j(x/t)$.

Part (b) follows as in [3] because the generators of the evolutions essentially coincide on these components of the states:

$$0 = [H - H(t)]\, F_j(x/t)\, f_j(p/m) \approx F_j(x/t)\, f_j(p/m)\, [H - H(t)].$$

The error in the last approximation decays faster than any inverse power of t.
$$\square$$

We have shown that uniformely in $t \ge \tau$

$$e^{-iHt}\, \Psi \approx \sum_j e^{-iHt}\, \Psi_j \approx \sum_j U(t,\tau)\, e^{-iH\tau}\, \Psi_j$$

$$\approx \sum_j F_j(p/m)\, U(t,\tau)\, e^{-iH\tau}\, \Psi_j. \tag{3.6}$$

The errors are arbitrarily small for suitably chosen large τ and the sum is finite.

4.   Asymptotic Completeness

From now on we assume a transversal gauge (1.1) and $\gamma,\, \mu > 1/2$. We shall show that each summand of (3.6) lies in the range of the Dollard wave operators. The Dollard modified free time evolution $U^D(t,T)$, generated by

$$H^D(t) := H_0 + V^{\mathrm{long}}(t\,p/m)), \tag{4.1}$$

$$U^D(t,T) = \exp -i\left\{ \int_T^t ds(H_0 + V^{\mathrm{long}}(s\,p/m)) \right\} \tag{4.2}$$

is an explicit multiplication operator in momentum space. V^{long} can be replaced by V_t in (4.1), (4.2) on the range of any $F_j(p/m)$. If

$$\lim_{t\to\infty} U^D(t,0)^* e^{-iHt} \Psi_j \approx \lim_{t\to\infty} U^D(t,0)^* F_j(p/m) e^{-iHt} \Psi_j$$

$$\approx \lim_{t\to\infty} F_j(p/m) U^D(t,0)^* U(t,\tau) e^{-iH\tau} \Psi_j \tag{4.3}$$

exists then Ψ_j lies in the range of

$$\Omega^D := s - \lim_{t\to\infty} e^{iHt} U^D(t,0). \tag{4.4}$$

Note that

$$e^{-iH\tau} \Psi_j \approx F_j(x/\tau) f_j(p/m) e^{-iH\tau} \Psi_j =: \Phi_j$$

belongs to $\mathcal{D}(x) \cap \mathcal{D}(p)$ and Proposition 2.1 can be applied. Convergence (4.3) follows from integrability of the time derivative:

$$\|(d/dt) F_j(p/m) U^D(t,0)^* U(t,\tau) \Phi_j\|$$

$$\leq \|(1/2m) \left\{A_t \cdot p + p \cdot A_t - A_t^2\right\} U(t,\tau) \Phi_j\|$$

$$+ \| \left\{V_t(x) - V_t(t\,p/m)\right\} U(t,\tau) \Phi_j\|. \tag{4.5}$$

$\|A_t^2\| \sim t^{-2\mu}$ is integrable for $\mu > 1/2$ and the integrability of the second norm in (4.5) is a standard estimate of long-range scattering theory with Proposition 2.1 for $\gamma, \mu > 1/2$. Moreover, $i p \cdot A_t - i A_t \cdot p = \text{div} A_t$ is integrable by assumption (1.5). It remains to estimate

$$\|A_t \cdot p\ U(t,\tau) \Phi_j\| = \|A_t \cdot (p - m\,x/t) U(t,\tau) \Phi_j\|$$

$$\leq \|A_t\|\ \|(p - m\,x/t) U(t,\tau) \Phi_j\|. \tag{4.6}$$

The first equality holds due to the transversality $A(x) \cdot x = 0$ of the chosen gauge. With (2.2) and (2.3) we have integrability of (4.6) for $\mu, \gamma > 1/2$. Thus we have shown that any vector from a dense set in $\mathcal{H}^{cont}(H)$ belongs to the range of Ω^D. Existence of Ω^D on all of $\mathcal{H}$ is easy to verify directly or with the estimates above. The range of Ω^D is well known to belong to the absolutely continuous spectral subspace of H. Summing up we have

Theorem 4.1. *Let the potentials in $H = (1/2m)\,[p - A(x)]^2 + V(x)$ satisfy (1.1), (1.4)-(1.7) with $\mu, \gamma > 1/2$. Then the Dollard wave operators exist and are asymptotically complete, i.e.*

$$\mathcal{H}^{cont}(H) = \mathcal{H}^{ac}(H) = \text{Ran}\ \Omega^D, \quad \mathcal{H}^{sc}(H) = \emptyset.$$

The modification of the free time evolution is needed only for the electrostatic potential $V(x)$. If the latter is of short range then the ordinary wave operators are complete as in [4] and [6]. The corresponding results for Dirac operators [5] can be obtained and generalized similarly along these lines.

References

[1] M. COMBESCURE: Spectral and scattering theory for a class of strongly oszillating potentials, Commun. Math. Phys. **73**, 43 - 62 (1980).

[2] V. ENSS: Asymptotic observables on scattering states, Commun. Math. Phys. **89**, 245 - 268 (1983); Section 7 of: Introduction to asymptotic observables for multiparticle quantum scattering, in: *Schrödinger Operators, Aarhus 1985*, E. Balslev ed., Springer LN Math. 1218, Berlin 1986, pp. 61 - 92.

[3] V. ENSS: Long-range scattering of two- and three-body quantum systems, in: *Journées "Equations aux dérivées partielles" Saint Jean de Monts 1989*, Publications Ecole Polytechnique, Palaiseau 1989, pp. 1 - 31.

[4] M. LOSS and B. THALLER: Scattering of particles by long-range magnetic fields, Ann. Phys. **176**, 159 - 180 (1987).

[5] M. LOSS and B. THALLER: Short-range scattering in long-range magnetic fields: the relativistic case, J. Diff. Equ. **73**, 225 - 236 (1988).

[6] F. NICOLEAU and D. ROBERT: Théorie de la diffusion quantique pour des perturbations à longue et courte portée du champ magnétique, Ann. Fac. Sci. de Toulouse **12**, n° 2, 1 - 10 (1991).

Volker Enss
Institut für Reine und Angewandte Mathematik
RWTH Aachen
Templergraben 55
D-5100 Aachen, Germany

iw010en@dacth11.Bitnet

Operator Theory:
Advances and Applications, Vol. 57
© 1992 Birkhäuser Verlag Basel

Spectral Invariance and Submultiplicativity for Fréchet Algebras with Applications to Pseudo-Differential Operators and Ψ^*-Quantization

Bernhard Gramsch, Johannes Ueberberg and Klaus Wagner

1. Introduction

Connected to the theory of pseudo-differential operators it turned out that for a perturbation theory [14] and for analytic operator functions [12] [17] [18] in these Fréchet algebras Ψ it is useful to consider the following classical properties

I) The group Ψ^{-1} of invertible elements is open.

II) Ψ is a countable projective limit of Banach algebras.

In this paper we show that the Hörmander classes $\Psi^0_{\rho,\delta}$, $0 \leq \delta \leq \rho \leq 1$, $\delta < 1$, of Fréchet operator algebras have the property II. The property I has been proved for various classes of pseudo-differential operators and for C^∞-algebras attached to C*-dynamical systems (e.g.[4], [6], [8], [10], [14]). The property I is not true already for the natural Fréchet algebra of all operators of order 0 on $C^\infty(S^1)$(cf. [14], 6.2).

The property I leads to a perturbation theory for algebras of pseudo-differential operators and to a rational (analytic) homogeneous structure for the set of semi-Fredholm operators [14] and for similarity orbits of Jordan operators [22], [22'] also on a Hilbert space. Moreover the property I implies a theory of analytic and meromorphic Fredholm functions [17] connected to the division problem for operator valued distributions. Furthermore, a slightly sharper formulation of I, especially the notion of a Ψ^*-algebra (cf.[14]) has with density arguments interesting implications in the K-theory of Fréchet algebras of operators, namely some isomorphisms with respect to the enveloping Banach algebra (cf. [8] [41]). The holomorphic functional calculus of Waelbroeck (cf. [13]) applies to Ψ^*-algebras.

The property II is essential for the development of the homotopy results of Davie [12] (cf. [3], [16], [18]) connected to the Arens-Royden theorem for noncommutative Fréchet algebras and for the Oka principle with Fréchet-Lie groups and with the attached analytic homogeneous spaces ([14], [17]).

In this paper we develop some general methods proving the properties I and II. The results of Cordes [9], [10], [11] imply that his Fréchet algebras of pseudo-differential operators have the properties I and II. For compactly supported $S_{1,0}^0$-operators a characterization by commutators with vector fields implies I.

It has been pointed out by Schulze [26] that the classical Boutet de Monvel algebra of boundary value problems is a Fréchet algebra with property I (cf. Grubb [19], p. 357); property II is very likely to hold in this case, too. The work of Ali Mehmeti [2], Lorentz [22], Schrohe [24], [25], [25'], Ueberberg [27], Gramsch [14] and Gramsch, Kalb [16] shows the significance and applicability of property I. Let us remark that it seems to be an open problem(cf.[23]) whether for Ψ^*-algebras (Def. 5.1 of [14]) the property II is always fulfilled; Ψ^*-algebras Ψ are symmetric unital Fréchet subalgebras of C^*-algebras B for which

$$\Psi \cap B^{-1} = \Psi^{-1} \quad \text{(spectral invariance, cf. also [16])}$$

is fulfilled, where B^{-1} resp. Ψ^{-1} denotes the group of invertible elements of B resp. Ψ. For commutative Fréchet algebras the property I implies essentially the property II (cf. [28], p. 123).

Let us try to explain that the theory of Ψ^*-algebras is very flexible tool also for the microlocal analysis on extremely singular spaces (Weyl-Lemma, subelliptic regularity, wave front). The operator calculus should indicate in some parts of the underlying space the locally given C^∞- or C^r-structure of the cotangent bundle in the neighborhood of good points. If one has constructed Ψ^*-algebras Ψ_k , k = 1,2,..., reflecting C^∞-structures on open sets W_k of a topological space Ω, then the countable intersection $\Psi = \bigcap_k \Psi_k$ is again a Ψ^*-algebra carrying the common C^∞-information.

For the local construction of Ψ^*-algebras one may use (§2, 2.10; §3, 3.12) compactly supported vector fields, abstract localized derivations and locally supported parameterized automorphisms ([14], 5.9 - 5.17), group representations and flows (cf. [38']). In the pseudodifferential analysis noncommutative Ψ^*-algebras with suitable properties are assigned to $\mathbb{R}^n$, to certain Riemannian manifolds with or without boundary, etc. ([9], [10], [11], [38], ch 2,[19], [26],[32],...). Since Ψ^*-algebras can be constructed also from local properties one may attach Ψ^*-algebras to manifolds with singularities [35], [42], to algebras with discontinuous symbols, to ramified spaces [2], and to more complicated spaces carrying generalized Laplacians (Cordes [11]) or a diffusion semigroup (cf. [2]). This is indicated in 2.12 - 2.14 and in 3.12.

A Ψ^*-quantization of a topological space Ω is an assignment (cf. 3.12) of a Hilbert space $H(\Omega)$ and a noncommutative Ψ^*-subalgebra Ψ of the C*-algebra $\mathcal{L}(H(\Omega))$ of bounded transformations of $H(\Omega)$ such that the local structure of Ω is reflected by Ψ.

$$\begin{array}{ccc} \Omega & \longmapsto & \Psi \\ \big\downarrow & & \cap \\ H(\Omega) & & \mathcal{L}(H(\Omega)) \end{array}$$

In certain situations when $\Psi/(\Psi \cap K)$ is commutative, where K denotes the ideal of compact operators of $H(\Omega)$, the Gelfand space of $\Psi/(\Psi \cap K)$ is a meaningful generalization of the cosphere bundle for a compact Ω (cf. [2]). An essential basis of our remarks is the localized Beals - Cordes - approach ([4], [9], [10], [11]) for algebras of pseudodifferential operators.

Some results of this paper are connected to [16] and to lectures on operator algebras of the first named author in 1985/86 (cf.[27],[29]). Let us remark that also the work of Lorentz on Jordan operators and of Schrohe on Boutet de Monvel's algebra (cf. [22], [22'], [24], [25], [25'], [32] and the references in these papers) show the applicability of the notion of a Ψ^*-algebra in connection with methods of topological algebras and differential geometry.

We thank E. Schrohe and K.G. Kalb for valuable remarks and discussions. The first named author is grateful for the hospitality of the Karl-Weierstrass-Institute at Berlin (Sept. - Nov. 1991) and of the Dept. of Math. of UC Berkeley (Febr. - April 1992). He thanks H.O. Cordes, M.A. Rieffel and B.-W. Schulze for interesting discussions.

2. Derivations, spectral invariance and closed operators

2.1 Definition. Let $(E, \| \ \|)$ be a Banach space and $\mathcal{X}$ a finite set of closed operators $Z: D(Z) \longrightarrow E$, where the domains $D(Z)$ are not necessarily dense in E. For $Z_1,...,Z_n \in \mathcal{X}$, the domain $D(Z_n...Z_1)$ of the product $Z_n...Z_1$ is

$$D(Z_n...Z_1) = \{x \in D(Z_1) : Z_j...Z_1 \, x \in D(Z_{j+1}), \ 1 \le j \le n-1\}.$$

a) $\quad \mathcal{D}_n(\mathcal{X}) := \{x \in E : x \in D(Z_j...Z_1), Z_\nu \in \mathcal{X}, 1 \le \nu \le j, 1 \le j \le n\}, \ n = 1,2,...;$

$\quad \mathcal{D}_0(\mathcal{X}) = E.$

b) $\qquad p_0(x) := \|x\|, \ x \in E,..., \ p_{n+1}(x) := p_n(x) + \sum_{Z \in \mathcal{X}} p_n(Zx).$

2.2 Remark. $(\mathcal{D}_n, p_n)$, $n = 1,2,...$, is a Banach space, $\mathcal{D}_{n+1} \subseteq \mathcal{D}_n$.

<u>Proof.</u> Use the closedness of $Z \in \mathcal{X}$ and induction with respect to n.

2.3 Definition. a) A subalgebra $\mathcal{A}$ of the algebra $\mathcal{B}$ with unit is called a full subalgebra of $\mathcal{B}$ if $\mathcal{A}$ has a unit e and if for the groups $\mathcal{A}^{-1}$ resp. $\mathcal{B}^{-1}$ of $\mathcal{A}$ respectively $\mathcal{B}$ the following relation holds (cf. Bourbaki 1967, Naimark 1973):

$$\mathcal{A} \cap \mathcal{B}^{-1} = \mathcal{A}^{-1}$$

(This is called "spectral invariance" for Fréchet algebras $\mathcal{A} \subseteq \mathcal{B}$, see [16]).

b) Let $\mathcal{A}$ be a subalgebra of an algebra $\mathcal{B}$ with unit e and $e \in \mathcal{A}$. A linear map $\delta \colon \mathcal{A} \longrightarrow \mathcal{B}$ is called a derivation if for $a,b \in \mathcal{A}$

$$\delta(ab) = \delta(a)b + a\delta(b).$$

If $\delta \colon D(\delta) = \mathcal{A} \longrightarrow \mathcal{B}$ is a closed operator from the domain $D(\delta)$ of the Banach algebra $\mathcal{B}$ into $\mathcal{B}$ then δ is called a closed derivation.

2.3' Remark. For $a, a^{-1} \in \mathcal{A}$ we have

$\qquad$ 1) $\delta(a^{-1}) = -a^{-1}\delta(a)\, a^{-1}, \quad \delta(e) = 0.$

$\qquad$ 2) $\delta(a^\nu) = \sum_{j=1}^{\nu} a^{j-1}\, \delta(a)\, a^{\nu-j}, \quad \nu = 1,2,... \ .$

$\qquad$ 3) $\delta(a_1 a_2 ... a_\nu) = \sum_{j=1}^{\nu} a_1 ... a_{j-1}\, \delta(a_j)\, a_{j+1} ... a_\nu.$

2.4 Proposition. Let $\delta \colon \mathcal{A} \longrightarrow \mathcal{B}$ be a closed derivation as in 2.3 for the Banach algebra $\mathcal{B}$ with e. Then the following properties are fulfilled.

$\qquad$ 1) $q(a) := \|a\| + \|\delta(a)\|$ is a norm on $\mathcal{A}$ with $q(ab) \leq q(a)q(b).$

$\qquad$ 2) $(\mathcal{A}, q)$ is a Banach algebra with the unit e.

$\qquad$ 3) For the closure $\overline{\mathcal{A}}$ of $\mathcal{A}$ in $(\mathcal{B}, \| \ \|)$ we have

$$\mathcal{A} \cap (\overline{\mathcal{A}})^{-1} = \mathcal{A}^{-1}.$$

<u>Proof.</u> 2) follows with 1) from 2.2. Ad 3): Using Lemma 5.3 of [14] it is enough to show that for $\|a\| < \varepsilon$

$$(e-a)^{-1} = e + a + a^2 + ...$$

is convergent with respect to the norm q. Using 2.3'.2) we get

$$\|\delta(a^\nu)\| \le \|a\|^{\nu-1} \, \nu \cdot \|\delta(a)\|.$$

This shows the convergence of the geometric series in $(\mathcal{A},q)$. For sake of completeness let us give the argument leading to 3): $\mathcal{A}$ is dense in $(\overline{\mathcal{A}},\|\ \|)$; if $a \in \mathcal{A}$ and $b = a^{-1} \in \overline{\mathcal{A}}$, then there exists a sequence $b_k \in \mathcal{A}$ with $b_k a = e - x_k$ with $x_k \in \mathcal{A}$ and $\|x_k\| < 1$; by the first part of the proof of 3), $(e-x_k)^{-1}$ exists in $\mathcal{A}$; by the uniqueness we have $a^{-1} \in \mathcal{A}$.

2.5 Theorem. Let $\mathcal{B}$ be a Banach algebra with a unit e, and let Δ be a finite set of closed derivations $\delta\colon D(\delta) \longrightarrow \mathcal{B}$. For $n = 0,1,2,\dots$ define $\mathcal{A}_0 := \mathcal{B}$,

$$\mathcal{A}_n := \mathcal{A}_n(\Delta) := \{a \in \mathcal{B}\colon a \in D(\delta_1\dots\delta_k), \ k \le n, \ \delta_i \in \Delta\},$$

and

$$q_0(a) := \|a\|_{\mathcal{B}},$$

and for $a \in \mathcal{A}_{n+1}$ define

$$q_{n+1}(a) := q_n(a) + \sum_{\delta\in\Delta} q_n(\delta(a)).$$

Then we have

1) $q_n(ab) \le q_n(a)\, q_n(b)$, $q_n(e) = 1$, $a,b \in \mathcal{A}_n$, $n = 0,1,\dots$ (submultiplicativity).

2) $(\mathcal{A}_n, q_n)$ is a Banach algebra with unit.

2') $\mathcal{A}_{n+1} \subseteq \mathcal{A}_n$, $(\delta_1\dots\delta_k)(\mathcal{A}_n) \subseteq \mathcal{A}_{n-k}$, $k \le n$.

3) For the closure $\overline{\mathcal{A}}_n$ of $\mathcal{A}_n$ in $\mathcal{B}$ the relation

$$\mathcal{A}_n \cap (\overline{\mathcal{A}}_n)^{-1} = \mathcal{A}_n^{-1}$$

is fulfilled for the groups $(\overline{\mathcal{A}}_n)^{-1}$ respectively $\mathcal{A}_n^{-1}$.

4) $\mathcal{A}_\infty := \displaystyle\bigcap_{n=0}^{\infty} \mathcal{A}_n$ is a Fréchet algebra with the system $(q_n)_{n\in\mathbb{N}}$ of norms.

5) For $a \in \mathcal{A}_\infty$ and $\|a\|_{\mathcal{B}} < 1$ the inverse $(e-a)^{-1}$ is contained in $\mathcal{A}_\infty$. $\mathcal{A}_\infty$ is a full subalgebra of the closure $\overline{\mathcal{A}}_\infty$ of $\mathcal{A}_\infty$ in $\mathcal{B}$; especially the group of invertible elements of $\mathcal{A}_\infty$ is open in $\mathcal{A}_\infty$ with respect to the topology induced by the norm of $\mathcal{B}$.

<u>Proof.</u> 1) $\mathcal{A}_n$ is an algebra for $n = 0$; assume that $\mathcal{A}_\nu$ is an algebra $\nu \le n$ and let $a,b \in \mathcal{A}_{n+1}$.

We mention a form of the Leibniz formula (cf. [6]):
For $\delta_1,\dots,\delta_k \in \Delta$, $a,b \in \mathcal{A}_k$, we have

$$\delta_1\dots\delta_k(ab) = \sum_{\substack{P\subseteq\{1,\dots,k\} \\ \overline{P}=\{1,\dots,k\}\backslash P}} \delta_P(a)\delta_{\overline{P}}(b)$$

where $P = \{j_1,\dots,j_r\}$, $1 \le j_1 < j_2 <\dots< j_r \le k$, $\overline{P} = \{1,\dots,k\}\backslash P =$ $\left\{\{k_1,\dots,k_s\} : k_1 < k_2 <\dots\le k_s\right\}$, $1 \le k_1 < k_2 <\dots< k_s$; $\delta_\emptyset(x) = x$, $\delta_P(a):= \delta_{j_1}\dots\delta_{j_r}(a)$ and $\delta_{\overline{P}}(b) =:\delta_{k_1}\dots\delta_{k_s}(b)$; this is immediate by induction using $\delta_P(a) \in D(\delta)$ for $\delta \in \Delta$ if $a \in A_k$ and the fact that P contains at most k elements.

Using the above Leibniz formula it follows that A_k is an algebra.

Let us prove the submultiplicativity of q_n by induction on n: Assume

$$q_n(ab) \le q_n(a)\, q_n(b);$$

$$q_{n+1}(ab) = q_n(ab) + \sum_{\delta\in\Delta} q_n(\delta(ab))$$

$$\le q_n(a)\, q_n(b) + \sum_{\delta\in\Delta} q_n(\delta(a)b + a\delta(b))$$

$$\le q_n(a)\, q_n(b) + \sum_{\delta\in\Delta} q_n(\delta(a))\, q_n(b) + q_n(a)q_n(\delta(b))$$

$$\le \left\{q_n(a) + \sum_{\delta\in\Delta} q_n(\delta(a))\right\} \left\{(q_n(b) + \sum_{\delta\in\Delta} q_n(\delta(b))\right\}$$

$$\le q_{n+1}(a)\, q_{n+1}(b).$$

$q_0(e) = 1$, assume $q_n(e) = 1$; then $q_{n+1}(e) = q_n(e) + \sum_{\delta\in\Delta} q_n(\delta(e)) = 1$, since $\delta(e) = 0$.

2) Completeness of (A_n, q_n) by induction: (A_0, q_0) is complete; let us assume that (A_n, q_n) is complete and that $\{a_k \in A_{n+1}, \ k \in \mathbb{N}\}$ is a Cauchy sequence in (A_{n+1}, q_{n+1}). Using the closedness of the derivations $\delta \in \Delta$ and 2.2 it follows that (A_{n+1}, q_{n+1}) is complete.

3) For $v, r \in \mathbb{N}$, $r \le v$, we have

(*) $\|(\delta_1\dots\delta_r)(a^v)\|_B \le v^r \|a\|_B^{v-r} q_r(a)^r$

This follows by induction (for large v) with respect to $r = 0,1,\dots,v$, from 2.3'.2), $\delta_j \in \Delta$, $a \in A_r$, and the product rule. With the estimate (*) we are able to show that for $\|a\|_B < 1$ and $a \in A_r$ the series $\sum_{v=0}^{\infty} a^v$ is absolutely convergent in (A_r, q_r). This implies that $s_k=$

$\sum\limits_{j=0}^{k} a^j$ is convergent to $(e-a)^{-1}$ which implies $(e-a)^{-1} \in \mathcal{A}_r$. Then we use Lemma 5.3

of [14] and conclude as in 2.4.3).

2.6 Definition. Let E be a Banach space and $X: D(X) \longrightarrow E$, $D(X) \subseteq E$, a densely defined closed linear operator. Let
$$\mathcal{T}(X) := \{a \in \mathcal{L}(E) : a: D(X) \longrightarrow D(X)\}.$$
For $a \in \mathcal{T}(X)$ we set $(\delta_X a)\varphi := Xa\varphi - aX\varphi,\ \varphi \in D(X)$.

1) Then $\delta := \delta_X : \mathcal{T}(X) \longrightarrow L(D(X),E)$, where $L(D(X),E)$ denotes the space of all linear operators from $D(X)$ to E. We will also use the following notation: $\delta_X a = (adX)(a) = [X,a]$.

2) Let $\mathcal{A}(X)$ denote the set of all $a \in \mathcal{T}(X)$ such that
$$(adX)(a) : D(X) \longrightarrow E$$
has a (unique, $D(X)$ is dense in E) extension to an element of $\mathcal{L}(E)$.

3) For $a \in \mathcal{A}(X)$ define $q(a) := \|a\|_{\mathcal{L}(E)} + \|\delta(a)\|_{\mathcal{L}(E)}$, where $\delta(a)$ denotes the unique extension of $(adX)(a)$.

2.7 Remark. 1) $\mathcal{T}(X)$ is a subalgebra of $\mathcal{L}(E)$.

2) $\delta_X(a): (D(X), \text{graph norm}) \longrightarrow (E,\|\ \|)$ is continuous for $a \in \mathcal{T}(X)$.

3) For $a,b \in \mathcal{T}(X)$ and $\delta := \delta_X$ we have
$$\delta(ab) = \delta(a)b + a\delta(b)\quad : D(X) \longrightarrow E.$$
4) $\mathcal{A}(X)$ is a subalgebra of $\mathcal{L}(E)$ with unit $e = \mathrm{Id}_E$.

<u>Proof.</u> 2) follows from the closed graph theorem. Ad 3):
$$[X,ab] = Xab - abX = Xab - aXb + aXb - abX$$
$$= [X,a]b + a[b,X] = \delta(a)b + a\delta(b)$$
4) Obviously, $\delta(ab)$ extends uniquely to an element of $\mathcal{L}(E)$ if a and b are contained in $\mathcal{A}(X)$. From $\delta(e) = \delta(e^2) = 2\delta(e)$ follows $\delta(e) = 0$ and $e \in \mathcal{A}(X)$.

2.8 Proposition. Let $X: D(X) \longrightarrow E$ be a densely defined closed linear operator on the Banach space E. Then the linear operator (cf. 2.6, 2.7)
$$\delta_X: \mathcal{A}(X) \longrightarrow \mathcal{L}(E), \qquad \delta_X(a) := (adX)(a),$$
is a closed derivation.

<u>Proof.</u> Let $a_k \in A(X)$ be a sequence converging to a in $\mathcal{L}(E)$ such that $\delta_X(a_k)$ is converging to b in $\mathcal{L}(E)$. For $\varphi \in D(X)$ we have $\delta_X(a_k)\varphi = (Xa_k - a_k X)\varphi = Xa_k\varphi - a_k X\varphi$.

We get

$$X(a_k\varphi) = (Xa_k - a_k X)\varphi + a_k X\varphi$$
$$= \delta_X(a_k)\varphi \quad + a_k X\varphi.$$

Therefore $Xa_k\varphi$ is convergent to $b\varphi + aX\varphi$ in E for $\varphi \in D(X)$. Since X is closed we obtain $a\varphi \in D(X)$; therefore $\delta(a_k)\varphi$ is convergent to $\delta(a)\varphi$ and we get $\delta(a) \in \mathcal{L}(E)$.

2.8' Corollary. 1) $A(X)$ equipped with the norm $q(a):= \|a\|_{\mathcal{L}(E)} + \|\delta(a)\|_{\mathcal{L}(E)}$ is a Banach algebra with unit $e = \mathrm{Id}_E$.

1) $q(ab) \leq q(a)q(b)$ for $a,b \in A(x)$, $q(e) = 1$.

2) For the Banach space $(D(X),p)$, $p(\varphi):= \|\varphi\| + \|X\varphi\|$, we have a continuous bilinear map

$$(D(X),p) \times (A(X),q) \longrightarrow (D(X),p)$$
$$(\varphi, a) \longrightarrow a\varphi,$$

with $p(a\varphi) \leq q(a)p(\varphi)$.

3) For the closure $\overline{A}$ of $A(X)$ in $\mathcal{L}(E)$ we have
$$(\overline{A})^{-1} \cap A(X) = A(X)^{-1} \quad \text{(spectral invariance)},$$
where $(\ )^{-1}$ denotes as usual the group of invertible elements.

<u>Proof.</u> 1) is obvious from 2.4. and 2.8. Ad 2):
$$\begin{aligned}
p(a\varphi): \quad &= \|a\varphi\| + \|Xa\varphi\|, \quad Xa = (Xa-aX) + aX, \\
&\leq \|a\varphi\| + \|\delta_X(a)\varphi\| + \|aX\varphi\| \\
&\leq \|a\|_{\mathcal{L}(E)}\|\varphi\| + \|\delta_X(a)\|_{\mathcal{L}(E)}\|\varphi\|_E + \|a\|_{\mathcal{L}(E)}\|X\varphi\| \\
&\leq (\|a\|_{\mathcal{L}(E)} + \|\delta_X(a)\|_{\mathcal{L}(E)})(\|\varphi\|_E + \|X\varphi\|_E) \\
&\leq q(a)\, p(\varphi).
\end{aligned}$$

Ad 3): This is contained in 2.4.

2.9 Remark. Let $X: D(X) \longrightarrow H$ be a densely defined symmetric closed operator on a Hilbert space H, and let $A := \{a \in A(X): a^* \in A(X)\}$.

1) Then for $a \in A$

$$\delta(a) := i\,(Xa - aX)$$

is a closed *-derivation on $\mathcal{A}$, namely

$$\delta(a)^* = \delta(a^*).$$

2) $(\mathcal{A},q)$ is a *-Banach subalgebra of $\mathcal{L}(H)$.

3) $\qquad \mathcal{L}(H)^{-1} \cap \mathcal{A} = \mathcal{A}^{-1}$ (spectral invariance).

<u>Proof.</u> 1) For $u,v \in D(X)$, $< \delta(a)\,u,v > = i < Xau,v > - i < aXu,v >,$

$$= i < au,Xv > - i < Xu,a^*v >$$

$$= < u,+i(Xa^*-a^*X)v > = < u,\delta(a^*)v >.$$

2) $(\mathcal{A},q)$ is a closed subalgebra of $(\mathcal{A}(X),q)$.

3) The closure $\overline{\mathcal{A}}$ of $\mathcal{A}$ in $\mathcal{L}(H)$ is a C^*-algebra. Therefore the assertion follows from 2.4.3) and the fact that C^*-subalgebras are always full in any C^*-algebra.

2.10 Theorem. Let $(E,\|\ \|)$ be a Banach space and $\mathcal{X}$ a finite set of densely defined closed linear operators $Z: D(Z) \longrightarrow E$, such that $(\mathcal{D}_n(\mathcal{X}),p_n)$ (cf.2.1) is dense in E, $n \in \mathbb{N}$.

Moreover let $\Delta := \{\delta_Z : Z \in \mathcal{X}\}$ be the set of closed derivations (cf. 2.8) attached to $\mathcal{X}$.

Let (cf. 2.5) $\Psi_0 := \mathcal{A}_0$ and $\Psi_n := \mathcal{A}_n$, $n \geq 1$, equipped with the norm q_n (cf. 2.5).

Then the following statements are fulfilled:

1) $\forall a \in \Psi_n,\ \varphi \in \mathcal{D}_n(\mathcal{X})$ we have $a\varphi \in \mathcal{D}_n(\mathcal{X})$ and $p_n(a\varphi) \leq q_n(a)\,p_n(\varphi)$.

2) (Ψ_n,q_n) is a Banach algebra with unit $e = \mathrm{Id}_E$ (cf. 2.5.2).

3) For $a \in \Psi_n$ and $\|a\|_{\mathcal{L}(E)} < 1$ we have $(e-a)^{-1} \in \Psi_n$; (cf. [14], Lemma 5.3).

3') For the closure $\overline{\Psi_n}$ of Ψ_n in $\mathcal{L}(E)$ the relation

$$(\overline{\Psi_n})^{-1} \cap \Psi_n = \Psi_n^{-1} \quad \text{(spectral invariance)}$$

is valid.

4) For the Fréchet algebra $\Psi_\infty = \bigcap_{n=0}^{\infty} \Psi_n$ and for the Fréchet space $\mathcal{D}_\infty = \bigcap_{n=0}^{\infty} \mathcal{D}(\mathcal{X})$ we

have the continuous bilinear map

$$\Psi_\infty \times \mathcal{D}_\infty \longrightarrow \mathcal{D}_\infty$$

induced by $(a,\varphi) \longrightarrow a\varphi$.

5) For the closure $\overline{\Psi_\infty}$ of Ψ_∞ in $\mathcal{L}(E)$ the relation

$$(\overline{\Psi_\infty})^{-1} \cap \Psi_\infty = (\Psi_\infty)^{-1}$$

is fulfilled.

<u>Proof.</u> 1) Induction, n=0, $\|a\varphi\| \le \|a\|_{\mathcal{L}(E)} \|\varphi\|$. Assume $a\varphi \in \mathcal{D}_\nu$ for

$a \in \varphi_\nu$, $\psi \in \mathcal{D}_\nu$ and $p_\nu(a\varphi) \le q_\nu(a) p_\nu(\varphi)$, $0 \le \nu \le$ n-1. From

$Za = (Za\text{-}aZ) + aZ$ and $\delta_Z(a) = Za\text{-}aZ \in \Psi_{n-1}$ (cf. 2.5.2')), $a \in \Psi_n$,

$\varphi \in \mathcal{D}_n$ we obtain $Za\varphi \in \mathcal{D}_{n-1}$. We get

$$p_n(a\varphi) = p_{n-1}(a\varphi) + \sum_{Z \in \mathcal{X}} p_{n-1}(Za\varphi)$$

$$\le q_{n-1}(a)\, p_{n-1}(\varphi) + \sum_{Z \in \mathcal{X}} [p_{n-1}(\delta_Z(a)\varphi) + p_{n-1}(aZ\varphi)]$$

$$\le q_{n-1}(a) p_{n-1}(\varphi) + \sum_{Z \in \mathcal{X}} [q_{n-1}(\delta_Z(a))p_{n-1}(\varphi) + q_{n-1}(a)p_{n-1}(Z\varphi)]$$

$$\le [q_{n-1}(a) + \sum_{Z \in \mathcal{X}} q_{n-1}(\delta_Z(a))][p_{n-1}(\varphi) + \sum_{Z \in \mathcal{X}} p_{n-1}(Z\varphi)]$$

$$= q_n(a)\, p_n(\varphi).$$

For 2) to 5) see 2.5.

2.10' Remark. If in 2.10 the space E is a Hilbert space and Ψ_∞ is a symmetric subalgebra of $\mathcal{L}(E)$, then Ψ_∞ is a Ψ^*-subalgebra of $\mathcal{L}(E)$ in the sense of Gramsch [14], namely $\Psi_\infty \cap \mathcal{L}(E)^{-1} = (\Psi_\infty)^{-1}$. In 2.11 and 2.12 we give an additional method to generate Ψ - and Ψ_0-algebras. Using 2.8' we obtain by induction the following result.

2.11 Lemma. Assume:

1) A scale of Banach spaces $E_0 \supset E_1 \supset ... \supset E_r$, with continuous and dense injections.

2) Bounded operators $\lambda_j : E_j \longrightarrow E_{j-1}$, j=1 ,..., r , such that the norm of E_j is equivalent to $\|x\|_{j-1} + \|\lambda_j x\|_{j-1}$, $x \in E_j$.

3) A Banach algebra $\mathcal{R}$ which is continuously embedded into $\mathcal{L}(E_0)$ and contains Id_{E_0} as a unit.

4) Define Ψ to be the algebra of all $a \in \mathcal{R} \cap \bigcap_{j=0}^{r} \mathcal{L}(E_j)$ such that for j=1,...,r, the mappings

$$\lambda_j a\text{-}a\lambda_j \in \mathcal{L}(E_j , E_{j-1})$$

can be extended to linear operators

$$\lambda_j a - a\lambda_j \in \mathcal{L}(E_{j-1}), \quad j=1,\dots, r.$$

Put
$$\|a\|_\Psi := \|a\|_\mathcal{R} + \sum_{j=1}^{r} \|\lambda_j a - a\lambda_j\|_{\mathcal{L}(E_{j-1})}.$$

Then α) $(\Psi, \| \ \|_\Psi)$ is a Banach algebra with

$$\|a\|_{\mathcal{L}(E_r)} \leq C \|a\|_\Psi$$

for a suitable constant C.

β) Ψ is a Ψ_0-algebra in $\mathcal{L}(E_0)$ (cf. [14], 5.1); this means that for the closure $\mathcal{B}$ of Ψ in the Banach algebra $\mathcal{L}(E_0)$

$$\Psi \cap \mathcal{B}^{-1} = \Psi^{-1} \ \text{(spectral invariance)}$$

is fulfilled.

Besides an application to interpolation scales of Banach spaces we get the following interesting consequence (cf. [2], 3.1, [27], 4.1)

2.12 Proposition. Let Λ be an unbounded selfadjoint operator on a Hilbert space E^0 with $\Lambda \geq I$.

Put $E^S := D(\Lambda^S)$ the domain of Λ^S, $s \geq 0$,

$$\Lambda^S := \int_0^\infty \lambda^S dP_\Lambda(\lambda)$$

with the norm
$$\|x\|_{E^S} = \|\Lambda^S x\|_{E^0}.$$

For a fixed $\varepsilon > 0$ define the algebra

$$\mathcal{A} = \left\{ a \in \bigcap_{s \geq 0} \mathcal{L}(E^S) : \Lambda^\varepsilon a - a\Lambda^\varepsilon \in \mathcal{L}(E^S), s \geq 0 \right\}$$

equipped with th system of norms

$$\|a\|_{\mathcal{L}(E^S)} + \|\Lambda^\varepsilon a - a\Lambda^\varepsilon\|_{\mathcal{L}(E^S)} , \ s \geq 0.$$

Then $\mathcal{A}$ is with this topology a submultiplicative Fréchet algebra, continuously embedded into $\mathcal{L}(E^0)$ such that $\mathcal{A}$ is a Ψ_0-algebra in $\mathcal{L}(E^0)$ (cf. [14], 5.1); $\mathcal{A} \cap \mathcal{B}^{-1} = \mathcal{A}^{-1}$ is true for the norm closure of $\mathcal{A}$ in $\mathcal{L}(E^0)$.

<u>Proof.</u> The interpolation procedure (cf. [11], I.6.3) yields a countable system of submultiplicative norms defining the complete topology of $\mathcal{A}$ (cf. 2.8'). The Ψ_0-property follows from 2.11.

We mention a method of Cordes (cf. [11], I.6.5) to construct Fréchet operator algebras with spectral invariance.

2.13 Theorem. Let $(E^s : s \in S)$ be a countable system of Banach spaces with linear continuous isomorphisms $\Lambda_s : E^{s_0} \longrightarrow E^s$. Assume that E^s is contained in E^{s_0} as a dense linear subspace for all $s \in S$. Consider on the algebra

$$\mathcal{L}_0 := \bigcap_{s \in S} \mathcal{L}(E^s)$$

the Fréchet topology τ defined by the family of operator norms on $\mathcal{L}(E^s)$. Let $\mathcal{K}(E^{s_0})$ be the ideal of compact operators on E^{s_0} and a_s the restriction of $a \in \mathcal{L}_0$ to E^s. Then the set

$$A := \left\{ a \in \mathcal{L}_0 : (\Lambda_s)^{-1} a_s \Lambda_s - a_{s_0} \in \mathcal{K}(E^{s_0}) \right\}$$

is a submultiplicative Fréchet algebra with respect to the topology induced by τ from $\mathcal{L}_0$. Furthermore we have

$$A \cap [\mathcal{L}(E^{s_0})]^{-1} = A^{-1}.$$

This follows from

2.13' Remark. Let D be a Banach space which is continuously embedded in the Banach space F and let $\lambda : D \longrightarrow F$ be a bounded linear isomorphism. Define

$$B := \left\{ a \in \mathcal{L}(D) \cap \mathcal{L}(F) : \lambda a_D - a_F \lambda \in \mathcal{K}(D,F) \right\},$$

where $\mathcal{K}(D,F)$ denotes the space of compact operators.

Then B is a closed subalgebra of the Banach algebra $\mathcal{L}(D) \cap \mathcal{L}(F)$ and we have

$$B \cap \mathcal{L}(F)^{-1} = B^{-1}.$$

<u>Proof.</u> One uses elementary Fredholm theory and the closedness of $\mathcal{K}(D,F)$ in $\mathcal{L}(D,F)$. Clearly, B is an algebra. If $a \in B$ has an inverse $(a_F)^{-1}$ in $\mathcal{L}(F)$ then the restriction a_D of a has an inverse $(a_D)^{-1}$ in $\mathcal{L}(D)$. From

$$(a_F)^{-1}\lambda - \lambda(a_D)^{-1} = (a_F)^{-1}(\lambda a_D - a_F \lambda)(a_D)^{-1}$$

follows the spectral invariance.

In 2.8' it was not possible to establish the equality

$$\delta_X(a^{-1}) = -a^{-1} \delta_X(a)a^{-1} \text{ for } a \in A(X),$$

because we did not know whether a^{-1} is in the domain of δ_X.

Using the Neumann series it could be shown that $(e - \mu a)^{-1}$ is contained in the domain

$D(\delta_X)$ if $a \in D(\delta_X)$ and $|\mu| < \|a\|^{-1}$ is fulfilled. Now, we present a situation where we can do better. For many closed operators $X: D(X) \longrightarrow E$ there exists the possibility of "approximation" by bounded operators, e.g. for generators of C_o-semigroups.

2.14 Lemma. Let E be a Banach space and $X: D(X) \longrightarrow E$ a closed, densely defined operator with a family $\{X_h : 0 < h < \varepsilon\}$, $\varepsilon > 0$ fixed, of bounded operators on E such that

$$D(X) = \left\{ \varphi \in E : \lim_{h \to 0} X_h \varphi \in E \right\} \quad \text{and}$$

$$X\varphi = \lim_{h \to 0} X_h \varphi, \, \varphi \in D(X).$$

Define $A_0(X)$ to be the set of all $a \in \mathcal{L}(E)$ such that

$$X_h a - a X_h \text{ is strongly convergent on } E \text{ as } h \longrightarrow 0.$$

Then we have

1) $a \in A_0(X)$ implies $a(D(X)) \subset D(X)$ and $[X,a]\,\varphi = \lim_{h \to 0} [X_h,a]\,\varphi, \, \varphi \in D(X)$.

2) $A_0(X)$ is an algebra which is full in $\mathcal{L}(E)$,

$$A_0(X) \cap \mathcal{L}(E)^{-1} = [A_0(X)]^{-1}.$$

2') $a \in A_0(X)$ and $a^{-1} \in \mathcal{L}(E)$ implies, $a^{-1}(D(X)) \subset D(X)$ and

$$Xa^{-1} - aX^{-1} = -a^{-1}(Xa - aX)a^{-1}.$$

3) The closure $\mathcal{A}$ of $A_0(X)$ in $A(X)$ (see 2.8') is a Banach algebra with $\mathcal{A} \cap \mathcal{L}(E)^{-1} = (\mathcal{A})^{-1}$.

Remark. The approximation of the closed operator $X: D(X) \longrightarrow E$ is (implicitely) used in [4], [25], [25'] and [27], 4.2, 4.3.

<u>Proof.</u> Using $X_h \in \mathcal{L}(E)$ we do not have to care about domains for algebraic calculations. Therefore we get

$$[X_h, ab] = [X_h, a]\,b + a\,[X_h, b]$$

$$[X_h, a^{-1}] = -a^{-1}\,[X_h, a]\,a^{-1}.$$

The assertion 1) follows from

$$X_h(a\varphi) = [X_h, a]\,\varphi + aX_h\,\varphi, \, \varphi \in D(X),$$

with $h \longrightarrow 0$. The assertions 2) and 2') follow with $h \longrightarrow 0$ from the above formulas.

3) is implied by the following simple observation.

2.15 Lemma. Let Ψ be a Ψ_0-algebra ([14],5.1) in the Banach algebra B and A_0 a subalgebra of Ψ with the property $A_0 \cap B^{-1} = A_0^{-1}$. Then we have for the closure A of A_0 in the Fréchet algebra Ψ the property $A \cap B^{-1} = A^{-1}$.

<u>Proof.</u> For $a \in A$ with $a^{-1} \in B$ there exists $c \in A_0$ such that $\|a - c\|_B < \varepsilon$. For a suitable $\delta > 0$ there exist a constant M with $\|b^{-1}\| \leq M$ for $\|b - a\|_B < \delta$.

From $a = c - (c - a)$ we get $c^{-1}a = e - c^{-1}(c - a)$ with $c^{-1}a \in A \subseteq \Psi$. For a small $\varepsilon > 0$ we have the inverse of $e - c^{-1}(c - a)$ in Ψ, and the corresponding power series is converging in A since $c^{-1}(c - a)$ is in A.

2.16 Remark. In view of 2.14 it makes sense to construct Ψ^*-algebras by considering unbounded sequences of bounded derivations; this especially suitable on very singular spaces.

3. Submultiplicativity and spectral invariance for derivation algebras with order shift

A series of Fréchet operator algebras A considered by Cordes [9], [10], [11] are submultiplicative, this means that the topology of A can be given by a family of seminorms $\{q_\mu\}$ such that for $a,b \in A$

$$(3.1) \qquad\qquad q_\mu(ab) \leq q_\mu(a)\, q_\mu(b)$$

is fulfilled. In many cases the norms q_μ are the operator norms on a set (scale) of (weighted) Sobolev spaces. If q is a norm on an algebra A with unit e such that there exists a constant C with

$$(3.2) \qquad\qquad q(ab) \leq C\, q(a)\, q(b)$$

for all $a,b \in A$, then the left regular representation of A gives a norm

$$(3.2') \qquad\qquad \tilde{q}(a) := \sup_{\substack{x \in A \\ x \neq 0}} \frac{q(ax)}{q(x)}$$

which fulfills

$$(3.3) \qquad\qquad \frac{q(a)}{q(e)} \leq \tilde{q}(a) \leq C\, q(a) \qquad \text{(equivalence)}$$

and

$$\tilde{q}(ab) \leq \tilde{q}(a)\, \tilde{q}(b), \qquad \tilde{q}(e) = 1.$$

For the classes of Hörmander $\Psi^0_{\rho,\delta}$, $0 \le \delta \le \rho \le 1$, $\delta < 1$ Beals [4] gave a characterisation by iterated commutators respectively derivations δ, where $\delta(a)$ has in general an order different from the order of a (we call this order shift).

3.1 Remark. Let $\{E^s : s \in \mathbb{R}\}$ be a family of Banach spaces with continuous and dense embeddings

$$\mathcal{D} \subseteq E^s \subseteq E^{s'} \subseteq \mathcal{V}, \quad s \ge s',$$

for separated locally convex vector spaces $\mathcal{D}$ and $\mathcal{V}$ such that "interpolation" is possible (Adams [1], 7.11-7.20, Cordes [11],I.6).

If $\mathcal{A}$ is a subalgebra of

$$\mathcal{L}_0 := \bigcap_{s \in \mathbb{R}} \mathcal{L}(E^s)$$

equipped with the topology τ given by the system

$$\{ \| \cdot \|_{\mathcal{L}(E^s)} : s \in \mathbb{R} \},$$

then this topology can be defined by a countable system of submultiplicative norms q_k, $k = 1,2,...$, using interpolation.

3.1'. Remark. The group of invertible elements of $\mathcal{L}_0$ is in general not open in $\mathcal{L}_0$ (cf.[14], 6.2).

In the following we give a (general) construction which leads to the result that the Hörmander classes $\Psi^0_{\rho,\delta}$, $0 \le \delta \le \rho \le 1$, $\delta < 1$, are countable intersections of Banach algebras.

3.2 Definition. Let $\{E^s : s \in \mathbb{R}^p\}$ be a family of Banach spaces contained in a separated topological vector space $\mathcal{V}$ with continuous embeddings $E^s \subseteq \mathcal{V}$, such that $\underline{E} := \bigcap_{s \in \mathbb{R}^p} E^s$ is dense in E^s for each s; $\underline{E}$ is equipped with the projective topology induced by the family $\{E^s : s \in \mathbb{R}^p\}$. $\mathcal{L}(\underline{E})$ denotes the algebra of continuous linear transformations of $\underline{E}$. An element $a \in \mathcal{L}(\underline{E})$ is called an operator of finite order, if there exists an element $m = m(a) \in \mathbb{R}^p$ such that for each $s \in \mathbb{R}^p$ the operator a has a (unique) extension $a \in \mathcal{L}(E^{s+m(a)}, E^s)$; a number m(a) with this property is called an order of a. The linear space $\mathcal{BO}$ of all operators of bounded order is a subalgebra of $\mathcal{L}(\underline{E})$.

3.2' Remark. Let $\tilde{\Delta}$ be a finite set $\tilde{\Delta} = \{Z_j : j = 1,...,r\}$ in the algebra $\mathcal{L}(\underline{E})$ (see 3.2).

Let

$$\delta_j := \text{ad } Z_j : \mathcal{L}(\underline{E}) \longrightarrow \mathcal{L}(\underline{E})$$

be defined by $(\text{ad } Z_j)(a) = Z_j a - a Z_j$, $j = 1,...,r$. $\Delta := \{\delta_j : 1 \leq j \leq r\}$ is in general a noncommuting set.

Since $\mathcal{L}(\underline{E})$ is an algebra and $Z_j \in \mathcal{L}(\underline{E})$ we get the Leibniz formula in the sense stated in the proof of 2.5 (cf. [6]) : for

$$\alpha = (\alpha_1,...,\alpha_r), \ \alpha' = (\alpha_1',...\alpha_r'), \ \delta^\alpha := \delta_{j_1}^{\alpha_{j_1}} ... \delta_{j_r}^{\alpha_{j_r}}, \ a,b \in \mathcal{L}(\underline{E}),$$

$$(3.4) \qquad\qquad \delta^\alpha(ab) = \sum_{\alpha'+\alpha''=\alpha} C_{\alpha',\alpha''} \, \delta^{\alpha'}(a) \, \delta^{\alpha''}(b)$$

in the sense of noncommuting derivations δ_j, where $C_{\alpha',\alpha''}$ are elements of $\mathbb{N}$.

3.3 Definition. Let $\tau : \mathbb{N}^r \longrightarrow \mathbb{R}^p$, $\mathbb{N} = \{0,1,...\}$, be an additive function and $\mathcal{L}_\tau$ be the subspace of those $a \in \mathcal{B}\mathcal{O}$ (cf. 3.2) such that $\delta^\alpha(a)$ is of order $m(a) - \tau(\alpha)$ for all $\alpha \in \mathbb{N}^r$ (this property is called order shift by derivation).

3.3' Remark. From the Leibniz formula in 3.2' it follows that $\mathcal{L}_\tau$ is an algebra; for the proof use the additivity of τ taking into account the possible noncommutativity of Δ .

3.4 Construction of submultiplicative norms. Let $M \neq \emptyset$ be a subset of $\mathbb{R}^p$; let us attach to every $s \in M$ a normed space E^s such that we have continuous embeddings $\mathcal{D} \subseteq E^s \subseteq \mathcal{V}$, $s \in M$ for the separated topological vector spaces $\mathcal{D}$ and $\mathcal{V}$, and the density of $\mathcal{D}$ in E^s. Let $\tau \colon \mathbb{N}^r \longrightarrow \mathbb{R}^p$ be defined as in 3.3.

1) Let $\mathcal{A}$ be a unital subalgebra of $\bigcap_{s \in M} \mathcal{L}(E^s, E^s)$, for which we assume that the linear mappings

$$\delta^\alpha : \mathcal{A} \longrightarrow \mathcal{L}(E^s, E^{s+\tau(\alpha)})$$

$(\alpha = (\alpha_1,...,\alpha_r) \in \mathbb{N}^r)$ are defined if $s+\tau(\alpha) \in M$ and $s \in M$.

2) Let $t_j := \{s \in \mathbb{R}^p : s = \tau(\alpha), |\alpha| = \alpha_1+...+\alpha_r = j\}$ and C_ν, $\nu = 0,1,...,k$, be a finite family of finite sets with

$$M \supset C_0 \supset C_1 \supset ... \supset C_k$$

such that the following properties are fulfilled.

$$i_1) \; C_1 \subseteq \{s \in C_0 : s+t_1 \subset C_0\} \, , \; s+t_1 := \{s+\tau(\alpha) : (\alpha) = 1\}$$

(s is an element of M and t_1 a finite subset of $\mathbb{R}^p$)

$$i_2) \; C_2 \subseteq \{s \in C_1 : s+t_1 \in C_1, \, s+t_2 \subset C_0\}$$

and for $0 \le \nu \le k$

$$i_\nu) \; C_\nu \subseteq \{s \in C_{\nu-1} : s+t_j \subset C_{\nu-j}, \, 1 \le j \le \nu\}.$$

3) We assume that the maps δ^α are chosen in such a way that for $s \in C_{|\alpha|}$, $|\alpha| \le k$, the Leibniz formula (3.4) is fulfilled with $a,b \in \mathcal{A}$ and

$$\delta^\alpha(ab) \in \mathcal{L}(E^{s-\tau(\alpha)}, E^s), \quad \delta^{\alpha'}(a) \in \mathcal{L}(E^{s-\tau(\alpha')}, E^s) \, ,$$
$$\delta^{\alpha''}(b) \in \mathcal{L}(E^{s-\tau(\alpha)}, E^{s-\tau(\alpha')}),$$

with an appropriate ordering where $\alpha = \alpha' + \alpha''$.

4) Corresponding to a finite family $\tilde{C}$ of sets C_j

$$M \supset C = C_0 \supset ... \supset C_k \quad \text{(as in 2))}$$

we define on $\mathcal{A}$ the norm, $0 \le \nu \le k$,

$$(3.5) \quad p_{\tilde{C},\nu}(a) := \sup \left\{ \|\delta_{j_1} ... \delta_{j_{|\alpha|}}(\alpha)\|_{\mathcal{L}(E^s, E^{s+\tau(\alpha)})} : \delta_{j_k} \in \Delta, \, |\alpha| \le \nu \right\}$$

3.5 Proposition. For the family $\tilde{C}$ and for $\nu \in \{0,...,k\}$, there exists a constant $K > 0$ such that for $a,b \in \mathcal{A}$

$$p_{\tilde{C},\nu}(ab) \le K \, p_{\tilde{C},\nu}(a) \, p_{\tilde{C},\nu}(b).$$

<u>Proof.</u> It is sufficient to show that for $|\alpha| \le \nu$ the quantities

$$\|\delta^\alpha(ab)\|_{\mathcal{L}(E^s, E^{s+\tau(\alpha)})}, \quad s \in C_{|\alpha|},$$

are bounded by $K \cdot p_{\tilde{C},\nu}(a) \, p_{\tilde{C},\nu}(b)$. The Leibniz formula (3.4) with the chain $C_0 \supset C_1 \supset ... \supset C_\nu \supset ... \supset C_k$ gives the required estimate since $s-\tau(\alpha') \in C_{|\alpha''|}$ for $\alpha = \alpha' + \alpha''$, $\tau(\alpha'+\alpha'') = \tau(\alpha') + \tau(\alpha'')$ using the operator norms of $\mathcal{L}(E^s, E^{s'})$.

3.6 Remark. 1) Considering (3.3) we get an equivalent submultiplicative norm from (3.5).

2) For $\tau(\alpha) \equiv 0$ (no shift of order) we may choose $C_\nu = C_0$, $\nu = 1,...,k$.

3) For $\tau(e_j) \neq 0$ we have $C_k = 0$ for large k since C_0 is finite.

4) In 3.4.2) we may assume the equality

$C_v := \{s \in C_{v-1} : s+t_j \subset C_{v-j}, 0 \leq j \leq v\}$.5) For a given finite set C_k and $M = \mathbb{R}^p$ we may choose with $T_k := \{\tau(\alpha) \in \mathbb{R}^p : |\alpha| \leq k\}$

$$C_{k-j} = C_{k-(j-1)} + T_k = \left\{x+y : x \in C_{k-(j-1)}, y \in T_k\right\}, j = 1,...,k,$$

to obtain a finite chain $C_0 \supset C_1 \supset ... \supset C_k$ with the properties i_v), $0 \leq v \leq k$, in 3.4.2).

3.7 Lemma. Let $\mathcal{D}$ and $\mathcal{V}$ be separated topological vector spaces with a continuous embedding $\mathcal{D} \subseteq \mathcal{V}$ and $Z \in \mathcal{L}(\mathcal{D}) \cap \mathcal{L}(\mathcal{V})$, where $\mathcal{L}$ denotes the space of continuous linear transformations. Assume E_j, $j=1,2,3,4$, to be normed spaces with continuous embeddings $\mathcal{D} \subseteq E_j \subseteq \mathcal{V}$, where $\mathcal{D}$ is dense in E_j, $j = 1,...,4$. For $a \in \mathcal{L}(E_1,E_2)$ define

$$\delta: \mathcal{L}(E_1,E_2) \longrightarrow \mathcal{L}(,\mathcal{D},\mathcal{V})$$

by

$$\delta(a) := (adZ)(a) := Za - aZ.$$

Let $D(\delta)$ be the space of those $a \in \mathcal{L}(E_1,E_2)$ for which $\delta(a)$ can be extended to an element of $\mathcal{L}(E_3,E_4)$.

Then

$$\delta : D(\delta) \longrightarrow \mathcal{L}(E_3,E_4)$$
$$\cap$$
$$\mathcal{L}(E_1,E_2)$$

is a closed map with respect to the operator norm topology.

<u>Proof.</u> For $\varphi \in \mathcal{D}$ we have $Za\varphi \in \mathcal{V}$ since $\mathcal{D} \subseteq E_1$, $a \in \mathcal{L}(E_1,E_2)$ and $E_2 \subseteq \mathcal{V}$, $Z \in \mathcal{L}(\mathcal{V})$; with $aZ\varphi \in E_2$ we conclude $\delta(a) \in \mathcal{L}(\mathcal{D},\mathcal{V})$. Let $a_n \in D(\delta)$, $n \in \mathbb{N}$, be a sequence converging in $\mathcal{L}(E_1,E_2)$ to $a \in \mathcal{L}(E_1,E_2)$ such that $\delta(a_n)$ is converging in $\mathcal{L}(E_3,E_4)$ to y. For $\varphi \in \mathcal{D}$ we get the convergence of

$$\delta(a_k)\varphi = Za_k\varphi - a_k Z\varphi$$

to an element

$$Za\varphi - aZ\varphi = (Za - aZ)\varphi \in \mathcal{V}$$

since we have $Z \in \mathcal{L}(\mathcal{D}) \cap \mathcal{L}(\mathcal{V})$; on the other hand $\delta(a_k) \longrightarrow y$ in $\mathcal{L}(E_3,E_4)$ implies $\delta(a_k)\varphi \longrightarrow y\varphi$ and therefore $\delta(a)\varphi = y\varphi$ $\forall \varphi \in \mathcal{D}$. Thus $\delta(a)$ has a continuous extension from $\mathcal{D}$ to an element of $\mathcal{L}(E_3, E_4)$. This implies $\bar{\delta}(a) = y$ for the unique extension $\bar{\delta}(a)$ of $\delta(a)$.

3.8 Theorem. Let us take the notations and assumptions of 3.4 with $\delta^\alpha = \delta_{j_1}^{\alpha_{j_1}} \ldots \delta_{j_r}^{\alpha_{j_r}}$ and $\delta_j(a) := (\mathrm{ad} Z_j)(a)$ for a finite family of linear maps $Z_j \in \mathcal{L}(\mathcal{D}) \cap \mathcal{L}(\mathcal{V})$, $j = 1, \ldots r$ Let E^s be Banach spaces. We assume furthermore that the mappings $\delta^\alpha(a) := (\mathrm{ad} Z_{j_1})^{\alpha_{j_1}} \ldots (\mathrm{ad} Z_{j_r})^{\alpha_{j_r}}(a) \in \mathcal{L}(\mathcal{D}, \mathcal{V})$ have unique extensions from $\mathcal{D}$ to elements of $\mathcal{L}(E^s, E^{s + \tau(\alpha)})$ with the properties as in 3.4.

Let $\mathcal{A}_{\widetilde{C}, \nu}$ be the completion of the normed (submultiplicative) algebra $(\mathcal{A}, p_{\widetilde{C}, \nu})$ (see 3.4.1).

Then the Banach algebra $\mathcal{A}_{\widetilde{C}, \nu}$ is a subalgebra of $\underset{s \in C}{\cap} \mathcal{L}(E^s)$ equipped with a stronger topology than that induced by $\underset{s \in C}{\cap} \mathcal{L}(E^s)$. The continuous maps

$$\delta^\alpha : (\mathcal{A}, p_{\widetilde{C}, \nu}) \longrightarrow \mathcal{L}(E^s, E^{s + \tau(\alpha)}), \; s \in C_{|\alpha|},$$

extend continuously to $\mathcal{A}_{\widetilde{C}, \nu}$, $|\alpha| \leq \nu$.

<u>Proof.</u> 1) First we remark that given $a \in \mathcal{A}$, $Z_j \in \mathcal{L}(\mathcal{D}) \cap \mathcal{L}(\mathcal{V})$, $j = 1, \ldots, r$,

$$(\mathrm{ad} Z_{j_1})^{\alpha_{j_1}} \ldots (\mathrm{ad} Z_{j_r})^{\alpha_{j_r}}(a) \in \mathcal{L}(\mathcal{D}, \mathcal{V})$$

are defined as mappings from $\mathcal{D}$ to $\mathcal{V}$. In 3.4.1) it is assumed that $\delta^\alpha(a) = (\mathrm{ad} Z_{j_1})^{a_{j_1}} \ldots (\mathrm{ad} Z_{j_r})^{a_{j_r}}(a)$, $\alpha = (\alpha_1, \ldots, \alpha_r)$, is in $\mathcal{L}(E^s, E^{s + \tau(\alpha)})$, $s \in C_{|\alpha|}$.

2) For $\nu = 0$, $\mathcal{A}_{\widetilde{C}, \nu}$ is a closed subalgebra of the Banach algebra $\underset{s \in C}{\cap} \mathcal{L}(E^s)$ since C is a finite set; $\mathcal{L}(E^s)$ is a Banach algebra since E^s is a Banach space.

3) Induction: Let us assume that the assertions are proved for $0, 1, \ldots, \nu - 1$. Given a Cauchy sequence $a_n \in \mathcal{A}$ with respect to $p_{\widetilde{C}, \nu}$. We have by assumption

$$\delta^\beta(a_n) \longrightarrow \delta^\beta(a) \quad \text{in } \mathcal{L}(E^s, E^{s + \tau(\beta)})$$

for $|\beta| \leq \nu - 1, s \in C_{|\beta|}$. Consider for $|\alpha| = \nu$ for example

$$\delta^\alpha(a_n) = \delta_j \, \delta^\beta(a_n) = Z_j \, \delta^\beta(a_n) - \delta^\beta(a_n) Z_j$$

for some Z_j. Using 3.7 we get $\forall \varphi \in \mathcal{D}$

$$\delta^{\alpha}(a_n)\varphi \text{ converges to } Z_j \, \delta^{\beta}(a)\varphi - \delta^{\beta}(a) \, Z_j \varphi.$$

Therefore we obtain that the limit d_{α} of $\delta^{\alpha}(a_n)$ in $\mathcal{L}(E^s, E^{s+\tau(\alpha)})$ fulfills $d_{\alpha}\varphi = Z_j\delta^{\beta}(a)\varphi - \delta^{\beta}(a)Z_j\varphi$. This leads to $d_{\alpha} = \delta^{\alpha}(a)$ in the sense of an unique extension from $\mathcal{D}$. This proves 3.8.

Clearly the Leibniz formula extends to $\mathcal{A}_{\mathbb{C},\nu}$ by continuity.

3.9 Lemma. With the notations and assumptions of 3.3. and 3.4. we suppose that for a (fixed) element a in the algebra $\mathcal{A}$ and $\forall s \in C \subseteq M$ the following condition is fulfilled

$$\lim_{n \to \infty} \left[\|a^n\|_{\mathcal{L}(E^s)} \right]^{1/n} < 1$$

1) Then we have $\forall \, \alpha = (\alpha_1,...,\alpha_n)$, $|\alpha| \le k$,

$$\overline{\lim} \left[\|\delta^{\alpha}(a^n)\|_{\mathcal{L}(E^s, E^{s+\tau(\alpha)})} \right]^{1/n} < 1$$

with $s \in C_j$, $|\alpha| = j$, $\delta^{\alpha} = \delta_1^{\alpha_1}...\delta_r^{\alpha_r}$.

2) $b_{\ell} := \sum_{n=0}^{\ell} a^n$ is a Cauchy sequence with respect to the norm $p_{\mathbb{C},\nu}$.

<u>Proof.</u> 1) Since C is finite there exists for a sufficiently small $\varepsilon > 0$ a natural number r such that for all $n \ge r$ and $\forall s \in C$

$$\left[\|a^n\|_{\mathcal{L}(E^s)} \right]^{1/n} \le 1-\varepsilon$$

Furthermore put

$$R = \max_{|\alpha| \le k} \, \max_{s \in C_{|\alpha|}} \, \|\delta^{(\alpha)}(a)\|_{\mathcal{L}(E^s, E^{s+\tau(\alpha)})}.$$

Computing $\delta^{(\alpha)}(a^n)$, n large in comparison with $|\alpha| \cdot r$, (cf. 2.3'.2)) we get a sum of $n^{|\alpha|}$ products of the form

$$a...a(\delta^{(\gamma_1)}(a)) \, a...a(\delta^{(\gamma_2)}(a)) \, a...a(\delta^{(\gamma_{\ell})}(a)) \, a...a$$

with $\sum \gamma_i = \alpha$, where the number of the factors is n; $\delta^{(\varphi_{\mu})}(a) \in \mathcal{L}(E^s, E^{s+\tau(\gamma_{\mu})})$, $s \in C_{|\gamma_{\mu}|}$, observe the special definition of the chain $C_0 \supset ... \supset C_{\nu}$ in 3.4.2). In the above finite product there are at most $|\alpha|$ intervals contained in $\{1,2,...,n\}$ of length less than r which have at the beginning and the end a derivation expression $\delta^{(\gamma)}(a)$ and no derivation expression in between. Therefore we obtain the following estimate, $n > |\alpha|r$,

$$\|\delta^{(\alpha)}(a^n)\|_{\mathcal{L}(E^s,E^{s+\tau(\alpha)})} \le n^{|\alpha|} R^{|\alpha|\cdot r} (1-\varepsilon)^{n-|\alpha|r}.$$

Obviously, this inequality proves 1).

Ad 2):

$$\|\delta^\alpha(b_\ell)\|_\sim = \sum_{n=n'+1}^{\ell} \|\delta^\alpha(a^n)\|_\sim \le \sum_{n=n'+1}^{\ell} n^{|\alpha|} R^{|\alpha|\cdot r}(1-\varepsilon)^{n-|\alpha|r}.$$

The sum on the right hand side is getting small for n' large enough since $|\alpha|$ and r are fixed. Following the definition of the norm $p_{\widetilde{C},v}$ as a finite sum of norms of $\delta^{(\alpha)}(a)$, $|\alpha| \le v$, we obtain the assertion 2).

3.10 Remark (product rule). Let $\mathcal{D}$ and $\mathcal{V}$ be separated topological vector spaces with a continuous embedding $\mathcal{D} \subseteq \mathcal{V}$ and assume $Z \in \mathcal{L}(\mathcal{D}) \cap \mathcal{L}(\mathcal{V})$, where $\mathcal{L}$ denotes the space of continuous linear maps. Let E_j, $j = 1,2,3$, be normed spaces with continuous embeddings $\mathcal{D} \subseteq E_j \subseteq \mathcal{V}$, where $\mathcal{D}$ is dense in E_j. Assume $a \in \mathcal{L}(E_2,E_3)$, $b \in \mathcal{L}(E_1,E_2)$. Then $(adZ)^k(a)$ and $(adZ)^k(b)$ are elements of $\mathcal{L}(\mathcal{D},\mathcal{V})$ for all $k = 0,1,2,...$. If we assume that the map $(Zb-bZ) \in \mathcal{L}(\mathcal{D},\mathcal{V})$ is an element of $\mathcal{L}(\mathcal{D},E_2)$ then Zb maps $\mathcal{D}$ into E_2. Furthermore let us assume that $Za-aZ$ can be extended to an element of $\mathcal{L}(E_2,\mathcal{V})$. Then we have for all $\varphi \in \mathcal{D}$, with $\delta(c):= Zc-cZ$.

$$(\delta(ab))\varphi = (\delta(a)b)\varphi + (a\delta(b))\varphi$$

This follows from

$$Zab-abZ = Zab - aZb + aZb - abZ$$
$$(adZ)(ab) = (Za-aZ)b + a(Zb-bZ).$$

Let us remark concerning 3.11 that the derivations $ad(-ix)^\alpha$ and $ad(D_x)^\beta$ commute in the characterization of the Hörmander classes (cf.[27]); $\Delta = \{\delta_j : 1 \le j \le 2n\}$ is a commuting set in this case.

3.11 Theorem. The Hörmander classes $\Psi^0_{\rho,\delta}$, $0 \le \delta \le \rho \le 1$, $\delta < 1$, (cf. [20], [21]) of Fréchet algebras of pseudo-differential operators are countable intersections of Banach algebras.

<u>Proof.</u> The original topology τ_1 on $\Psi^0_{\rho,\delta}$ (cf. Hörmander [20], Kumanogo [21]) is given via the symbols $p(x,\xi)$ on $\mathbb{R}^{2n}$ (cf. [27]):

$$\|p\|_{\ell,\ell} := \max_{\substack{|\alpha|\le\ell\\|\beta|\le\ell}} \sup_{(x,\xi)} | \partial_\xi^\alpha D_x^\beta p(x,\xi)| \, \lambda(\xi)^{\rho|\alpha|-\delta|\beta|}$$

with $D_{x_j} := - i\partial_{x_j}$, $\lambda(\xi):= (1 + |\xi|^2)^{\frac{1}{2}}$; $\| \cdot \|_{\ell,\ell}$ is a countable system of norms inducing a Fréchet topology on the operator algebras $\Psi^0_{\rho,\delta}$, $0 \le \delta \le \rho \le 1$, $\delta < 1$, stronger than the topology of $\mathcal{L}(L^2(\mathbb{R}^n))$. Therefore $\Psi^0_{\rho,\delta}$ is a Fréchet subalgebra of $\mathcal{L}(L^2(\mathbb{R}^n))$. Following the characterization of Beals [4] (cf. [27], 1.7 Def., 1.8) we have an equivalent topology τ_2 on $\Psi^0_{\rho,\delta}$ defined by the norms

$$\|P\|_{s,\ell,\ell} := \sup_{\substack{|\alpha|\le\ell\\|\beta|\le\ell}} \|\mathrm{ad}(-ix)^\alpha \, \mathrm{ad}(D_x)^\beta P\|_{\mathcal{L}(H^{s-\rho|\alpha|+\delta|\beta|},H^s)} \, ,$$

where $\mathcal{L}(H^{s'},H^s)$ denotes the Banach space of all linear operators from the Hilbert-Sobolev space $H^{s'} = H^{s'}(\mathbb{R}^n)$ into H^s. Since we can use interpolation of operator norms with respect to the spaces H^s (Adams [1], 7.11 - 7.20), we get a topology equivalent to τ_2 considering the above norms only for discrete value of s. Therefore τ_2 defines a Fréchet topology on $\Psi^0_{\rho,\delta}$. From the H^s-continuity properties of $\Psi^0_{\rho,\delta}$ (cf. [27]) follows that τ_2 is weaker than τ_1 by an application of the closed graph theorem. Now choose in 3.4 the function $\tau\colon \mathbb{N}^r \longrightarrow \mathbb{R}^p$ in the following way, $r = 2n$, $p = 1$, $\tau(\alpha_1,...,\alpha_n; \beta_1,...,\beta_n):= \rho|\alpha| - \delta|\beta|$, $|\alpha| = \alpha_1+...+\alpha_n$, $|\beta| = \beta_1+...+\beta_n$. The norms $p_{\widetilde{C},v}$ (cf. (3.5)) constructed with $\| \cdot \|_{s,\ell,\ell}$ as in 3.4 are essentially submultiplicative (3.5. Prop.). The corresponding Banach algebra (Prop. 3.8) is a subalgebra of $\mathcal{L}(L^2(\mathbb{R}^n))$ with a finer topology than that induced by the operator norm of $\mathcal{L}(L^2(\mathbb{R}^n))$. Therefore (cf. 3.6.5)) the Fréchet algebra $\Psi^0_{\rho,\delta}$ is algebraically and topologically a countable intersection of Banach algebras.

3.12. Special Ψ^*-quantizations, general Hörmander classes $\Psi_{\rho,\delta}$.

Let Ω be a toplogical space (e.g. locally compact) with a measure μ. Furthermore we consider a diffusion semigroup on the Hilbert space $H:= L^2(\Omega,\mu)$ with a local infinitesimal generator $- A \geq 0$; put $\Lambda = (I - A)^{\frac{1}{2}}$. With the real powers Λ^s, $s \in \mathbb{R}$, (cf. 2.11, 2.12, 2.13, [11]), we generate a scale $\{E^s : s \in \mathbb{R}\}$ of Hilbert spaces and a Ψ^*-subalgebra $\mathcal{A}$ of $\mathcal{L}(E^0)$, $E^0 = H$, which is contained in the Fréchet algebra $\cap \, \mathcal{L}(E^s)$. We denote $S^{-\infty} := \underset{s,k}{\cap} \, \mathcal{L}(E^{s-k}, E^s)$; with this we can define pseudolocal operators in $\mathcal{L}(\underset{\sim}{E})$,

$\underset{\sim}{E} = \underset{s}{\cap} E^s$ of bounded order. Assume

    1) $m_1,\dots,m_r$ local or pseudolocal operators of order 0 in $\mathcal{L}(\underset{\sim}{E})$.

    2) $d_1,\dots,d_r$ local or pseudolocal operators of order 1 in $\mathcal{L}(\underset{\sim}{E})$.

Put $\delta_j := \mathrm{ad}(m_j)$, $\delta_{r+j} := \mathrm{ad}(d_j)$, $1 \leq j \leq r$.

These derivations are defined on $\mathcal{A}$ with values in $\mathcal{B}\mathcal{O}$ (cf. 3.2 Def.) Assume for $\alpha = (\alpha_1,\dots,\alpha_r,\dots,\alpha_{2r})$ and $0 < \delta \leq \rho \leq 1$, $\delta < 1$, in 3.3 Def.

$$\tau(\alpha) := \rho\left(\sum_{j=1}^{r} \alpha_j \right) - \delta \left(\sum_{j=r+1}^{2r} \alpha_j \right)$$

(consider also $\rho = 1$, $\delta = 0$).

Beginning with the Ψ^*-subalgebra $\mathcal{A}$ of $\mathcal{L}(E^0)$ the construction of § 3 leads to a submultiplicative Ψ^*-subalgebra $\Psi_{\rho,\delta}$ of $\mathcal{L}(E^0)$ contained in $\mathcal{A}$. This algebra may be called a general Hörmander class assigned to (Ω, μ). Of course this class depends heavily on the choice of Λ, m_j and d_j.

3.12'. Let us specialize to the situation where Ω is a submanifold of $\mathbb{R}^r$ with the induced (Riemannian) metric; furthermore assume for the sake of simplicity that the closure of the set Ω in $\mathbb{R}^r$ is compact (in many cases the set $\Omega\backslash K$ contains the singularities). For the selfadjoint positive operator $- A$ we take the Friedrichs extension of the induced Laplacian of the Riemannian submanifold Ω with respect to $C_c^\infty(\Omega)$. (One may take other Friedrichs extensions starting with other dense subspaces of $L^2(\Omega,\mu)$, where μ is the canonical measure of the Riemannian manifold Ω). Put $\Lambda = (I-A)^{\frac{1}{2}}$ (cf. 3.12, 2.12). Take as m_j the multiplication with the induced coordinate functions x_j of $\mathbb{R}^r, j=1,\dots,r$, and for d_j the (local) differential operator $\mathrm{grad}\,(m_j)$ relative to the Riemannian submanifold Ω of $\mathbb{R}^r$. In view of well known embedding theorems this procedures applies to Riemannian manifolds, but the resulting Fréchet algebra depends on the embedding.

4. C^{∞} - extensions of the holomorphic functional calculus

For a complex Banach (or Ψ-algebra) B with unit let $C^{\infty}E(B)$ be the set of all $b \in B$ such that $\sigma(b) \subseteq \mathbb{R}$ and that the holomorphic functional calculus extends to a continuous homomorphism

$$(4.1) \qquad\qquad \phi_b : C^{\infty}(\mathbb{R}) \longrightarrow B.$$

For certain Ψ-algebras A in a Banach algebra B we are interested in the equality

$$4.I \qquad\qquad C^{\infty}E(B) \cap A = C^{\infty}E(A)$$
$$(C^{\infty}\text{-extension invariance of } A \text{ in } B)$$

(cf. [6], [16]). Clearly, the property (4.I) is stable with respect to (countable) intersections in A.

Algebras of ultradifferentiable functions show that (4.I) is in general not true.

We are going to prove that the relation (4.I) holds in case A is one of the following algebras

$$\alpha) \quad \Psi := \langle B, C^{\infty}, \alpha \rangle \qquad\qquad (\text{cf. } [14], 5.10. - 5.12)$$
$$\beta) \quad A_{\infty} := \bigcap_{n=0}^{\infty} A_n \qquad\qquad (2.5. \text{ Theorem } 4)$$
$$\gamma) \quad \Psi^0_{\rho,\delta} \qquad\qquad (\text{general Hörmander classes}).$$

We also use ideas of Köthe and Tillmann (cf. [13]) treating vector distributions (see also [6], [15], [16]).

<u>Proof</u> of 4.I α) : Let Ω be a finite dimensional σ-compact C^{∞}-manifold and

$$\Omega \in t \longrightarrow \alpha_t \in \mathrm{Aut}(B).$$

a parametrizied (locally bounded) family of automorphisms α_t of B. It is well known that the property

$$a \in C^{\infty}E(B)$$

is equivalent to the resolvent estimate (for $\lambda \in \mathbb{C} \smallsetminus \sigma(a)$)

$$(4.2) \qquad\qquad |\mathrm{Im}\, \lambda|^m \|(\lambda e - a)^{-1}\|_B \leq C$$

for some $m > 0$. We have to show that if

$$[\Omega \ni t \longrightarrow \alpha_t(a)] \in C^{\infty}(\Omega, B)$$

then for each defining norm q_v of $A := \langle B, C^{\infty}, \alpha \rangle$ there exist constants $m_v > 0$ and $C_v > 0$ such that

$$(4.3) \qquad\qquad q_v((\lambda e - a)^{-1}) \leq C_v |\mathrm{Im}\, \lambda|^{-m} v.$$

For the sake of simplicity let us first consider the case where Ω is an interval in $\mathbb{R}$. For

$$a(t,\lambda) := \alpha_t(\lambda e-a) = \lambda e - \alpha_t(a)$$

$$a(t,\lambda)^{-1} = \alpha_t((\lambda e-a)^{-1})$$

$$\frac{1}{\Delta t}\left(a(t+\Delta t,\lambda)^{-1} - a(t,\lambda)^{-1}\right) =$$

$$(4.4)\qquad = - a(t+\Delta t,\lambda)^{-1}\left\{\frac{1}{\Delta t}\left(a(t+\Delta t,\lambda) - a(t,\lambda)\right)\right\} a(t,\lambda)^{-1}.$$

Since $a(t,\lambda)$ is a C^∞-function in t and the inversion is continuous in B we obtain with $\Delta t \longrightarrow 0$

$$(4.5)\qquad \partial_t\left(\alpha_t((\lambda e-a)^{-1})\right) = - \alpha_t((\lambda e-a)^{-1})\left(\partial_t(\alpha_t(a))\,\alpha_t(\lambda e-a)^{-1}\right).$$

Using (4.2) and $a \in \,<B,C^\infty,\alpha>$ it follows

$$(4.6)\qquad \sup_{t\in K}\left\|\partial_t\,\alpha_t((\lambda e-a)^{-1}))\right\|_B \leq C\,|\operatorname{Im}\lambda|^{-m},$$

where K denotes a compact subset of Ω.

These procedures apply to all partial derivatives on charts of the manifold Ω. This implies that the property (4.3) follows from (4.2). Therefore the assertion 4.I.α is proved.

<u>Proof</u> of 4.I.β): We have to show that the estimate

$$(4.7)\qquad \left\|\delta^\alpha((\lambda e-a)^{-1})\right\|_B \leq C_\alpha\,|\operatorname{Im}\lambda|^{-m_\alpha}$$

holds for $\alpha = (\alpha_1,...,\,\alpha_\nu)$ with constants $m_\alpha >0$ and $C_\alpha >0$ where $a \in C^\infty E(B)$ fulfills (4.2). The estimate (4.7) follows by iterated use of 2.3'.1) applying (4.2).

Let us indicate another extension method using [6], 2.3, and [15], §4. We apply the following (well known)

4.1 Lemma. Let $A \subset B$ be submultiplicative Fréchet algebras with a unit over $\mathbb{C}$ and A continuously embedded into B. For a continuous derivation $\delta : A \longrightarrow B$ $(a \in A,\ \xi \in \mathbb{R})$ the formula holds

$$(4.8)\qquad \delta(e^{i\xi a}) = i\,\xi \int_0^1 \left\{e^{\lambda i\xi a}\,\delta(a)\,e^{i(1-\lambda)\xi a}\right\} d\lambda.$$

<u>Proof.</u> For a, b $\in B$ we have

$$(4.9)\qquad \int_0^1 e^{\lambda i\xi a}\,(i\xi b)\,e^{(1-\lambda)\,i\xi a}\,d\lambda = \sum_{n=1}^\infty \frac{(i\xi)^n}{n!}\,C_n$$

with $C_n := \sum_{j=1}^n a^{j-1}ba^{n-j}$; this follows from the use of the Beta-function

$$\int_0^1 \lambda^j (1-\lambda)^k d\lambda = \frac{j!\ k!}{(j+k+1)!}$$

The assertion (4.8) follows from (4.9) applying 2.3'.2).

The combination of 4.1 with [15], 4.4. Theorem leads also to 4.1. α) and β).

Considering the extension procedure for the holomorphic functional calculus in several commuting elements as in 4.4. Theorem of [15] and in the paper of Droste [13] it is clear the above procedures leading to 4.I. α) and 4.I.β) qualify for more general situations (totally real submanifolds of $\mathbb{C}^n$ and several commuting elements). This proceeds along the lines indicated by Dunau (cf. [16], [27]).

4.2 Theorem. The property 4.I. holds for the Hörmander classes $A := \Psi^0_{\rho,\delta}$, $0 \le \delta \le \rho \le 1$, $\delta < 1$, on $L^2(\mathbb{R}^n)$ with $B = \mathcal{L}(L^2(\mathbb{R}^n))$. 4.I. is fulfilled for the general Hörmander classes $\Psi^0_{\rho,\delta}$ too.

<u>Proof.</u> For $a \in C^\infty E(\mathcal{L}(L^2(\mathbb{R}^n))) \cap \Psi^0_{\rho,\delta}$ we have for $\Lambda := \lambda(D_x)$, $\lambda(\xi) = (1+|\xi|^2)^{\frac{1}{2}}$,

the property

(4.10) $\Lambda^\varepsilon a - a\Lambda^\varepsilon \in \mathcal{L}(H^0(\mathbb{R}^n))$, $\varepsilon = 1-\delta$(cf.[27],4.1) ,

implies inductively (see [4], [27], §4)

(4.11) $\|a\|_{\mathcal{L}(H^s)} \le C_s \|a\|_{\mathcal{L}(H^0)}$

using (e.g.) interpolation and

$$\Lambda^\varepsilon a\varphi = [\Lambda^\varepsilon, a]\varphi + a\Lambda^\varepsilon\varphi \ , \ \varphi \in H^\varepsilon.$$

For the resolvent $(\lambda e - a)^{-1}$ of a we obtain from (4.2)

(4.12) $\|(\lambda e - a)^{-1}\|_{\mathcal{L}(H^s)} \le M_s |\operatorname{Im} \lambda|^m$

with constants M_s. For all $\alpha = (\alpha_1, ..., \alpha_n)$ and $\beta = (\beta_1, ..., \beta_n)$ we have to show

$$\|(\operatorname{ad}(-ix))^\alpha (\operatorname{ad}(D_x))^\beta (\lambda e - a)^{-1}\|_{\mathcal{L}(H^{s-\rho|\alpha|+\delta|\beta|}, H^s)} \le$$

(4.13) $\le N_{s,\alpha,\beta} |\operatorname{Im} \lambda|^{m_{s,\alpha,\beta}}$

for some constants $N_{s,\alpha,\beta}$ and $m_{s,\alpha,\beta}$. Using the product rule with order shift and (4.12) we obtain (e.g. see (4.5)) inductively the inequality (4.13).

References

1.   Adams, R.A.: Sobolev Spaces. Academic Press 1975
2.   Ali Mehmeti, F.: Regular solutions of transmission and interaction problems for wave equations. Mathematical Methods in Applied Sciences 11, 665 - 685 (1989)
3.   Arens, R.: Weierstrass products for inverse limits of Banach algebras. Math.Ann. 163, 155-160 (1966)
4.   Beals, R.: Characterizations of pseudodifferential operators and applications. Duke Math.J. 44, 45-57 (1977), ibid. 46, 215 (1979)
5.   Bratteli, O.: Derivations, dissipations and group actions on C*-algebras. Lect.Notes Math. 1229, Springer Verlag 1986
6.   Bratteli, O., Elliot, G.A., Jorgensen, P.E.T.: Decomposition of unbounded derivations into invariant and approximately inner parts. J. reine angewandte Math. 346, 166-193 (1984)
7.   Coifman, R., Meyer, Y.: Au delà des opérateurs pseudo-différentiels, Astérisques 57 (1978)
8.   Connes, A.: Géométrie non commutative. InterEditions Paris 1990
9.   Cordes, H.O.: Elliptic pseudo-differential operators - an abstract theory. Lect. Notes Math. 756, Springer 1979
10   Cordes, H.O.: On some C*-algebras and Fréchet-*-algebras of pseudodifferential operators. Proc.Symp.Pure Math. AMS 43, 79-104 (1985)
11.   Cordes. H.O.: Spectral theory of linear differential operators and comparison algebras. Lond.Math.Soc.Lect. Notes Ser.76 (1987)
12.   Davie, A.M.: Homotopy in Fréchet algebras. Proc.Lond.Math.Soc. 23, 31-52 (1971)
13.   Droste, B.: Extension of analytic functional calculus mappings and duality by $\bar{\partial}$-closed forms with growth. Math.Ann. 261, 185-200 (1982)
14.   Gramsch, B.: Relative Inversion in der Störungstheorie von Operatoren und Ψ-Algebren. Math.Ann. 269, 27-71 (1984)
15.   Gramsch, B.: An extension method of the duality theory of locally convex spaces with applications to extension kernels and the operational calculus. Functional Analysis: Surveys and Recent Results (Ed. K.D. Bierstedt, B. Fuchssteiner) North Holland, 131-147 (1977)
16.   Gramsch, B., K.G. Kalb: Pseudo-locality and hypoellipticity in operator algebras. Semesterbericht Funktionalanalysis 51-61, Tübingen, Sommersemester 1985
17.   Gramsch, B., Kaballo, W.: Decompositions of meromorphic Fredholm resolvents and Ψ^*-algebras. Integral Eq. and Operator Th. 12, 23 - 41 (1989)
18.   Gramsch, B.: Analytische Bündel mit Fréchet-Faser in der Störungstheorie von Fredholmfunktionen zur Anwendung des Oka-Prinzips in F-Algebren von Pseudo-Differentialoperatoren. Arbeitsgruppe Funktionalanalysis, Univ. Mainz, 120 S. (1990)
19.   Grubb, G.: Functional calculus of pseudo-differential boundary problems. Birkhäuser 1986
20.   Hörmander, L.: Pseudo-differential operators and hypoelliptic equations. Proc.Symp. Pure Math. AMS 10, 138-183 (1967)
21.   Kumano-go, H.: Pseudo-differential operators. MIT-Press 1981
22.   Lorentz, K.: On the local structure of the similarity orbits of Jordan elements in operator algebras. Annales Universitatis Saraviensis, vol. 2, No. 3, 159 - 189 (1989)
22'. Lorentz, K.: Ähnlichkeitsbahnen von Jordanoperatoren in topologischen Algebren als unendlichdimensionale homogene Mannigfaltigkeiten. Dissertation U. Mainz 1991.
23.   Michael, E.A.: Locally multiplicatively-convex topological algebras. Memoirs AMS 11 (1952)
24.   Schrohe, E.: The symbols of an algebra of pseudodifferential operators. Pac.J.Math. 211-224 (1986)

25.   Schrohe, E.: A Ψ^*-algebra of pseudodifferential operators on noncompact manifolds. Arch. Math. 51, 81-86 (1988)

25'.   Schrohe, E: Boundedness and spectral invariance for standard pseudodifferential operators on anisotropically weighted L^p-Sobolev spaces. Integr. Equations and Operator Th. 13, 271-284 (1990).

26.   Schulze, B.-W.: Topologies and invertibility in operator spaces with symbolic structures. Proc. 9, TMP, Karl-Marx-Stadt, Teubner Texte zur Math. Bd. 111, 257 - 270 (1989)

27.   Ueberberg, J.: Zur Spektralinvarianz von Algebren von Pseudodifferentialoperatoren in der L^p-Theorie. Manuscripta math. 61, 459-475 (1988)

28.   Waelbroeck, L.: Topological vector spaces and algebras. Springer Lect. Notes in Math. 230 (1971)

29.   Wagner, K.: Kommutatoren in der Theorie der Pseudodifferentialoperatoren mit Anwendungen auf die Submultiplikativität der Klassen $S^0_{1/2,1/2}$. Diplomarbeit, FB Math., Univ. Mainz 1987

30.   Widom, H.: A complete symbolic calculus for pseudo-differential operators Bull.Sc.Math., 2e série, 104, 19-63 (1980).

31.   Traute, M.: Spezielle Fréchetalgebren für singuläre Integraloperatoren mit Unstetigkeiten in den Symbolen. Diplomarbeit F.B. Math., U. Mainz 1989

32.   Schrohe, E.: A pseudodifferential calculus for weighted symbols and a Fredholm criterion for boundary value problems on noncompact manifolds. Arbeitsgruppe Funktionalanalysis, U. Mainz, 190 S.(1991) preprint, Habilitationsschrift.

33.   Schäfer, K.-J.: Klassen von Pseudodifferentialoperatoren in Hilberträumen auf dem $\mathbb{R}^n$ mit invarianten und singulären Symbolen. Dissertation U. Kaiserslautern 1980

34.   Payne, K. R.: Smooth tame Fréchet algebras and Lie groups of pseudodifferential operators. Comm Pure and Appl. Math. 44, 309 - 337 (1991)

35.   Schulze, B.-W.: Pseudo-Differential Operators on Manifolds with Singularities. Akademie-Verlag, North-Holland 1991

36.   Alvarez, J., Hounie, J.: Spectral invariance and tameness of pseudo-differential operators on weighted Sobolev spaces. Preprint.

36'.   Alvarez, J.: Functional calculi for pseudodifferential operators, III. Studia Math. 95, 155 - 173 (1989)

37.   Blackadar, B., J. Cuntz: Differential algebra norms and smooth subalgebras of C*-algebras, preprint 1991

38.   Folland, G.B.: Harmonic analysis in phase space. Princeton Univ. Press 1989

38'.   Sakai, S.: Operator algebras in dynamical systems. Cambridge Univ. Press 1991

39.   Rieffel, M.A.: Deformation quantization for actions of $\mathbb{R}^d$. 89 p., preprint 1991

40.   Schweitzer, L.B.: Dense m-convex subalgebras of operator algebra crossed products by Lie groups, preprint 1991

41.   Phillips, N.C.: K-theory for Fréchet algebras. International J. of Math. 2, 77 - 129 (1991)

42.   Unterberger A., H. Upmeier: Pseudodifferential analysis on symmetric cones. 193 p., preprint 1992

F.B. Mathematik
Universität Mainz
6500 Mainz, Germany

F.B. Mathematik
Universität Giessen
6300 Giessen, Germany

Operator Theory:
Advances and Applications, Vol. 57

Décroissance exponentielle des fonctions propres pour l'opérateur de Kac:
le cas de la dimension >1

par

Bernard Helffer

Abstract

The study of ferromagnetic is reduced by Kac (1966) to the study of spectral properties of an integral operator (that we call the Kac operator) of the following type :

$$(0.1) \quad K_V(h) = \exp(-V/2) . \exp (h^2 \Delta) . \exp (-V/2)$$

In order to study the mechanism of transition of phase, it is interesting to study the splitting between the two first eigenvalues of this operator. As in the case of the Schrodinger operator this is a consequence of the study of the decay of the eigenfunctions. This article is devoted to the proof by pseudo-differential techniques of the analog of the Agmon estimates for the Kac operator. Another application to the study of a double well problem for the Klein-Gordon operator is sketched.

§0 Introduction

On s'intéresse ici aux propriétés de l'opérateur de Kac introduit dans Kac (1966). On a vu dans Brunaud–Helffer (1991), en reprenant les arguments de Kac, que cet opérateur est très proche par sa structure de l'exponentielle d'un opérateur de Schrödinger. Un des problèmes posés par la mécanique statistique dans ce contexte est alors l'étude d'un double puits pour l'opérateur de Kac. Comme on l'a vu dans Brunaud–Helffer (1991), on peut considérer l'opérateur de Kac comme un opérateur pseudodifférentiel h–admissible au sens de Helffer–Robert (1983) (cf Robert 1986) qu'on peut écrire comme le composé de trois opérateurs pseudodifférentiels :
$$(0.1) \quad K_V(h) = \exp(-V/2).\,\exp(h^2\Delta).\,\exp(-V/2)$$
où V est une fonction C^∞ positive sur $\mathbb{R}^n$ telle que :
$$(0.2) \quad \text{Lim inf}_{|x|\to\infty} V > \text{Min } V = 0.$$
Un exemple de potentiel apparaissant en mécanique statistique (cf Kac 1966) est :
$$V(x) = (1/4)\,\Sigma_{k=1}^{n}\,x_k^{\,2} - \Sigma_{k=1}^{n}\log \text{ch}(\sqrt{v/2}\,(x_k+x_{k+1})) + c_v(n)$$
(en convenant que $x_{n+1}=x_1$, et en ajustant la constante $c_v(n)$ pour que le minimum soit 0).

Comme on l'a déjà dit, cet opérateur de Kac a beaucoup de similitudes avec l'opérateur $\exp(-P(h))$ où $P(h)$ est un opérateur de Schrödinger:
$$(0.3) \quad P(h) = -h^2\Delta + V(x).$$
On s'intéresse alors aux valeurs propres contenues à $O(h)$ près dans un intervalle I fermé inclus dans [MinV, Lim inf$_{|x|\to\infty}$ V[et plus précisément aux propriétés de décroissance des fonctions propres.

Un des outils de base dans l'étude spectrale fine du splitting est la bonne connaissance a priori de la décroissance des fonctions propres du problème considéré. Même si l'étude de ce problème a une longue histoire, des progrès décisifs ont été accomplis grâce à ce qui est appelé dans la littérature les inégalités d'Agmon (1980), du nom de celui qui les a introduit dans un autre contexte (décroissance des fonctions propres à l'∞). La situation est un peu plus délicate dans le cas pseudo–différentiel et n'a été traité qu'au coup par coup. Des cas en dimension 1 sont traités dans Helffer–Sjöstrand (1988) et on a expliqué dans Brunaud–Helffer (1991) comment les techniques de cet article permettait d'étudier la décroissance des fonctions propres de l'opérateur de Kac, dans le cas de la dimension 1.

Il se trouve que l'on rencontre le même type de problèmes pseudodifférentiels dans l'étude de l'opérateur de Klein–Gordon

$(0.4)\ D = \sqrt{(1-\Delta)} + V,$

dont le symbole est

$(0.5)\ (x,\xi) \rightarrow \sqrt{1+\xi^2} + V(x),$

 et c'est la deuxième motivation de ce travail.

Dans ce deuxième cadre des techniques probabilistes ont été utilisées par Nardini(1986), Carmona(1988) et Carmona–Masters– Simon (1990) pour l'étude de la décroissance à l'∞ des fonctions propres.

On se propose ici d'adapter certains résultats de Helffer–Sjöstrand (1984) (établis dans le cadre de l'équation de Schrödinger) ou de Helffer–Sjöstrand (1988) (établis pour des opérateurs pseudodifférentiels mais en dimension 1 et dans des cas assez particuliers) au cas pseudodifférentiel en dimension n. Certes, dans le cas qui nous intéresse ici, on peut envisager de travailler directement avec des opérateurs intégraux et utiliser le caractère très explicite du noyau de l'opérateur, et il faudra peut être revenir à cette approche pour contrôler le problème de la limite quand la dimension tend vers l'∞, problème qui apparaît comme naturel si on pense aux motivations statistiques (cf Sjöstrand (1990) et (1991) , Helffer–Sjöstrand (1991)), mais l'intéret de l'approche présentée ici est qu'elle s'applique telle quelle à l'étude d'autres problèmes de splitting pseudo-différentiels comme ceux posés pour l'équation de Klein–Gordon déjà mentionnée (cf §4).

Pour contrôler la décroissance exponentielle des fonctions propres, on va de manière cruciale utiliser que les symboles $p(x,\xi;h)$ de tous les opérateurs pseudo-différentiels $p^w(x,hD_x;h)$ (où p^w fait référence à la quantification de Weyl) entrant en jeu admettent des extensions holomorphes en ξ dans des bandes :

$(0.6)\ |\mathrm{Im}\ \xi| \leqslant C_0$ avec C_0 assez grand .

On a déjà vu dans Helffer–Sjöstrand (1988) (cf aussi Brunaud–Helffer 1991) que ces opérateurs pseudo-différentiels opèrent alors continûment non seulement dans $L^2(\mathbb{R}^n)$ mais aussi dans des espaces

$(0.7)\ L^2_{\phi,h}(\mathbb{R}^n)$ (où $L^2_{\phi,h}(\mathbb{R}^n) = \{ u \in L^2_{\ell oc}(\mathbb{R}^n);\ \exp(\phi/h)u \in L^2(\mathbb{R}^n)\}$

tant qu'on a la propriété :

$(0.8)\ |D_x^\alpha \phi(x)| \leqslant D_k$

uniformément sur $\mathbb{R}$ et pour tout multiindice α avec $|\alpha| \geqslant 1$, avec de plus :

$(0.9)\ \quad D_0 < C_0.$

La démonstration s'appuie sur la propriété que l'opérateur

$\exp(\phi/h).P(x,hD_x;h).\exp(-\phi/h)$

défini a priori sur $C_0^\infty(\mathbb{R}^n)$ peut être réalisé comme opérateur h–pseudodifférentiel.

Ici, on munit l'espace $L^2_{\bullet,h}$ de sa norme naturelle, h est un paramètre dans un intervalle de la forme $]0,h_0]$ et, quand on parle de continuité (d'un espace normé dans un autre) pour une famille (par rapport au paramètre h) d'opérateurs, on sous-entend que la famille d'opérateurs continus est bornée indépendamment de h pour la norme naturelle associée. Pour simplifier l'exposition, on va surtout se concentrer sur le cas du simple puits, le cas du double puits ou de plusieurs puits ne posant pas de problèmes réellement nouveaux par rapport au cas de l'équation de Schrödinger.

Le plan de cet article est le suivant :

Au §1 nous étudions la décroissance des fonctions propres que nous mesurons à l'aide de la distance d'Agmon à l'ensemble des puits. Il s'agit d'obtenir l'estimation grossière en ce sens qu'on accepte une erreur (en facteur) de $O_\varepsilon(\exp(\varepsilon/h))$ avec $\varepsilon > 0$ arbitrairement petit.

Au §2 nous raffinons les estimations pour transformer dans l'estimation le terme $O_\varepsilon(\exp(\varepsilon/h))$ avec $\varepsilon > 0$ arbitrairement petit en $O(h^{-N})$ pour N convenable. Ceci est obtenu sous la condition habituelle que le potentiel a des minima non dégénérés.

Au §3, on donne une application à l'approximation dans certains domaines de la fonction propre par une solution BKW. Cette approximation joue un rôle essentiel dans l'étude fine de l'effet tunnel (cf Helffer-Sjöstrand 1984, Brunaud-Helffer 1991).

Enfin au § 4, nous esquissons une application possible à l'étude de l'équation de Klein-Gordon (cf Helffer-Parisse (1991) pour des développements).

§1 Décroissance des fonctions propres : estimations grossières.

On suppose que V atteint son minimum (égal à 0) en un unique point 0 de $\mathbb{R}^n$:

(1.1) $\text{Min } V = V(0) = 0$.

On se propose de démontrer dans ce paragraphe le

Theorème 1.1 :

On suppose que les hypothèses (0.2) et (1.1) sont vérifiées.

Soit u_h une (famille de) fonction(s) propre(s) normalisée(s) de

(1.2) $K_V(h)\, u_h = \mu(h).\, u_h$ avec $\mu(h) \to 1$ lorsque $h \to 0$.

Alors, pour tout $\varepsilon > 0$, et tout compact K, il existe une constante $C_{\varepsilon,K}$ telle que, pour h assez petit, on ait

(1.3) $\| \exp d/h \cdot u_h \|_{L^2(K)} \leqslant C_{K,\varepsilon} \cdot e^{\varepsilon/h}$

où $x \to d(x)$ désigne la distance d'Agmon du point x au point 0 attachée à la métrique $(V - \min V)dx^2$.

On a de plus des estimations du même type pour les dérivées.

Démonstration

Bien évidemment, la démonstration suit étroitement celle du cas de l'équation de Schrödinger (cf Helffer–Sjöstrand 1984) mais nous insisterons sur les difficultés spécifiques rencontrées dans le cas pseudodifférentiel.

On va comme dans le cas de Schrödinger introduire deux familles de fonctions χ_ε et ψ_ε. La fonction de χ_ε est de boucher le puits tandis que ψ_ε est une fonction régulière approchant la fonction poids optimale mesurant la décroissance des solutions. Plus précisément, soit $\varepsilon > 0$; soit χ une fonction C_0^∞ positive telle que $\chi(0) = 1$ et dont le support est contenu dans la boule de rayon 1.

Pour tout ε, soit $\psi_\varepsilon(x)$ une fonction C^∞ réelle telle que :

$(1.4)_a$ $\psi_\varepsilon(0) = 0$,

$(1.4)_b$ $|D_x^\alpha \psi_\varepsilon(x)| \leqslant C_k$ pour $|\alpha| \geqslant 1$,

$(1.4)_c$ $\varepsilon\chi(x/\varepsilon) + V(x) - |\psi_\varepsilon{}'(x)|^2 \geqslant C\varepsilon^3 > 0$,

où $\varepsilon \in \,]0,\varepsilon_0]$ avec ε_0 assez petit et C et C_k sont des constantes > 0 indépendantes de ε.

On posera :

(1.5) $\chi_\varepsilon(x) = \varepsilon\chi(x/\varepsilon)$

Typiquement, on peut partir, en dimension 1, de la fonction $\phi(x)$ solution de $|\phi'(x)|^2 = V(x)$ et définie pour $x > 0$ par $\phi(x) = \int_0^x V^{1/2}(t)\, dt$

puis poser $\psi_\varepsilon(x) = (1-\varepsilon)\,\phi(x)$.

On obtient alors :

$\varepsilon\chi(x/\varepsilon) + V(x) - |\psi_\varepsilon{}'(x)|^2 = \varepsilon\chi(x/\varepsilon) + \varepsilon\,(2-\varepsilon)V(x) \geqslant C\varepsilon^3$

Dans le cas général on peut prendre comme fonction ϕ une expression régularisée (en dehors d'un voisinage de 0) de la distance d'Agmon $d(x)$ de x au puits.

Loin de la région qui nous intéresse, on peut prendre $\psi_\varepsilon = $ Cste ou $\psi_\varepsilon = $ $C|x|$.

Considérons

(1.6) $(\exp(\psi_\varepsilon(x)/h)(K^\varepsilon(x,hD_x,h))\exp(-\psi_\varepsilon(x)/h)$.

K^ε est obtenu en remplaçant V par $V + \chi_\varepsilon$ dans la définition de K_V.

On va plutôt regarder :

(1.7) $P^\varepsilon = I - K^\varepsilon = I - e^{-(\chi_\varepsilon + V)/2} \exp(h^2\Delta)\, e^{-(\chi_\varepsilon + V)/2}$
$\qquad\quad = I - e^{-(\chi_\varepsilon/2)} \cdot K_V \cdot e^{-(\chi_\varepsilon/2)}$

et

(1.8) $P_\#^\varepsilon = (\exp(\psi_\varepsilon/h) \cdot (P^\varepsilon(x,hD_x,h)) \cdot \exp(-\psi_\varepsilon/h))$.

D'après Helffer–Sjöstrand (1988) (cf aussi la proposition(3.1) de Brunaud–
Helffer (1991)), on obtient ainsi une famille d'opérateurs pseudo–différentiels
$P_{\#}^{\varepsilon}$ dont le symbole principal est donné par :

$$(1.9) \qquad (x,\xi) \to p_{\#,0}^{\varepsilon}(x,\xi) = p_0(x,\xi + i\psi_{\varepsilon}'(x))$$

Dans notre cas,

$$p_0(x,\xi + i\psi_{\varepsilon}'(x)) = 1 - \exp(i\xi\,\psi_{\varepsilon}'(x))\exp\,-(\xi^2 + \chi_{\varepsilon} + V - |\psi_{\varepsilon}'(x)|^2)$$

et on observe que, compte–tenu de (1.2), le symbole est elliptique :

$$(1.10) \quad |p_{\#,0}^{\varepsilon}(x,\xi)| \geqslant (1 - \exp(-C\varepsilon^3))\,.$$

Soit maintenant $u_h(x)$ la fonction propre dans $L^2(\mathbb{R}^n)$ (qu'on suppose de
norme 1) attachée à la valeur propre $\mu(h)$ de K_V.

On pose

$$P(x,hD_x,h) = I - K_V \quad \text{et} \quad \lambda(h) = 1 - \mu(h)\,.$$

On a, en posant :

$$(1.11) \qquad v_{\varepsilon}(x,h) = \exp(e^{\psi_{\varepsilon}(x)/h}).u_h(x) :$$

$$(1.12) \qquad (p_{\#}^{\varepsilon}(x,hD_x,h) - \lambda(h))\,v_{\varepsilon}(x,h) = w_{\varepsilon}(x,h)$$

avec :

$$w_{\varepsilon}(.,h) = \exp(\psi_{\varepsilon}/h)\,[p^{\varepsilon}(x,hD_x,h) - p(x,hD_x,h)]u_h$$
$$= \exp(\psi_{\varepsilon}/h)\,[\,e^{-(\chi_{\varepsilon}/2)}K_V.\,e^{-(\chi_{\varepsilon}/2)} - K_V]\,u_h$$
$$= [\exp(\psi_{\varepsilon}/h)\,(e^{-(\chi_{\varepsilon}/2)} - I)].[\,K_V.\,e^{-(\chi_{\varepsilon}/2)}]u_h$$
$$+ [\exp(\psi_{\varepsilon}/h).K_V.\exp(-\psi_{\varepsilon}/h)]\,[\,e^{-(\chi_{\varepsilon}/2)} - I).\,\exp(\psi_{\varepsilon}/h)u_h]$$

On observe maintenant que :

$$(1.13) \quad \|[\exp(\psi_{\varepsilon}/h)\,(e^{-(\chi_{\varepsilon}/2)} - 1)]\|_{L^{\infty}} \leqslant C\,e^{C\varepsilon/h}$$

Par ailleurs K_{V_2} et $[\exp(\psi_{\varepsilon}/h).K_V.\exp(-\psi_{\varepsilon}/h)]$ sont des opérateurs
continus dans L^2 de norme majorée indépendamment de h. On a donc:

$$(1.14) \quad \|w_{\varepsilon}(.,h)\| \leqslant C_{\varepsilon}\,e^{C.\varepsilon/h}\,.$$

Remarquons maintenant que $(p_{\#}^{\varepsilon}(x,hD_x,h) - \lambda(h))$ est uniformément inversible
(compte–tenu de (1.10) et du calcul pseudodifférentiel) comme opérateur de
L^2 dans L^2 (pour $h < h(\varepsilon)$) et ceci implique :

$$(1.15) \quad \|\exp(\psi_{\varepsilon}/h).u_h\| \leqslant C_{\varepsilon}.\,e^{C.\varepsilon/h}\,.$$
$$\#\#\#\#\#\#\#\#$$

Remarque 1.2

Dans le cas où il y a plusieurs puits, si on désigne par $d(x)$ la distance d'Agmon
aux différents puits (convenablement modifiée à l'∞, pour éviter certains
contrôles à l'∞), le résultat ci–dessus reste valable.

Il faudrait pouvoir raffiner pour avoir h^{-N} (pour un N convenable) dans le membre de droite de l'inégalité (1.3) à la place de $C_{K\varepsilon}$. $e^{\varepsilon/h}$ pour tout ε dans les domaines où la distance d'Agmon est C^∞. Ceci est démontré dans Helffer–Sjöstrand (1984) dans le cas de l'équation de Schrödinger (en toute dimension) sous l'hypothèse que V admet un minimum non–dégénéré et en dimension 1 (dans Helffer–Sjöstrand 1988) pour des opérateurs pseudo-différentiels dont le symbole est "voisin" du symbole
$(x,\xi) \to \cos x + \lambda \cos \xi$,
mais la démonstration reste générale pour des o.p.d plus généraux en dimension 1 et s'applique ici (comme nous l'avons utilisé dans Brunaud–Helffer (1991)) si $n = 1$. On se propose dans le prochain paragraphe de traiter le cas de la dimension > 1.

§2 Décroissance des fonctions propres dans le cas du simple puits : raffinements dans le cas d'un puits non–dégénéré.

On suppose désormais en plus des hypothèses du paragraphe précédent que :
(2.1) V admet un minimum *non–dégénéré* en 0.
On se propose de démontrer le

Théorème 2.1
Soit u_h une (famille de) fonction(s) propre(s) normalisée(s) de
(2.2) $K_V(h)u_h = \mu(h)u_h$ avec $\mu(h) \to 1$ lorsque $h \to 0$.
Alors sous les hypothèses (1.2) et (2.1), on a :
Pour tout compact K contenu dans un ouvert où $x \to d(x)$ est C^∞, il existe une constante C_K telle que, pour h assez petit, on ait
(2.3) $\| \exp d/h . u_h\|_{L^2(K)} \leq C_K . h^{-c_K}$
On a de plus des estimations du même type pour les dérivées.
#######

Comme déjà indiqué au paragraphe précédent, ce théorème était déjà démontré en dimension 1 (cf Helffer–Sjöstrand 1988 , Brunaud–Helffer 1991). On va maintenant présenter une démonstration valable en toute dimension pour des opérateurs pseudodifférentiels dont les symboles sont holomorphes en ξ dans une bande contenant l'axe réel. Dans un premier temps, on travaillera près d'un minimum non dégénéré du symbole qu'on supposera atteint en $(x,\xi) = (0.0)$.

On introduit, comme dans le cas de l'équation de Schrödinger, une distance modifiée. L'idée essentielle est de modifier la distance d'Agmon au puits 0 dans la région $|x| \leq C\sqrt{h}$.

On reprend bien entendu les idées développées pour le cas de Schrödinger dans Helffer-Sjöstrand (1984) en essayant d'absorber les difficultés nouvelles produites par le caractère non local des opérateurs pseudo-différentiels.

On cherche donc de nouveau une phase ψ telle que pour une petite modification de $(P-\lambda(h))$, l'opérateur

$\exp(\psi/h).(P(x,hD_x,h)-\lambda(h))\exp(-\psi/h)$

soit elliptique (en fait, on remplacera cette propriété d'ellipticité par un contrôle du type de celui donné par l'inégalité de Garding).

On serait tenter de prendre $\varepsilon = h$, dans la démarche qui a conduit au Théorème 1.1. Il y a cependant différents problèmes de contrôle des paramètres qui ne semblent pas permettre de faire ceci aussi simplement. On va donc plutôt prendre $\varepsilon = Dh$ avec D à déterminer. Les paramètres C, D étant à choisir convenablement, on cherche une fonction $\psi_{h,C}(x)$ C^∞ (avec contrôle uniforme des dérivées par rapport à h) telle que :

$$(2.4)\ P^{\psi_{h,C}\varepsilon} = (\exp(\psi_{h,C}/h)(P^\varepsilon(x,hD_x,h))\exp(-\psi_{h,C}/h))$$

vérifie :

$$(2.5)\ P^{\psi_{h,C};\varepsilon} + (P^{\psi_{h,C};\varepsilon})^* \geq E\,h \quad \text{avec } \varepsilon = Dh$$

pour h assez petit.

Plus précisément, on cherche à trouver pour tout $E \geq 0$, des constantes C (qui apparaîtra dans la définition de la phase $\psi_{h,C}(x)$) et D (qui apparait dans la modification de l'opérateur P) telles que (2.5) soit satisfaite.

La phase est construite (comme dans Helffer-Sjöstrand 1984, §5) à partir de la distance d'Agmon. On va la supposer C^∞, ce qui est toujours possible en la modifiant en dehors de l'origine. On s'intéresse ici surtout à améliorer l'estimation dans un voisinage du minimum.

On suppose, par conséquent, que $d(x)$[1] est une fonction C^∞ telle que:

$(2.6) \quad |\nabla d|^2 \leq V$, avec égalité dans un voisinage de 0

$(2.7) \quad d \geq 0,\ d(0) = 0$

---

[1] Dans l'étude de l'équation de Schrodinger, on a beaucoup utilisé des fonctions Lipschitziennes qui fournissaient un cadre naturel pour les différentes distances qui apparaissaient. Compte-tenu des opérateurs pseudodifférentiels qui interviennent ici, il vaut mieux travailler ici avec des fonctions régulières.

(2.8) $d \geqslant \varepsilon_0$, en dehors d'un voisinage de 0,

(2.9)   Pour tout α tel que $|\alpha| \geqslant 1$, $|D_x^\alpha d| \leqslant L_\alpha$ partout (où L_α est une constante)

On pose :

(2.10) $\psi_{h,C}(x) = d(x) - Ch \, Log \, (C \, \theta(d(x)/Ch))$

où $\theta(x)$ est une fonction C^∞, paire, qui satisfait à :

(2.11) $\theta = 3/2$ sur $[-1, 1]$

(2.12) $\theta(t) = t$ si $|t| \geqslant 2$

et

(2.13) $0 \leqslant \theta'/\theta < 1$ sur $[0, \infty)$.

Observons que :

(2.14)$_1$ $\psi_{h,C}(x) = d(x) - Ch \, Log \, (3C/2)$ si $d(x) \leqslant Ch$

et

(2.14)$_2$ $\psi_{h,C}(x) = d(x) - Ch \, Log \, (d(x)/h)$ si $d(x) \geqslant 2Ch$.

Ici, C est une constante $\geqslant 2$, qui sera choisie assez grande ultérieurement.

Il est important pour la suite de suivre très soigneusement la dépendance par rapport à h et C. On remarque d'abord que :

(2.15) $\nabla \psi_{h,C}(x) = \nabla d(x) [1 - (\theta'/\theta)(d(x)/Ch)]$

On a donc :

(2.16) $|\nabla \psi_{h,C}(x)|^2 \leqslant |\nabla d(x)|^2 \leqslant V$

pour toute valeur de C et h.

Regardons maintenant les dérivées d'ordre supérieur. On obtient :

Pour tout α t.q. $|\alpha| \geqslant 2$, il existe C_α tel que, pour tout $C \geqslant 2$, tout $h \in]0, h_0]$, on ait :

(2.17) $|D_x^\alpha \psi_{h,C}(x)| \leqslant C_\alpha . h^{1-(|\alpha|/2)}$.

Ici, on utilise que :

(2.18) $|\nabla d|^2 \leqslant A \, d$

(lié à la positivité de d au voisinage de 0, conséquence de (2.7),(2.8),(2.9)), et on vérifie séparément (2.17) dans les régions $\{(d/h) \leqslant C\}$, $\{ (d/h) \geqslant 2C\}$ et $\{C \leqslant (d/h) \leqslant 2C\}$.

Considérons maintenant le symbole de l'opérateur

$P^{\psi_{h,C} ; \varepsilon} + (P^{\psi_{h,C} ; \varepsilon})^*$

 Pour des raisons techniques, on modifie un peu la définition de la fonction "bouche-trou" introduite en (1.5). Ici, on va prendre plutôt :

(2.19) $\tilde{\chi}_\varepsilon = \varepsilon \, \chi(d(x)/\varepsilon)$

Observons pour plus tard que, grâce à (2.16), on a :

(2.20) $|\nabla \tilde{\chi}_\varepsilon| \leqslant C_0$ et, pour tout α t.q. $|\alpha| \geqslant 2$, $|D_x^\alpha \tilde{\chi}_\varepsilon(x)| \leqslant C_\alpha . \varepsilon^{1-(|\alpha|/2)}$

où les constantes C_0 et C_α sont indépendantes de $\varepsilon > 0$.

Pour séparer les différentes "fonctions" du paramètre h, on rend indépendants les paramètres et on va prendre comme phase $\psi_{\rho,C}$. Les choix de ρ et de ε comme fonctions de h seront précisés après.

Le symbole " principal" *(mais attention, le sous-principal ne pourra plus être négligé !)* est donné par :

$$(2.21) \quad p_0^\varepsilon(.\,,\xi+i\nabla\psi_{\rho,C}) = 1 - \exp(i\,\xi\,.\nabla\psi_{\rho,C})\exp - (\xi^2 + \tilde{\chi}_\varepsilon + V - |\nabla\psi_{\rho,C}|^2)$$

On est intéressé en fait par :

$$\mathrm{Re}\,p_0^\varepsilon(.\,,\xi+i\nabla\psi_{\rho,C}) = 1 - \cos(\xi\,.\nabla\psi_{\rho,C})\,.\,\exp - (\xi^2 + \tilde{\chi}_\varepsilon + V - |\nabla\psi_{\rho,C}|^2)$$

D'où (dans l'application, on prendra ε = Dh et ρ = h) :

$$(2.22) \quad \mathrm{Re}\,p_0^\varepsilon(.\,,\xi+i\nabla\psi_{\rho,C}) \geq 1 - \exp - (\tilde{\chi}_\varepsilon + V - |\nabla\psi_{\rho,C}|^2)$$

Or on a :

$$(\tilde{\chi}_\varepsilon + V - |\nabla\psi_{\rho,C}|^2) = \tilde{\chi}_\varepsilon + V - |\nabla\,d|^2\,[1 - (\,\theta'/\theta)(d/C\rho)\,]^2$$

Comme $|\nabla d|^2 \leq V$, cela donne :

$$(\tilde{\chi}_\varepsilon + V - \nabla\psi_{\rho,C}^2) \geq \chi_\rho + V\,.\,(\,\theta'/\theta)(d/C\rho)\,)$$

Lorsque d(x) $\leq$ 1/2 ε, on a :

$$(\tilde{\chi}_\varepsilon(x) + V(x) - |\nabla\psi_{\rho,C}(x)|^2) \geq (1/4)\,\varepsilon$$

Donc en choisissant ε = Dh, avec D assez grand, on contrôle cette région.

Si d(x) $\geq$ (1/2)ε, en prenant ρ = h, on peut en choisissant D = Cα avec α convenable (indépendant des autres paramètres) avoir:

$$V(x)\,.\,(\theta'/\theta)(d(x)/C\rho)\,) = C\rho\,V(x)/d(x) \geq \beta\,C\,\rho$$

pour $\beta > 0$, convenable.

On a donc la propriété :

Il existe α, $\beta > 0$ tel que, pour C$\geq$2, D = Cα, ε = Dh, ρ = h,

$$(2.23) \quad \mathrm{Re}\,p_0^\varepsilon(x,\xi+i\nabla\psi_{\rho,C}(x)) \geq \beta\,C\,h$$

On pourra choisir C ultérieurement assez grand. Tout ceci est vrai pour h assez petit.

Nous avons maintenant à vérifier qu'on est dans les conditions d'application d'un théorème de positivité pour une classe d'opérateurs différentiels et on pense naturellement à l'une des formes des inégalités de Garding. On observe que, lorsque D = αC, et ε = Dh , l'opérateur

$$Q = P^{\psi_{h,C};\varepsilon} + (P^{\psi_{h,C};\varepsilon})^*$$

est un opérateur pseudodifférentiel dont le symbole de Weyl total q_h (usuel, pas h-quantifié, c'est à dire que $Q = q_h(x,D_x)$) est réel (car Q est autoadjoint) et vérifie :

$$(2.24) \quad q_h(x,\xi) \geq \delta(C)\,.\,h \quad \text{avec } \delta(C) \to +\infty \text{ lorsque } C \to +\infty.$$

En effet, on a contrôlé le symbole principal mais les termes d'ordre inférieur sont de l'ordre de h avec des majorations indépendantes de C (C$\geq$2). Par ailleurs :

(2.25) $|D_x^\alpha D_\xi^\beta q_h(x,\xi)| \leqslant C_{\alpha,\beta} \, h^{-|\alpha|/2 \, + |\beta|}$

(où les constantes $C_{\alpha,\beta}$ peuvent être choisies indépendamment de C, cf (2.17)). C'est donc très bon en ξ mais moins bon en x. Pour appliquer un théorème de positivité à la Fefferman avec un contrôle par rapport aux paramètres suffisamment précis, il est plus agréable de faire le changement de variable :

(2.26) $x = h^{3/4} y$

Dans ces nouvelles coordonnées, le symbole du nouvel opérateur $\tilde{q}_h(y,\eta)$ vérifie:

(2.27) $|D_y^\alpha D_\eta^\beta \tilde{q}_h(y,\eta)| \leqslant C_{\alpha,\beta} \, h^{(|\alpha|+|\beta|)/4}$

et

(2.28) $\tilde{q}_h \geqslant \delta(C).h$

En divisant par h et une constante indépendante de C et h, on se retrouve exactement dans les conditions d'applications du lemme 18.6.10 [2] de Hörmander (1985) qui va permettre le contrôle des paramètres. Soyons un peu plus précis. Compte tenu de (2.24) et (2.28), on voudrait appliquer le lemme (avec $\lambda = h^{1/2}$) à l'opérateur (après changement d'échelle) $(\tilde{Q} - \delta(C)h))/h$, pour montrer qu'il est semiborné avec une constante indépendante du paramètre C et de h. Le seul problème est que le nouvel opérateur ne satisfait les conditions du théorème que si $\delta(C)h \leqslant 1$. On travaille donc avec $(\tilde{Q} - (\delta(C)h).\chi(\delta(C)h))/h$. Finalement, on a :

$\tilde{Q} \geqslant (\delta(C)h).\chi(\delta(C)h) - C_0 h$, avec C_0 indépendant de $C \geqslant 2$ et $h \in]0,h_0]$,

et retournant aux coordonnées initiales, on a la même propriété pour Q. On a donc démontré la proposition :

Proposition 2.2

Pour tout E, il existe D et C et une fonction $\psi_h(x) = \psi_{h,C}(x)$ introduite en (2.10) tels que :

$P^{\psi_h \varepsilon} = (\exp(\psi_h(x)/h)(P^\varepsilon(x,hD_x,h))\exp(-\psi_h(x)/h)$

vérifie :

(2.29) $P^{\psi_h \varepsilon} + (P^{\psi_h \varepsilon})^* \geqslant E h$

pour $\varepsilon = Dh$, avec h assez petit.

Ici $P^{\psi_h \varepsilon} = P_*$ est défini par :

(2.30) $I - \exp(-(\tilde{\chi}_\varepsilon + V)/2) \exp(h^2 \Delta) \exp(-(\tilde{\chi}_\varepsilon + V)/2)$

---

[2] Le lemme 18.6.10 dit :

Soit $0 \leqslant a \in C^\infty(\mathbb{R}^{2n},\mathbb{R})$, et $0 \leqslant \lambda \leqslant 1$ et supposons que :

$|D_{x,\xi}^\alpha a| \leqslant \lambda^{-2 + (|\alpha|/2)}$ pour $|\alpha| \leqslant N$

Si N est suffisamment grand : $(a^w(x,D) u, u) \geqslant - C_0 \|u\|^2, u \in \mathscr{S}$

où C_0 est indépendant de λ et de a .

avec : $\tilde{\chi}_\varepsilon(x) = \varepsilon\, \chi(d(x)/\varepsilon)$

Fin de la démonstration

On reprend la démonstration du cas où on travaillait à un exponentiellement petit près. Soit maintenant u_h une fonction propre dans $L^2(\mathbb{R}^n)$ attachée à une valeur propre $\lambda(h)$ de $P(x,hD_x,h)$ qui tend vers 0, qu'on suppose de norme 1. On a, en posant

$(2.31)\ v_h(x) = \exp(e^{\psi_h(x)/h})u_h(x)$,

$(2.32)\ (p_*(x,hD_x,h) - \lambda(h))\, v_h = w_h$

avec :

$$w_h = \exp(\psi_h/h)\,[p^\varepsilon(x,hD_x,h) - p(x,hD_x,h)]\, u_h$$

et $\varepsilon = Dh$.

Comme au paragraphe précédent (cf $(1.13)-(1.15)$), on observe d'abord que :

$$\|[\exp(\psi_h/h)\,(\exp(-(\tilde{\chi}_{Dh}/2)) - I)]\|_{L^\infty} \leqslant C_1$$

Par ailleurs K_V et $[\exp(\psi_h/h).K_V.\exp(-\psi_h/h)]$ sont des opérateurs continus dans L^2 de norme majorée indépendamment de h. Pour montrer le deuxième point on peut ou bien travailler directement sur le noyau ou bien procéder ainsi. On pose :

$(2.33)_a\ \ x = h^{1/2}y,\ \tilde{h} = h^{1/2}$,

$(2.33)_b\ \ \tilde{\psi}_h(y) = h^{-1/2}\psi_h(h^{1/2}y)$

$(2.33)_c\ \ \tilde{p}(y,hD_y,\tilde{h}) = p(hy,hD_y,h)$

et on observe maintenant que ψ_h vérifie (cf $(2.16)-(2.17)$) :

Pour tout α t.q. $|\alpha| \geqslant 2$, il existe $\tilde{C}_\alpha$ tel que, pour tout $C \geqslant 2$, tout $h \in\,]0,h_0]$, on ait :

$(2.34)\ \ |D_y^\alpha\tilde{\psi}_h(y)| \leqslant \tilde{C}_\alpha$.

On est alors ramené à l'étude de la nouvelle famille :

$$\exp(-\tilde{\psi}_h/\tilde{h}).\tilde{p}(y,hD_y,\tilde{h}).\exp(\tilde{\psi}_h/\tilde{h})$$

avec cette fois-ci des estimations uniformes.

On obtient ainsi l'estimation :

$(2.35)\ \|w_h\| \leqslant C_2$

pour une constante C_2 indépendante de h .

Maintenant, retournant à $(2.31)-(2.32)$, on observe qu'on peut choisir C (et $D = \alpha C$), telle que : $\mathrm{Re}(p_*(x,hD_x,h) - \lambda(h)) \geqslant h$ et on obtient que : $\|\exp(\psi_h/h)u_h\|^2 \leqslant h^{-1}\|w_h\| \leqslant C_3\, h^{-1}$

Si on revient à la définition de ψ_h, on a ainsi démontré le théorème .

§ 3 Comparaison entre la fonction propre et la solution BKW.

On a vu dans Helffer–Sjöstrand (1984) et Brunaud–Helffer (1991) comment on pouvait construire une solution BKW correspondant au niveau fondamental de $P(x,hD_x,h)$ lorsque le minimum est non–dégénéré. Le problème est en effet résolu sans restriction sur la dimension, pour ce type d'opérateurs pseudodifférentiels. Le cas considéré ici est d'autant plus sympathique que le symbole principal de $P = I-K_V$ est

$$(x,\xi) \to 1-\exp- (\xi^2 +V(x))$$

et que l'on peut utiliser tel quel tous les résultats de Helffer–Sjöstrand (1984) concernant l'existence, si V admet un minimum non–dégénéré en 0 tel que $V(0) = 0$, d'une fonction C^∞ ϕ définie au voisinage de 0 telle que :

$$(3.1) \quad \phi(0) = 0; \ \nabla\phi(0) = 0, \ \phi \geqslant 0 \text{ et } |\nabla\phi|^2 = V.$$

On obtient ainsi la :

Proposition 3.1 :
On suppose que les hypothèses (1.1)–(2.1) sont satisfaites. Soit Ω un voisinage ouvert de 0 assez petit. Alors, si Ω_1 est un ouvert relativement compact dans Ω, on peut trouver

$$(3.2) \quad \lambda^{bkw}(h) \approx \Sigma_{j=1}^{\infty} \lambda_j h^j$$

(avec $\lambda_1 .h$ première valeur propre de l'oscillateur harmonique en fond de puits)
et un symbole défini dans Ω

$$a(x,h) \approx \Sigma_j a_j(x). h^j$$

tel que l' on ait :

$$(3.2) \quad (e^{\phi(x)/h}(P(x,hD_x,h)-\lambda(h))e^{-\phi(x)/h})\chi(x).a(x,h) = 0(h^\infty) \text{ dans } L^2(\Omega_1)$$

où χ est une fonction à support compact dans Ω et égale à 1 sur un voisinage de Ω_1 .

La fonction $a(x,h)e^{-\phi(x)/h}$ est le candidat comme quasi–mode. Rappelons que a est obtenu comme la réalisation (définie modulo $0(h^\infty)$) d'un symbole formel $\Sigma a_j(x)h^j$ obtenu en résolvant successivement des équations de transport.

De la construction précédente, on déduit l'existence d'une valeur propre égale à $\lambda(h)$ modulo $0(h^\infty)$ sous réserve que le spectre soit discret au voisinage de 0 (par exemple sous l'hypothèse $\liminf_{|x|\to\infty} V >0$).

On souhaite maintenant comparer la solution BKW ainsi construite au moins au voisinage de 0 avec la fonction propre correspondante. Il reste donc à adapter la procédure utilisée pour Schrödinger dans le cas pseudo–différentiel. Rappelons qu'en dimension 1, il est donné dans Helffer–Sjöstrand (1988) une autre démonstration utilisant très explicitement le fait que l'on est en dimension 1 et que dans ce cas les fonctions propres elles–même sont des solutions BKW. Dans le cas où la dimension est quelconque, la propriété n'est plus vraie partout mais seulement dans certains ouverts où la distance d'Agmon est C^∞. Il apparait cependant que dans les applications la connaissance de la fonction propre dans ces régions est en fait suffisante (cf Helffer–Sjöstrand 1984).

On va suivre ici étroitement (et on renvoie pour certains détails à) la démonstration de Helffer–Sjöstrand (1984) (p.394) en éliminant le problème des fonctions poids irrégulières.

Soit donc Ω un ouvert assez bon (i.e.satisfaisant[3] aux hypothèses (5.11) et (5.12) de Helffer–Sjöstrand (1984)). On notera tout d'abord que les solutions BKW, construites dans la proposition 3.1 au voisinage de 0, se prolongent dans un tel ouvert . On se donne un compact K dans Ω et $\eta_1(x)$ une fonction dans $C_0^\infty(\Omega)$ égale à 1 dans un voisinage de la réunion $\hat{K}$ de toutes les géodésiques minimales pour la métrique d'Agmon allant d'un point de K à 0. Soit u_h la fonction propre et u_h^{bkw} la construction BKW correspondante qui existe dans un voisinage de $\hat{K}$. Il n'est pas trop difficile de voir que :

$$(3.3) \quad \|\eta_1 u_h - \eta_1 u_h^{bkw}\|_{L^2} = O(h^\infty)$$

et que par ailleurs :

$$(3.4) \quad \|\exp(d(x)/h)[\eta_1 u_h - \eta_1 u_h^{bkw}]\|_{L^2} \leqslant C\, h^{-C}$$

pour C assez grand.

On souhaite montrer que ces deux propriétés impliquent :

$$(3.5) \quad \|\exp(d(x)/h)[u_h - u_h^{bkw}]\|_{L^2\,(K)} \leqslant C_N\, h^N$$

pour tout N.

Il s'agit d'un résultat de propagation en ce sens que l'inégalité qu'on souhaite démontrer dans K est vraie dans un voisinage de 0 de taille $\sqrt{h}$ et qu'on souhaite propager ce résultat jusqu'à K.

Dans Helffer–Sjöstrand (1984), il est introduit la fonction :

$$(3.6)$$

$$\phi_{N,h}(x) = \text{Min}\,(\,\psi_h(x) + Nh.\,\text{Log}(1/h),\ \text{Inf}_{y\,\in\,\text{Supp}\chi}\,[\psi_h(y) + (1-\varepsilon)\,d(x,y)]$$

(Ici, on prend la fonction $\psi_h = \psi_{h,C}$ introduite au §2)

L'idée sous-jacente à l'introduction de cette fonction est qu'on peut traduire (3.3) et (3.4) en la propriété que la norme dans L^2 de $\exp(\phi_{N,h}/h)v_h$ (où on a posé $v_h = \eta_1\,[u_h - u_h^{bkw}]$) est contrôlée par $C_0\,h^{-C}$ dans le complémentaire d'un voisinage de $\hat{K}$ et dans une boule de rayon $C_1\,\sqrt{h}$.

Compte-tenu des techniques développées aux paragraphes précédents (cf pour les détails la prépublication à l'ENS de ce travail), la démonstration suit les grandes lignes de Helffer-Sjöstrand (1984).

§4 Remarques sur l'équation de Klein-Gordon

Les propositions qui précèdent s'appliquent sans problème à l'équation de Klein-Gordon, c'est à dire à l'opérateur pseudo-différentiel de symbole de Weyl :
$$(4.1)\quad \sqrt{(1+\xi^2)}^{-} + V(x)$$
où $V(x)$ vérifie les hypothèses (0.2),(1.1) (et (2.1) pour les estimations les plus fines) où on étudie le spectre près de 1. L'opérateur pseudodifférentiel a en effet un symbole holomorphe en ξ dans la bande : $|\text{Im}\xi| < 1$.
On peut alors mesurer de la décroissance exponentielle par rapport à une fonction d Lipschitzienne satisfaisant :
$$(4.2)\ |\nabla d|^2 \leqslant \text{Inf}\,\big[1 - (1-V)^2,\ 1-\varepsilon\big]\quad \text{avec } \varepsilon > 0 \text{ et}$$
$$(4.3)\ d(0) = 0 \text{ et } d \geqslant 0$$
Les résultats de Carmona (1988), Nardini (1986) et Carmona-Masters-Simon (1990) qui recoupent ceux-ci (mais ne sont pas semi-classiques) suggèrent que le résultat est presqu'optimal et que le point singulier en $\xi^2 = -1$, peut influer effectivement sur la décroissance de la fonction propre si V atteint la valeur 1. En particulier, l'étude du problème du double puits pour l'équation de Klein-Gordon devrait conduire à des phénomènes particuliers, si V prend la valeur 1. Des problèmes de ce genre sont évoqués dans Carmona (1988) (cf une des dernières remarques) et seront examinés dans Helffer-.Parisse (1991) en comparaison avec le même problème pour l'équation de Dirac.

Remerciements
Ce travail a été réalisé lors du séjour de l'auteur au WissenschaftsKolleg zu

Berlin. Je tiens à remercier cette institution pour les bonnes conditions de travail que j'y ai trouvées et également V.Enss, R.Seiler et R.Schrader qui ont été à l'origine de cette invitation.

<u>Références</u>

S.Agmon (1980), Lectures on exponential decay of solutions of second order elliptic equations, Math. Notes, t.29, Princeton University Press.

M.Brunaud – Helffer (1991), Un problème de double puits provenant de la théorie statistico-mécanique des changements de phase (ou relecture d'un cours de M.Kac), Prépublication de l'ENS.

R.Carmona (1988),Path integrals for relativistic Schrödinger operators dans "Schrödinger operators", Proceedings, SØnderborg, Denmark, 1988, Lecture Notes in Physics n°345, Springer Verlag.

R.Carmona,W.C.Masters et B.Simon (1990), Relativistic Schrödinger operators : Asymptotic behavior of   the eigenfunctions, J. Functional Anal. , 91, n°1, 117 – 142.

C.Fefferman – D. Phong (1981), The uncertainty principle and sharp Garding inequalities, Comm. in Pure Appl. Math. 34 , p.285 – 331.

B.Helffer – .B.Parisse (1991), Remarques sur l'équation de Klein – Gordon (en préparation).

B.Helffer et D.Robert (1983), Calcul fonctionnel par la transformée de Mellin et opérateurs admissibles, Journal of Functional Analysis 53 (1983) p.246 – 268

B.Helffer et D.Robert (1984), Puits de potentiels généralisés et asymptotique semi – classique, Annales de l'IHP, vol.41, n°3, p.291 – 331.

B.Helffer – J.Sjöstrand (1984), Multiple wells in the semi – classical limit I, Comm. in PDE, 9(4), 337 – 408 .

B.Helffer – J.Sjöstrand (1988), Analyse semi – classique pour l'équation de Harper,

Mémoire n°34, Supplément au Bulletin de la SMF, Tome 116, Fasc.4, (113 p.).

B.Helffer-J.Sjöstrand (1991),Semi-classical expansions of the thermodynamic limit for a Schrödinger equation, I. the one well case, à paraître.

L.Hörmander (1985), The Analysis of Linear Partial differential operators III, Springer Verlag

M.Kac (1966) Mathematical mechanisms of phase transitions, Cours à Brandeis (1966)

F.Nardini (1986), Exponential decay for the eigenfunctions of the 2-body relativistic hamiltonian, J.Analyse Mathématique 47, p 87-109

D.Robert (1986) , Autour de l'approximation semi-classique, Progress in Mathematics n°68, Birkhaüser (1986).

B.Simon (1983a), Instantons,double wells and large deviations
Bull.AMS.,t.8, 1983 p.323-326

B.Simon (1983b), Semi-classical analysis of low lying eigenvalues I
Ann. Inst. Poincaré, t.38, 1983, p.295-307

B.Simon (1984) Semi-classical analysis of low lying eigenvalues
II tunneling, Ann. of Math. 120, p.89-118 (1984)

 J.Sjöstrand (1990), Potential wells in high dimensions I, preprint et à paraître aux annales de l'IHP section Physique théorique.

 J.Sjöstrand (1991), Potential wells in high dimensions II, more about the one well case, preprint et à paraître aux annales de l'IHP section Physique théorique.

B.Helffer
DMI-ENS
45 rue d'Ulm
F75230 Paris Cedex 05

Operator Theory:
Advances and Applications, Vol. 57
© 1992 Birkhäuser Verlag Basel

Second Order Perturbations
of Divergence Type Operators with a Spectral Gap.

Rainer Hempel

Abstract. While it follows from Floquet theory that the spectrum of periodic elliptic operators $A = -\sum \partial_j a_{ij} \partial_i$, acting in the Hilbert space $L_2(\mathbf{R}^\nu)$, has band structure, we construct operators of this type with a spectral gap.

We also report on some recent results of Alama et al. (1992) concerning spectral properties of divergence form operators $A + \lambda B$, where $B = -\sum \partial_j b_{ij} \partial_i$ is a non-negative operator whose coefficients tend to zero at ∞. Here we ask for eigenvalues of $A + \lambda B$, $\lambda > 0$, in a spectral gap of A.

1. Introduction.

In mathematical physics, one frequently has to analyze the effect of localized perturbations in a homogeneous reference medium. An important example is provided by solid state physics where impurities in insulators or semi-conductors lead to new energy levels in spectral gaps, socalled *impurity levels* (cf., e. g., Deift and Hempel 1986); such impurity levels are basic for the quantum mechanical theory of the color of crystals and of the conductivity of doped semi-conductors. In an entirely different context, eigenvalues in spectral gaps also occur in certain control problems for the wave equation in perturbed media (see Avellaneda et al. 1992). In the present note, we are going to report on some of the recent results obtained jointly with S. Alama, M. Avellaneda and P. A. Deift (cf. Alama et al. 1992) concerning the existence of eigenvalues of divergence type operators $A + \lambda B = -\sum \partial_j (a_{ij} + \lambda b_{ij}) \partial_i$ in a spectral gap of $A = -\sum \partial_j a_{ij} \partial_i$, where it is assumed that the matrix functions $a := (a_{ij})$ and $b := (b_{ij})$ are uniformly Lipschitz–continuous and positive definite, with $(a_{ij}) \geq I$, while b_{ij} decays at ∞. Here a major difficulty arises form the fact that the perturbation is of the same order as the unperturbed operator A; nonetheless, it is still true that $\sigma_{\text{ess}}(A + \lambda B) = \sigma_{\text{ess}}(A)$, $\lambda \geq 0$, so that the spectrum of $A + \lambda B$ will be purely discrete inside any given spectral gap of A. Introducing the counting function

$$N(\lambda, b, E) := \sum_{0 < \mu < \lambda} \dim \ker (A + \mu B - E)$$

for a fixed E in the spectral gap of A, our main results are as follows:

(**A**) If the matrix function b has compact support in a ball of radius R, with $R \geq 1$, then $N(\lambda, b, E)$ is not only finite, but we have a constant C such that $N(\lambda, b, E) \leq C \cdot R^{\nu}$, for all perturbations b of the given type with support inside the ball B_R, and for all $\lambda > 0$.

Furthermore, in dimension $\nu \geq 2$, there exists $\rho_0 = \rho_0(E) > 0$ such that $N(\lambda, b, E) = 0$ for all b having support in the ball B_{ρ_0}, and for all $\lambda > 0$.

Finally, if the lowest eigenvalue of the matrix $(b_{ij}(x))$ is strictly positive, for all x in a sufficiently large ball, then $N(\lambda, b, E)$ will be non-zero, for λ large.

(**B**) In the case where b satisfies an upper polynomial bound of the type

$$\beta_{\nu}(x) \leq c(1 + |x|)^{-\alpha}, \quad x \in \mathbf{R}^{\nu}, \tag{1.1}$$

with $\beta_{\nu}(x)$ denoting the greatest eigenvalue of the matrix $b(x)$, we obtain an upper bound for the counting function,

$$N(\lambda, b, E) \leq C \cdot \lambda^{\nu/\alpha}, \quad \lambda \geq \lambda_0. \tag{1.2}$$

In Section 2, below, we shall give a more detailed exposition of a corresponding lower bound for $N(\lambda, b, E)$; cf. Theorem 2.1.

(**C**) Under suitable assumptions on the asymptotic behavior of b, it is possible to carry over the phase space analysis of Alama et al. (1989) for the Schrödinger operator case to the present situation and to derive the first order asymptotics of $N(\lambda, b, E)$, $\lambda \to \infty$.

Precise statements of these results are given in Theorem I-V in Alama et al. (1992). Their proof, while relying on some of the basic ideas developed for the Schrödinger case $H - \lambda W$ in the work of Alama, Deift and Hempel (Deift and Hempel 1986, Alama et al. 1989, Hempel 1987, 1989, 1990), has to cope with several difficulties: First of all, we are dealing with a *repulsive* perturbation which shifts eigenvalue branches from the lower band upwards into the gap; in the Schrödinger case, this corresponds to the analysis of $H + |\lambda W|$, where it is known that some subtle phenomena occur. Second, while many questions concerning the Schrödinger analog with W of constant sign are rather easily answered by looking at the associated (compact and symmetric) Birman-Schwinger kernel $|W|^{1/2}(H-E)^{-1}|W|^{1/2}$, one of the main difficulties in the present situation is related to the fact that the corresponding kernel is no longer compact—and hence virtually useless.

In Section 3, finally, we construct examples of divergence type operators with a spectral gap, in any dimension. This complements some recent work of Davies and Harrell (1987) on Laplace-Beltrami-operators with periodic metrics. Here we start with a periodic o. d. e. operator having a narrow first band, followed by a wide gap. Then it is clear that we can produce periodic operators $A = -\sum \partial_j a_{ij} \partial_i$ with a spectral gap in higher dimensions by summing copies of such 1-dimensional operators.

2. Eigenvalues of $A + \lambda B$ in a gap of $\sigma(A)$.

We now give a more precise description of one of the main results of Alama et al. (1992) and, at the same time, we shall try to convey some idea of the strategy of proof. We begin with the general setup:

For $i, j = 1, \ldots, \nu$, let a_{ij} and b_{ij} be bounded, uniformly Lipschitz continuous functions on $\mathbf{R}^\nu$, and suppose that, for each $x \in \mathbf{R}^\nu$, the matrices (a_{ij}) and (b_{ij}) are symmetric and positive (semi-) definite. Note that Lipschitz functions have a distributional gradient in $L_{\infty,loc}$. Furthermore, let us assume that there exists a constant M such that

$$|\xi|^2 \leq \sum a_{ij}(x)\xi_i\bar{\xi}_j \leq M|\xi|^2, \qquad \xi \in \mathbf{C}^\nu, \quad x \in \mathbf{R}^\nu. \tag{2.1}$$

The quadratic form

$$\mathbf{h}[\varphi, \psi] := \int \sum \partial_j \varphi\, a_{ij} \overline{\partial_i \psi}\, dx, \qquad \varphi, \psi \in \mathcal{H}^1(\mathbf{R}^\nu), \tag{2.2}$$

is closed and positive and hence there exists a self-adjoint operator $A \geq 0$ with domain $\mathcal{D}(A) \subset \mathcal{H}^1(\mathbf{R}^\nu)$ such that

$$(Au, v) = \mathbf{h}[u, v], \quad u \in \mathcal{D}(A), \quad v \in \mathcal{H}^1(\mathbf{R}^\nu). \tag{2.3}$$

Elliptic regularity theory (cf., e. g., Theorem 8.8 in Gilbarg and Trudinger 1977) now implies that, more strongly, $\mathcal{D}(A) \subset \mathcal{H}^2(\mathbf{R}^\nu)$ and that there exists a constant c such that

$$\|u\|_{\mathcal{H}^2} \leq c \cdot (\|Au\| + \|u\|), \quad u \in \mathcal{D}(A). \tag{2.4}$$

The estimate (2.4) may also be derived by perturbational arguments; cf. the Appendix in Alama et al. (1992). From (2.4) one easily derives the well known fact (cf. Stetkær-Hansen 1966) that the operator A is essentially self-adjoint on $C_0^\infty(\mathbf{R}^\nu)$. Our main additional assumption on A is the existence of a *spectral gap*: for the rest of this section, we shall always assume that there exist $E_1 < E_2$ such that $E_1 > \inf \sigma(A) = \inf \sigma_{\mathrm{ess}}(A) = 0$ and

$$(E_1, E_2) \cap \sigma(A) = \emptyset. \tag{2.5}$$

In Section 3 of the present paper it will be shown that one may in fact construct periodic operators $A = -\sum \partial_j a_{ij} \partial_i$ with a spectral gap. For the perturbation b_{ij}, let us assume that

$$b_{ij}(x) \to 0, \quad \nabla b_{ij}(x) \to 0, \quad |x| \to \infty. \tag{2.6}$$

Finally, let $\beta_1(x)$ (resp., $\beta_\nu(x)$) denote the smallest (resp., largest) eigenvalue of the matrix $(b_{ij}(x))$, so that, in particular, $0 \le \beta_1 \le \beta_\nu$. As before, we now may construct the (unique) self-adjoint operator $A + \lambda B$, associated with the matrix function $a + \lambda b$, for $\lambda \ge 0$. Our assumptions and eq. (2.4) imply that the perturbation λB is relatively bounded with respect to A and that the domain of the operators $A + \lambda B$ is independent of λ; furthermore, the estimate (2.4) holds for $A + \lambda B$ with a constant $c(\lambda)$ which may be chosen uniformly for λ in any bounded interval $[0, \Lambda]$. Therefore, the family $(A + \lambda B; \lambda \ge 0)$ can be extended to a self-adjoint holomorphic family of type (A) in the sense of Kato (cf. Reed and Simon 1978) and it follows that the eigenvalues of $A + \lambda B$ occuring away from the essential spectrum depend analytically on the parameter λ. But, clearly, the above estimates and the second resolvent equation imply that $(A + 1)^{-1} - (A + \lambda B + 1)^{-1}$ is compact, so that

$$\sigma_{\mathrm{ess}}(A + \lambda B) = \sigma_{\mathrm{ess}}(A), \quad \lambda \ge 0. \tag{2.7}$$

Therefore, the interval (E_1, E_2) will be a gap in the essential spectrum of $A + \lambda B$, $\lambda \ge 0$, and, fixing a control point $E \in (E_1, E_2)$, we can define the counting function

$$N(\lambda, b, E) := \sum_{0 < \mu < \lambda} \dim \ker (A + \mu B - E) \tag{2.8}$$

which indicates how many eigenvalue branches of $A + \mu B$ cross the level E while the coupling constant μ increases from 0 to λ, counting multiplicities. As an example for the results obtained in Alama et al. (1992), we discuss the following lower bound for $N(\lambda, b, E)$.

Theorem 2.1 *Suppose that (2.6) holds and that the smallest eigenvalue $\beta_1(x)$ of the matrix $b(x)$ satisfies a lower bound*

$$\beta_1(x) \ge c \cdot (1 + |x|)^{-\alpha}, \quad x \in \mathbf{R}^\nu \tag{2.9}$$

for some positive constants c and α. Then there exists a constant $C > 0$ such that

$$N(\lambda, b, E) \geq C \cdot \lambda^{\nu/\alpha}, \quad \lambda \geq 1. \tag{2.10}$$

As discussed in the Introduction, there is also a similar upper bound. Furthermore, one can obtain related results in the cases where B is of compact support, and, under somewhat stronger assumptions on A and B, it is possible to derive the precise first order asymptotics of the counting function $N(\lambda, b, E)$; see Alama et al. (1992).

We are now going to give a very brief introduction to the strategy of proof used in obtaining Theorem 2.1. Following the scheme developed by Alama, Deift and Hempel, we split the proof into three main steps:
1. The construction of a sequence of approximating problems on large balls;
2. Estimating the counting function for the approximating problems;
3. The "convergence step": estimates for the approximating problems yield estimates for $N(\lambda, b, E)$.

As for the first step, we proceed as in Deift and Hempel (1986) and Hempel (1987): Letting A_n denote the operator $-\sum \partial_j a_{ij} \partial_i$ acting in $L_2(B_n)$, B_n the ball of radius n, with Dirichlet boundary conditions on ∂B_n, we are again confronted with the difficulty that A_n may have eigenvalues inside the gap which are produced by the boundary condition on ∂B_n. If we now focus our attention on a compact subset K of the gap (E_1, E_2), the usual techniques proving exponential decay of eigenfunctions make it possible to show that the eigenfunctions associated with eigenvalues of A_n in the compact set K are exponentially small on the ball $B_{3n/4}$, for n large. Since, on the other hand, the number of these eigenvalues can grow at most like n^ν, by Weyl's Law, we can project these eigenfunctions out, at exponentially small cost. To be more precise, we can construct operators $\tilde{A}_n$ with the following main properties: $\tilde{A}_n$ coincides with A on $C_0^\infty(B_{n/2})$, and the interval $((E_1 + E)/2, (E + E_2)/2)$ is a spectral gap for $\tilde{A}_n$, for n large. We shall also have to consider the perturbed operator

$$\tilde{A}_{\lambda;n} := \tilde{A}_n - \lambda \sum \partial_j b_{ij} \partial_i, \tag{2.11}$$

acting in $L_2(B_n)$; note that $\tilde{A}_{\lambda;n}$ may have eigenvalues inside the gap.

The second step involves a refined eigenvalue counting for the unperturbed operator $\tilde{A}_n$ as compared with $\tilde{A}_{\lambda;n}$. More precisely, we have to establish a lower bound

$$\dim P_{(-\infty, E)}(\tilde{A}_n) - \dim P_{(-\infty, E]}(\tilde{A}_{\lambda;n}) \geq c \cdot \lambda^{\nu/\alpha}, \tag{2.12}$$

for λ large. Here a direct application of Dirichlet-Neumann-bracketing (Reed and Simon 1978) runs into the difficulty that the lower bound (2.12) should be independent of n, as $n \to \infty$. Therefore, we first single out a ball of radius $R = R(\lambda)$, with $R(\lambda) = c \cdot \lambda^{1/\alpha}$ and introduce Neumann boundary conditions on ∂B_R. This decouples the ball B_R into a region where the perturbation is active and a spherical shell $B_n - B_R$ where the lower bound (2.9) is small. The control of the additional boundary condition of ∂B_R is perhaps the most cumbersome technical problem in the whole setup; for, while it is immediate by min–max that adding Dirichlet boundary conditions *decreases* the number of eigenvalues below E, and that adding a Neumann b. c. *increases* the number of eigenvalues, we need precise estimates for the change in the number of eigenvalues. Of course, one will expect that the overall change is bounded by the surface volume of ∂B_R, which is proportional to $R^{\nu-1}$. Making use of the fact that we are in a spectral gap of $\tilde{A}_n$, this expectation can be substantiated by mimicking certain trace ideal estimates in Hempel (1990).

In the third step, finally, we assume that we are given solutions of the approximating eigenvalue problems and we try to preserve multiplicities as we let n tend to ∞. We refrain from indicating the full argument (which requires some astute notation) and merely discuss the central estimate in the most simple situation: Suppose we have sequences $(f_n) \subset \mathcal{D}(\tilde{A}_n)$ and $(\lambda_n) \subset (0, \infty)$ such that $\|f_n\| = 1$, $\lambda_n \to \hat{\lambda}$, as $n \to \infty$, and

$$(\tilde{A}_{\lambda_n;n} - E)f_n = 0, \quad n = 1, 2, \ldots.$$

Then we wish to prove that some subsequence of (f_n) converges weakly to a non-zero $f \in \mathcal{D}(A)$ satisfying $(A + \hat{\lambda}B - E)f = 0$. In the following, we shall not distinguish in the notation between the function f_n and its extension by 0 to all of $\mathbf{R}^\nu$. We start from the following a-priori estimates: By simple quadratic form arguments, we have a constant C such that

$$\|f_n\|_{\mathcal{H}^1} \leq C, \tag{2.13}$$

while arguments similar to the ones used in obtaining eq. (2.4) lead to a uniform bound for the second derivatives,

$$\|f_n\|_{\mathcal{H}^2(B_n)} \leq C. \tag{2.14}$$

From this, we immediately see that there exists a $f \in \mathcal{H}^1(\mathbf{R}^\nu) = \mathcal{Q}(A)$ (where $\mathcal{Q}$ denotes the form domain), and a subsequence $(f_{n_j}) \subset (f_n)$ such that

$$f_{n_j} \to f, \quad j \to \infty, \quad \text{weakly in } \mathcal{H}^1(\mathbf{R}^\nu).$$

Then it is easy to conclude that $\left(f, (A + \hat{\lambda}B - E)\varphi\right) = 0$, for all $\varphi \in C_0^\infty(\mathbf{R}^\nu)$, and hence, by self-adjointness, $f \in \mathcal{D}(A)$, $(A + \hat{\lambda}B - E)f = 0$. In order to prove that

$f \neq 0$, we now assume for a contradiction that

$$f_{n_j} \to 0, \quad j \to \infty, \quad \text{weakly in } \mathcal{H}^1(\mathbf{R}^\nu). \tag{2.15}$$

For simplicity, we'll write (f_n) instead of making subsequences explicit, in the sequel. By Rellich's compactness theorem and (2.14), (2.15) we may also assume that

$$\varphi f_n \to 0, \quad \text{strongly in } \mathcal{H}^1(\mathbf{R}^\nu) \tag{2.16}$$

for any fixed $\varphi \in C_0^\infty(\mathbf{R}^\nu)$. By the construction of the approximating operators $\tilde{A}_n$ in Step 1, it follows that there exists a $\gamma > 0$ such that $\mathrm{dist}(E, \sigma(\tilde{A}_n)) \geq \gamma$, $n \geq n_0$, whence

$$\begin{aligned}
\gamma \|f_n\| &\leq \left\| (\tilde{A}_n - E) f_n \right\| \\
&\leq \left\| (\tilde{A}_{\lambda_n;n} - E) f_n \right\| + \lambda_n \left\| \sum \partial_j b_{ij} \partial_i f_n \right\| \\
&= \lambda_n \left\| \sum \partial_j b_{ij} \partial_i f_n \right\|, \quad n \geq n_0.
\end{aligned}$$

As the sequence λ_n is bounded we will arrive at a contradiction if we can show that

$$\sum \partial_j b_{ij} \partial_i f_n \to 0, \quad n \to \infty. \tag{2.17}$$

This involves the *second* derivatives of the f_n. However, since b_{ij} and also ∇b_{ij} decay at infinity, it is clear from eq. (2.14) that it is sufficient to show that

$$\|\varphi f_n\|_{\mathcal{H}^2} \to 0, \quad n \to \infty, \tag{2.18}$$

for any fixed cut-off function $\varphi \in C_0^\infty$: using the bound (2.4) and the boundedness of the sequence λ_n, we obtain

$$\|\varphi f_n\|_{\mathcal{H}^2} \leq C \cdot \left(\left\| \tilde{A}_{\lambda_n;n}(\varphi f_n) \right\| + \|\varphi f_n\| \right),$$

where $\varphi f_n \to 0$, by (2.16), and

$$\begin{aligned}
\tilde{A}_{\lambda_n;n}(\varphi f_n) &= \varphi \tilde{A}_{\lambda_n;n} f_n + \left[\tilde{A}_{\lambda_n;n}, \varphi \right] f_n \\
&= E\varphi f_n + \left[\tilde{A}_{\lambda_n;n}, \varphi \right] f_n \to 0,
\end{aligned}$$

since $\left[\tilde{A}_{\lambda_n;n}, \varphi \right] f_n$ involves only f_n and its first derivatives on the support of φ; this concludes the proof of (2.17).

3. On the existence of gaps for operators of divergence type.

Here we consider 1–dimensional operators of the type

$$h_a = -\frac{d}{dx} a(x) \frac{d}{dx}$$

where $1 \le a : \mathbf{R} \to \mathbf{R}$ is bounded, periodic, with period $L > 0$, and *symmetric*, i. e., $a(-x) = a(x)$, for $0 \le x \le L/2$. Our aim is to find examples of such operators with a spectral gap. Here Floquet theory (cf., e. g., Eastham 1973) describes the edges of the spectral bands in terms of the eigenvalues of the periodic and anti-periodic eigenvalue problems on the interval $(-L/2, L/2)$, and it is a simple consequence that for symmetric a the edges of the bands are equivalently given by the eigenvalues of the Dirichlet and Neumann boundary value problems. More precisely, letting $\lambda_{k;D}$ and $\lambda_{k;N}$, $k = 1, 2, \ldots$, denote the eigenvalues of the Dirichlet (resp. Neumann) problem on the interval $(-L/2, L/2)$, it follows that the first spectral band of h_a is given by the interval $[0, \lambda_{1;D}]$, while the second band is given by the interval $[\lambda_{2;N}, \lambda_{2;D}]$; note that here $\lambda_{1;N} = 0$. This setup is quite similar to the discussion in Section 1 of Davies and Harrell (1987) where periodic Laplace-Beltrami-operators are considered.

Lemma 3.1. *For $n = 1, 2, \ldots$, let a_n be defined on the interval $(-n - 1, n + 1)$ by*

$$a_n(x) := e^n, \quad |x| \le n, \quad \text{and} \quad a_n(x) := 1, \quad n < |x| \le n + 1.$$

Letting $\lambda_{k;D}(n)$, $\lambda_{k;N}(n)$, $k = 1, 2, \ldots$, denote the Dirichlet (resp., Neumann) eigenvalues of the operator $-\frac{d}{dx} a_n(x) \frac{d}{dx}$ on the interval $(-n - 1, n + 1)$, we have

$$\lambda_{1;D}(n) \to 0, \quad n \to \infty, \tag{3.1}$$

and

$$\lambda_{2;N}(n) \to \pi^2/4, \quad n \to \infty. \tag{3.2}$$

In particular, the periodic operator h_{a_n} constructed from (the periodic continuation of) a_n has the following property: For any $\epsilon > 0$, there exists n_ϵ such that the first band of h_{a_n} is contained in the interval $[0, \epsilon)$, while the interval $[\epsilon, \pi^2/4 - \epsilon]$ is contained in a spectral gap, for $n \ge n_\epsilon$. Hence the operators

$$h_{a_n(x)} \otimes 1_{\{y\}} + 1_{\{x\}} \otimes h_{a_n(y)} = -\frac{d}{dx} a_n(x) \frac{d}{dx} - \frac{d}{dy} a_n(y) \frac{d}{dy}$$

provide examples of periodic divergence type operators in 2 dimensions with a spectral gap, for n large. Of course, the functions a_n can be smoothed out without affecting our main results.

Proof of Lemma 3.1. While min-max implies that

$$\lambda_{1;D}(n) = \inf\left\{ \int a_n(x)\, |\varphi'(x)|^2 \, dx \; ; \; \varphi \in C_0^\infty(-n-1, n+1),\ \|\varphi\| = 1 \right\},$$

it is easy to see that we may construct functions φ_n which are of constant height $(2n)^{-1/2}$ over the interval $(-n, n)$ and decrease monotonically to zero over the intervals $(-n-1, -n)$ and $(n, n+1)$ in such a way that $\|\varphi_n\| \to 1$ and $\int a_n |\varphi_n'|^2 \to 0$, as $n \to \infty$; this proves (3.1).

The proof of (3.2) is less evident. Letting $u_n := u_{2;N}^{(n)}$ denote the (normalized) eigenfunction of the Neumann problem for h_{a_n} on $(-n-1, n+1)$, associated with the eigenvalue $\lambda_{2;N}(n)$. As u_n is orthogonal to the constants, it has to change sign. More precisely, since u_n minimizes the value of the quadratic form amongst all normalized functions which are orthogonal to the constants, it is easily shown that u_n is anti-symmetric; hence $u_n(0) = 0$. (Of course, u_n has precisely one zero, by Sturm oscillation theory.) Integrating u_n' from 0 to $\pm t$, we conclude that

$$|u_n(\pm t)| \le t^{1/2} e^{-n/2} (\lambda_{2;N}(n))^{1/2}, \quad |t| \le n.$$

As $\lambda_{2;N}(n)$ is clearly bounded, it follows that

$$|u_n(\pm n)| \to 0, \quad \int_{-n}^{n} |u_n(x)|^2 \, dx \to 0, \quad n \to \infty.$$

This implies that for any fixed $\epsilon > 0$, we can find $n_0 \in \mathbf{N}$ such that $|u_n(n)| \le \epsilon$ while $\int_n^{n+1} |u_n(x)|^2 dx \ge 1/2 - \epsilon$, for $n \ge n_0$; therefore, the restriction of u_n to the interval $(n, n+1)$ satisfies mixed Dirichlet-Neumann boundary conditions, up to small errors. Since the lowest eigenvalue of the Dirichlet-Neumann problem for the operator $-d^2/dx^2$ on the interval $(n, n+1)$ is just $\pi^2/4$, our claim follows.

Acknowledgements. It is a pleasure to thank the organizers, Dr. M. Demuth and Prof. Dr. H.–W. Schulze, for the kind invitation to this Symposium.

References.

Alama S. et al. (1989), Eigenvalue branches of the Schrdinger operator $H - \lambda W$ in a gap of $\sigma(H)$. Commun. Math. Phys. **121**, 291–321.

Alama S. et al. (1992), On the existence of eigenvalues of a divergence form operator $A + \lambda B$ in a gap of $\sigma(A)$. Preprint.

Avellaneda M. et al. (1992), Controllabilité exacte, homogénéisation et localization d'ondes dans un milieu non-homogène. To appear in Asymptotic Analysis.

Birman M. Sh. (1990), Discrete spectrum in the gaps of the continuous one in the large coupling constant limit. In: Operator Theory, Advances and Applications, Vol. **46**, Birkhuser, Basel.

Davies E. B. and Harrell E. (1987), Conformally flat Riemannian metrics, Schrdinger operators, and semiclassical approximation. J. Differential Equations **66**, 165–188.

Deift P. A. and Hempel R. (1986), On the existence of eigenvalues of the Schrdinger operator $H - \lambda W$ in a gap of $\sigma(H)$. Commun. Math. Phys. **103**, 461–490.

Eastham M. S. P. (1973), The spectral theory of periodic differential equations. Scottish Academic Press, Edinburgh.

Gesztesy F. and Simon B. (1988), On a theorem of Deift and Hempel. Commun. Math. Phys. **116**, 503–505.

Gilbarg D. and Trudinger N. S. (1977), Elliptic partial differential equations of second order. Grundlehren der Mathematischen Wissenschaften **224**, Springer, Berlin etc.

Hempel R. (1987), A left-indefinite generalized eigenvalue problem for Schrdinger operators. Habilitationsschrift, Univ. Mnchen 1987.

Hempel R. (1989), On the asymptotic distribution of eigenvalue branches of the Schrdinger operator $H - \lambda W$ in a spectral gap of H. J. Reine Angew. Math. **399**, 38–59.

Hempel R. (1990), Eigenvalues in gaps and decoupling by Neumann boundary conditions. To appear in J. Math. Anal. Appl.

Reed R. and Simon B. (1978), Methods of modern mathematical physics, Vol. IV: Analysis of operators. Academic Press, New York.

Stetkær-Hansen H. (1966), A generalization of a theorem of Wienholtz concerning essential selfadjointness of singular elliptic operators. Math. Scand. **19**, 108–112.

Author's address:

Math. Inst. der Universitt Mnchen
Theresienstr. 39
W–8000 Mnchen 2
Germany

Rainer.Hempel@Mathematik.Uni–Muenchen.DBP.DE

Operator Theory:
Advances and Applications, Vol. 57
© 1992 Birkhäuser Verlag Basel

On the Weyl Quantized Relativistic Hamiltonian

Takashi Ichinose

Dedicated to Professor Hisanao Ogura on his sixtieth birthday

Abstract. This note discusses a general condition on essential selfadjointness of the Weyl quantized relativistic Hamiltonian and the path integral representation for its semigroup.

1. Introduction

Consider the imaginary-time relativistic Schrödinger equation

$$(1) \qquad \begin{aligned} (\partial/\partial t)u(t,x) &= -[H - m]u(t,x), \quad t > 0, x \in \mathbf{R}^d, \\ u(0,x) &= g(x), \end{aligned}$$

for a spinless particle in an electromagnetic field. The operator $H = H_A + \Phi$ is the *Weyl quantized relativistic Hamiltonian* or *Weyl quantized relativistic Schrödinger operator* corresponding to the classical symbol

$$(2) \qquad \sqrt{(p - A(x))^2 + m^2} + \Phi(x), \quad (p,x) \in \mathbf{R}^d \times \mathbf{R}^d,$$

dependent on the mass $m \geq 0$, with the vector and scalar potentials $A\colon \mathbf{R}^d \to \mathbf{R}^d$ and $\Phi\colon \mathbf{R}^d \to \mathbf{R}$ (e.g. Landau-Lifschitz 1975), where the light velocity c is taken to be equal to 1.

We have studied the problems about this Hamiltonian H in Ichinose-Tamura (1986), Ichinose (1987, 1989, 1991) and Ichinose-Tsuchida (1991 a, b). In this lecture we should like to give some recent results, mainly based on our work Ichinose-Tsuchida (1991a, b).

For the time being we assume that

$$(3) \qquad A \in \mathcal{B}^\infty(\mathbf{R}^d) \text{ and } \Phi \in \mathcal{B}^\infty(\mathbf{R}^d),$$

and define the operator $\mathcal{H} = \mathcal{H}_A + \Phi$ as follows: $\mathcal{H}_A$ is the Weyl pseudo-differential operator defined by the oscillatory integral

$$(\mathcal{H}_A u)(x)$$

$$\text{(4)} \qquad = (2\pi)^{-d} \int\!\!\int_{\mathbf{R}^d \times \mathbf{R}^d} e^{i(x-y)p} \sqrt{(p - A(\tfrac{x+y}{2}))^2 + m^2}\, u(y)\,dy\,dp,$$

$$u \in \mathcal{S}(\mathbf{R}^d),$$

while Φ is the multiplication operator by the function $\Phi(x)$. Then it can be seen that $\mathcal{H}$ defines a symmetric operator in $L^2(\mathbf{R}^d)$ with domain $C_0^\infty(\mathbf{R}^d)$. Then using ellipticity of the symbol $\sqrt{(p - A(x))^2 + m^2}$ we can show (e.g. Shubin 1987) that $\mathcal{H}$ is essentially selfadjoint on $C_0^\infty(\mathbf{R}^d)$. The unique selfadjoint extension of $\mathcal{H}$ (resp. $\mathcal{H}_A$), denoted by H (resp. H_A), is called the *Weyl quantized relativistic Hamiltonian*. If $A(x) \equiv 0$, H_A is nothing but the free relativistic Hamiltonian $H_0 \equiv \sqrt{-\Delta + m^2}$. Further, in Ichinose-Tamura (1986) it was proved for $m > 0$ that the semigroup $e^{-t[H-m]}$ resolving Eq. (1) admits the following path integral representation:

$$\text{(5a)} \qquad (e^{-t[H-m]}g)(x) = \int_{D_x} e^{-S(t,X)} g(X(t))\,d\lambda_x(X),$$

for $g \in L^2(\mathbf{R}^d)$, with

$$\text{(5b)} \qquad \begin{aligned}
S(t,X) = &\,i \int_0^{t+}\!\!\int_{|y|\geq 1} A(X(s-) + y/2)y\, N_X(ds\,dy) \\
&+ i \int_0^{t+}\!\!\int_{0<|y|<1} A(X(s-) + y/2)y\, \tilde{N}_X(ds\,dy) \\
&+ i \int_0^{t}\!\!\int_{0<|y|<1} (A(X(s) + y/2) - A(X(s)))y\, ds\, n(dy) \\
&+ \int_0^{t} \Phi(X(s))\,ds.
\end{aligned}$$

Here λ_x is a probability measure on the space $D_x \equiv D_x([0,\infty) \to \mathbf{R}^d)$ of the right-continuous paths $X \colon [0,\infty) \to \mathbf{R}^d$ with left-hand limits and $X(0) = x$ such that

$$\text{(6)}$$

$$\exp[-t(\sqrt{p^2 + m^2} - m)] = \int_{D_x} e^{ip(X(t)-x)}\,d\lambda_x(X), \quad t \geq 0,\; p \in \mathbf{R}^d.$$

The measure $n(dy)$, called the Lévy measure, is a σ-finite measure with rotationally invariant density $n(y)$ with respect to the Lebesgue measure dy : $n(dy) = n(y)dy$. It behaves as $O(|y|^{-(d+1)})dy$ near $y = 0$ and is a bounded measure for large $|y|$ (For its explicit expression see Ichinose 1989). By the Lévy-Khinchin formula (e.g. Ikeda-Watanabe 1981, Reed-Simon 1975)

$$(7) \quad \sqrt{p^2 + m^2} - m = - \int_{|y|>0} [e^{ipy} - 1 - ipyI_{\{|y|<1\}}]n(dy), \quad p \in \mathbf{R}^d,$$

where $I_{\{|y|<1\}}$ is the indicator function for the set $\{|y| < 1\}$. For each $X \in D_x$, $\tilde{N}_X(dsdy) = N_X(dsdy) - dsn(dy)$ by definition, where $N_X(dsdy)$ is a counting measure on $(0,\infty) \times (\mathbf{R}^d \setminus \{0\})$ defined by

$$N_X((t,t'] \times U) = \#\{s \in (t,t']; 0 \neq X(s) - X(s-) \in U\}$$

with $0 < t < t'$ and U a Borel set in $\mathbf{R}^d \setminus \{0\}$ and $\int N_X(dsdy)d\lambda_x(X) = dsn(dy)$ (e.g. Ikeda-Watanabe 1981). The Lévy-Itô theorem gives the representation of X in D_x:

$$(8) \quad X(t) = x + \int_0^{t+} \int_{|y|\geq 1} yN_X(dsdy) + \int_0^{t+} \int_{0<|y|<1} y\tilde{N}_X(dsdy).$$

$X(s)$ has at most finitely many points s at which the jump $|X(s) - X(s-)|$ exceeds a given positive constant. In particular, $X(s)$ has at most countably many discontinuities. It is bounded on every finite interval $[0, t]$, but not of bounded variation there (e.g. Bilingsley 1968).

The representation (5a,b) has a close analogy with the Feynman-Kac-Itô formula (see Simon 1979) for the nonrelativistic Schrödinger semigroup.

If we observe this path integral representation (5a,b), we shall see it will also be valid for less regular or singular and not necessarily bounded potentials $A(x)$ and $\Phi(x)$. However, how about the definition of $\mathcal{H} = \mathcal{H}_A + \Phi$, in particular, of $\mathcal{H}_A$? The previous definition in terms of a pseudo-differential operator does not make sense, because the usual theory of pseudo-differential operators needs sufficient smoothness of the symbols and boundedness of their higher derivatives, that is, the vector potential $A(x)$ should be sufficiently smooth and bounded. Here it is mentioned that Nagase-Umeda (1987,1990) proved essential self-adjointness of $\mathcal{H}_A$ in (4) under a milder assumption than (3) that $A(x)$ is sufficiently smooth and has bounded derivatives: $|\partial^\alpha A(x)| \leq C_\alpha$ for $|\alpha| \geq 1$ with constants C_α depending on the multi-indices α.

The aim of this note is to give a meaning to both the sides of this formula (5a,b) for general, singular and not necessarily bounded vector and scalar potentials $A(x)$ and $\Phi(x)$.

2. Definition of the Weyl quantized relativistic Hamiltonian

In this section we shall give an equivalent definition of the Weyl quantized relativistic Hamiltonian corresponding to the classical symbol (2). First, for

$$(9) \quad A \in L_{loc}^{2+\delta}(\mathbf{R}^d) \text{ for some } \delta > 0 \text{ and } \Phi \in L_{loc}^2(\mathbf{R}^d), \Phi(x) \geq 0 \ a.e.,$$

we want to define the operator $\mathcal{H} = \mathcal{H}_A + \Phi$ as follows: $\mathcal{H}_A$ is the singular integral operator

$$
(10) \quad
\begin{aligned}
&(\mathcal{H}_A u)(x) - mu(x) \\
&= -\lim_{r \downarrow 0} \int_{|y| \geq r} [e^{-iy A(x+y/2)} u(x+y) - u(x)] n(dy),
\end{aligned}
$$

$$u \in C_0^\infty(\mathbf{R}^d),$$

while Φ is the multiplication operator by the function $\Phi(x)$ as before.

Then we can show (see Ichinose-Tsuchida 1991b) with the Calderon-Zygmund theorem that the limit $\lim_{r \downarrow 0}$ in (10) for $u \in C_0^\infty(\mathbf{R}^d)$ exists pointwise a.e. in x as well as in the L^2 norm, so that $\mathcal{H} = \mathcal{H}_A + \Phi$ defines a symmetric operator in $L^2(\mathbf{R}^d)$ with domain $C_0^\infty(\mathbf{R}^d)$. This kind of definition of $\mathcal{H}_A$ was given first in Ichinose (1989), though for less general $A(x)$ than in (9), that is, for $A(x)$ satisfying that

$$(11) \quad A(x) \text{ and } \int_{0<|y|<1} |A(x+y/2) - A(x)| |y| n(dy) \text{ are in } L_{loc}^\infty(\mathbf{R}^d),$$

as the integral operator

$$
(12) \quad
\begin{aligned}
(\mathcal{H}_A u)(x) - mu(x) = &-\int_{|y|>0} [e^{-iy A(x+y/2)} u(x+y) - u(x) \\
&- I_{\{|y|<1\}} y(\nabla - iA(x)) u(x)] n(dy),
\end{aligned}
$$

$$u \in \mathcal{S}(\mathbf{R}^d),$$

because the integrand on the right of (12) is integrable with respect to $n(dy)$. This expression for $\mathcal{H}_A$ can also be obtained by calculating with the Itô formula the generator of the semigroup $e^{-t[H-m]}$ represented by the path integral (5a,b). Of course, $\mathcal{H}_A$ in (10) as well as in (12) coincides with the pseudo-differential operator $\mathcal{H}_A$ in (4) if $A(x)$ is sufficiently smooth and has bounded derivatives up to sufficiently higher order (see Ichinose 1989).

Further, we introduce, for

(13)
$$A \in L^{1+\delta}_{loc}(\mathbf{R}^d) \text{ for some } \delta > 0 \text{ and } \Phi \in L^1_{loc}(\mathbf{R}^d), \Phi(x) \geq 0 \quad a.e.,$$

the maximal symmetric form $h = h_{A,\Phi}$:

$$h[u] - m\|u\|^2 \equiv h[u,u] - m\|u\|^2 =$$

$$(14) \qquad \iint_{|x-y|>0} |e^{-i(x-y)A(\frac{1}{2}(x+y))}u(x) - u(y)|^2 n(x-y)dxdy$$

$$+ \int \Phi(x)|u(x)|^2 dx$$

with form domain $Q(h) = \{u \in L^2(\mathbf{R}^d); h[u] < \infty\}$. We can see that $h[u,v] = (\mathcal{H}u,v)$ for $u,v \in C_0^\infty(\mathbf{R}^d)$.

Remark. For the scalar potential $\Phi(x)$ having negative part: $\Phi(x) = \Phi_+(x) - \Phi_-(x)$, where $\Phi_\pm(x)$ satisfy the condition of Φ in (9) or (13), it will be possible to show the theorems of this note, if only $\Phi_-(x)$ is small in a certain sense, but we content ourselves with such Φ as in (9) or (13). The main point of this note is in treating the Hamiltonian with vector potential $A(x)$ as general as possible.

THEOREM 1. (i)*If $A(x)$ and $\Phi(x)$ satisfy (13), h is a closed form with form domain $Q(h)$ including $C_0^\infty(\mathbf{R}^d)$ as a form core, so that the minimal symmetric form $h_{\min}$ defined as the form closure of $h|C_0^\infty(\mathbf{R}^d) \times C_0^\infty(\mathbf{R}^d)$ coincides with h. Therefore there exists a unique selfadjoint operator H with domain $D(H)$ corresponding to the form h such that $h[u,v] = (Hu,v)$ for $u \in D(H), v \in Q(h)$.*

(ii) *If $A(x)$ and $\Phi(x)$ satisfy (9), $\mathcal{H} = \mathcal{H}_A + \Phi$ is essentially selfadjoint on $C_0^\infty(\mathbf{R}^d)$ and the closure H of $\mathcal{H}$ is bounded from below by m.*

We can show Theorem 1,mimicking the arguments used in Leinfelder-Simader (1981), who proved a definitive result that the nonrelativistic Schrödinger operator $(2m)^{-1}(-i\nabla - A(x))^2 + \Phi(x)$ is essentially

selfadjoint on $C_0^\infty(\mathbf{R}^d)$ when
(15)
$$A \in L_{loc}^4(\mathbf{R}^d), \operatorname{div} A \in L_{loc}^2(\mathbf{R}^d) \text{ and } \Phi \in L_{loc}^2(\mathbf{R}^d), \ \Phi(x) \geq 0 \quad a.e.$$

The idea of proof of the statement (ii) consists in showing that, when $A(x)$ and $\Phi(x)$ satisfy (9), $C_0^\infty(\mathbf{R}^d)$ is an operator core of the selfadjoint operator H obtained through the form h in the statement (i). The details of the proof will be referred to Ichinose-Tsuchida (1991b).

We want to call the selfadjoint operator H obtained in Theorem 1 (i) and (ii) the *Weyl quantized relativistic Hamiltonian* corresponding to the classical symbol (2).

Remarks. 1) Ichinose (1989) proved that $\mathcal{H}_A$ in (12) is essentially selfadjoint on $C_0^\infty(\mathbf{R}^d)$ under the assumption (11), by establishing what is called *Kato's inequality* for $\mathcal{H}_A$: if $u \in L^2(\mathbf{R}^d)$ with $\mathcal{H}_A u \in L_{loc}^1(\mathbf{R}^d)$, then

$$(16) \qquad \operatorname{Re}((\operatorname{sgn} u)\mathcal{H}_A u) \geq \sqrt{-\Delta + m^2}|u|$$

in the sense of distributions, where $(\operatorname{sgn} u)(x) = \overline{u(x)}/|u(x)|$ for $u(x) \neq 0; = 0$ for $u(x) = 0$. This result has recently been improved in Ichinose-Tsuchida (1991a) by extending (16) to the case where $A(x)$ is $L_{loc}^{2+\delta}(\mathbf{R}^d)$ and $\int_{0<|y|<1} |y(A(x+y/2) - A(x))|n(dy)$ is in $L_{loc}^2(\mathbf{R}^d)$.

2) We can take the zero-mass limit and nonrelativistic limit of our Hamiltonian (see Ichinose 1991, cf. Ichinose 1987). Namely, for its semigroup and unitary group, we have, if $A(x)$ and $\Phi(x)$ satisfy (9),

$$\exp[-t(H^{m,c} - mc^2)] \xrightarrow{s} \exp[-tH^{0,c}], \quad m \downarrow 0, \ t \geq 0,$$

$$\exp[-it(H^{m,c} - mc^2)] \xrightarrow{s} \exp[-itH^{0,c}], \quad m \downarrow 0, \ t \in \mathbf{R},$$

and, if $A(x)$ and $\Phi(x)$ satisfy (15),

$$\exp[-t(H^{m,c} - mc^2)] \xrightarrow{s} \exp[-tH^{NR}], \quad c \to \infty, \ t \geq 0,$$

$$\exp[-it(H^{m,c} - mc^2)] \xrightarrow{s} \exp[-itH^{NR}], \quad c \to \infty, \ t \in \mathbf{R},$$

all the convergences being uniform on each finite t-interval. Here $H^{m,c}$ is the Weyl quantized relativistic Hamiltonian, with the light velocity c restored, that is, corresponding to the classical symbol
$$\sqrt{c^2(p - A(x))^2 + m^2c^4} + \Phi(x), \text{ where the usual factor } 1/c \text{ in front of}$$

$A(x)$ is omitted so that it can be kept fixed in the limit, and H^{NR} is the selfadjoint nonrelativistic Schrödinger operator $(2m)^{-1}(-i\nabla - A(x))^2 + \Phi(x)$.

3. Path integral representation

We want to extend the path integral formula (5a,b) for the semigroup $e^{-t[H-m]}$ to the case where $A(x)$ and $\Phi(x)$ satisfy (13). H is the selfadjoint operator associated with the form h in (14). However, in this note we content ourselves with the case that

$$(17) \qquad A \in L^2_{loc}(\mathbf{R}^d) \text{ and } \Phi \in L^1_{loc}(\mathbf{R}^d), \quad \Phi(x) \geq 0 \quad a.e.,$$

to show for $m > 0$ the following path integral representation for the semigroup $e^{-t[H-m]}$ resolving Eq. (1).

THEOREM 2. *For* $g \in L^2(\mathbf{R}^d)$,

$$(18a) \qquad (e^{-t[H-m]}g)(x) = \int_{D_x} e^{-S(t,X)} g(X(t))d\lambda_x(X),$$

with

$$
(18b) \quad
\begin{aligned}
S(t,X) = {}& i \int_0^{t+} \int_{|y|\geq 1} A(X(s-) + y/2)y N_X(dsdy) \\
& + i \int_0^{t+} \int_{0<|y|<1} A(X(s-) + y/2)y \tilde{N}_X(dsdy) \\
& + i \int_0^{t} ds \text{ p.v.} \int_{0<|y|<1} A(X(s) + y/2)y n(dy) \\
& + \int_0^{t} \Phi(X(s))ds.
\end{aligned}
$$

Proof. I. Let $A(x)$ and $\Phi(x)$ be as in (17). Choose a sequence $\{A_k\}$ in $\mathcal{B}^\infty(\mathbf{R}^d)$ with $|A_k(x)| \leq |A(x)|$ which is convergent to $A(x)$ in L^2_{loc} and a.e., and a sequence $\{\Phi_k\}$ in $\mathcal{B}^\infty(\mathbf{R}^d)$ with $0 \leq \Phi_k(x) \leq \Phi(x)$ which is convergent to $\Phi(x)$ in L^1_{loc} and a.e., as $k \to \infty$. Then by (5a,b) we have

$$(19) \qquad (e^{-t[H^k-m]}g)(x) = \int_{D_x} e^{-S_k(t,X)} g(X(t))d\lambda_x(X),$$

where $S_k(t, X)$ is the $S(t, X)$ in (5b) with A_k and Φ_k in place of A and Φ, and H_k is the selfadjoint operator associated with the form $h_k \equiv h_{A_k,\Phi_k}$ in (14). We shall show both the sides of (19) converge to those of (18a) as $k \to \infty$.

As far as the left-hand side is concerned, by Ichinose-Tsuchida(1991b) [Lemma 3.6], H_k converges to H in the strong resolvent sense, and so by Kato (1976) [Chap. IX, Theorem 2.16, p.504], $\exp[-t(H_k - m)]g$ converges to $\exp[-t(H - m)]g$, uniformly on each finite t-interval in $[0, \infty)$, in L^2 and, if a subsequence is taken, a.e.

II. To see convergence of the right-hand side, we shall show that as $k \to \infty$, $\exp[-S_k(t, X)]$ converges for a.e. x and λ_x-a.e. X and its limit can be written as $e^{-S(t,X)}$. Put $S_k(t, X) = \sum_{j=1}^4 S_k^{(j)}(t, X)$ and $S(t, X) = \sum_{j=1}^4 S^{(j)}(t, X)$. We show each $\exp[-S_k^{(j)}(t, X)]$ converges for a.e. x and λ_x-a.e. X.

(i) There exists a Borel set K_1 in $\mathbf{R}^d$ of Lebesgue measure zero such that $|A(x)|$ is finite and $A_k(x) \to A(x)$ for $x \notin K_1$. Then for each (s, y) with $0 < s \le t$ and $|y| \ge 1$, $G_1(s, y) = \{X \in D_x; X(s-) + y/2 \in K_1\}$ has λ_x-measure zero, because $\int_{G(s,y)} d\lambda_x(X) = \int_{K_1 - y/2} k_0(s, x - z)dz$, where $k_0(t, x)$ is the kernel of the operator $\exp[-t(\sqrt{-\Delta + m^2} - m)]$, i.e. the inverse Fourier transform of (6) (For its explicit expression see Ichinose 1989). Therefore by the Fubini theorem

$$G_1 = \{(X, s, y) \in D_x \times (0, t] \times \{|y| \ge 1\}; \quad X(s-) + y/2 \in K_1\}$$

has $d\lambda_x \otimes dsn(dy)$-measure zero:

$$\iiint I_{G_1}(X, s, y)d\lambda_x(X)dsn(dy)$$
$$= \int_0^t \int_{|y| \ge 1} dsn(dy) \int_{G_1(s,y)} d\lambda_x(X) = 0,$$

if $I_{G_1}(X, s, y)$ is the indicator function for the set G_1. It follows again by the Fubini theorem that for λ_x-a.e. X, $G_1(X) = \{(s, y) \in (0, t] \times \{|y| \ge 1\}; X(s-) + y/2 \in K_1\}$ has $N_X(dsdy)$-measure zero, since

$$\int_{D_x} d\lambda_x(X) \iint_{G_1(X)} N_X(dsdy) = \int_{D_x} d\lambda_x(X) \iint_{G_1(X)} dsn(dy).$$

Therefore for λ_x-a,e. X, as $k \to \infty$, $A_k(X(s-) + y/2) \to A(X(s-) + y/2), N_X(dsdy)$-a.e., and the integral $S^{(1)}(t, X)$ exists, being a finite sum because $X(s)$ has at most finitely many discontinuities s with the jump $|X(s) - X(s-)|$ exceeding a given constant. By the Lebesgue dominated convergence theorem, for λ_x-a.e. X, $S_k^{(1)}(t, X) \to S^{(1)}(t, X)$ and hence $\exp[-S_k^{(1)}(t, X)] \to \exp[-S^{(1)}(t, X)]$.

(ii) For $n > 0$ let $\sigma_n(X) = \inf\{s > 0; |X(s-)| > n\}$. Then for λ_x-a.e. X, $\lim_{n\to\infty} \sigma_n(X) = \infty$. We have, for n fixed,

$$\int_{\mathbf{R}^d} dx \int_{D_x} |S_k^{(2)}(t \wedge \sigma_n(X), X) - S_\ell^{(2)}(t \wedge \sigma_n(X), X)|^2 d\lambda_x(X)$$

$$= \int dx \int d\lambda_x(X) \int_0^t \int_{0<|y|<1} I_{[0,\sigma_n(X)]}(s)$$
$$\times |[A_k(X(s) + y/2) - A_\ell(X(s) + y/2)]y|^2 dsn(dy)$$
$$\leq \int dx \int_0^t \int_{0<|y|<1} dsn(dy)$$
$$\times \int_{|z|\leq n} |(A_k(z + y/2) - A_\ell(z + y/2))y|^2 k_0(s, x - z) dz$$
$$\leq Ct \int_{|z|\leq n+1} |(A_k(x) - A_\ell(x)|^2 dx \to 0, \quad k, \ell \to \infty,$$

with a constant C. By passing to a subsequence, we see that for a.e. x and for λ_x-a.e. X, $\{S_k^{(2)}(t \wedge \sigma_n(X), X)\}$ converges to $S^{(2)}(t \wedge \sigma_n(X), X)$ as $k \to \infty$, so that $S_k^{(2)}(t, X) \to S^{(2)}(t, X)$ and hence $\exp[-S_k^{(2)}(t, X)] \to \exp[-S^{(2)}(t, X)]$, because $\lim_{n\to\infty} \sigma_n(X) = \infty$ for λ_x-a.e. X.

(iii) We can show with the theory of singular integrals that $a(x) \equiv$ p.v. $\int_{0<|y|<1} A(x + y/2)yn(dy)$ exists pointwise a.e. in x, while $a_k(x) \equiv \int_{0<|y|<1}(A_k(x + y/2) - A_k(x))yn(dy) = $ p.v. $\int_{0<|y|<1} A_k(x + y/2)yn(dy)$ exists for every x, and that as $k \to \infty$, $a_k(x)$ converges to $a(x)$ in L_{loc}^2. Similarly to (ii) with the same $\sigma_n(X)$ as there, we have, for n fixed,

$$\int_{\mathbf{R}^d} dx \int_{D_x} |S_k^{(3)}(t \wedge \sigma_n(X), X) - S_\ell^{(3)}(t \wedge \sigma_n(X), X)|^2 d\lambda_x(X)$$

$$= \int dx \int d\lambda_x(X)$$

$$\times \left| \int_0^t I_{[0,\sigma_n(X)]}(s)[a_k(X(s)) - a_\ell(X(s))]ds \right|^2$$

$$\leq \int dx \int_0^t ds \int_{|z| \leq n} |(a_k(z) - a_\ell(z))|^2 k_0(s, x - z)dz$$

$$\leq t \int_{|z| \leq n} |(a_k(x) - a_\ell(x))|^2 dx \to 0, \quad k, \ell \to \infty.$$

By passing to a subsequence, it can be seen that for a.e. x and λ_x-a.e. X, $\{S_k^{(3)}(t \wedge \sigma_n(X), X)\}$ converges to $S^{(3)}(t \wedge \sigma_n(X), X)$, as $k \to \infty$, so that $S_k^{(3)}(t, X) \to S^{(3)}(t, X)$ and hence $\exp[-S_k^{(3)}(t, X)] \to \exp[-S^{(3)}(t, X)]$.

(iv) The proof will proceed in the same way as the proof of Simon (1979)[Chap.II, Theorem 6,2, p.51]. We may suppose that $\Phi_k(x) \uparrow \Phi(x)$ a.e. There exists a Borel set K_4 in $\mathbf{R}^d$ of Lebesgue measure zero such that $\Phi_k(x) \uparrow \Phi(x)$ for $x \notin K_4$. Then for $0 < s \leq t$, $G_4(s, y) = \{X \in D_x; X(s) \in K_4\}$ has λ_x-measure zero. Therefore by the Fubini theorem $G_4 = \{(X, s) \in D_x \times (0, t]; X(s) \in K_4\}$ has $d\lambda_x \otimes ds$-measure zero, so that for λ_x-a.e. X, $G_4(X) = \{s \in (0, t]; X(s) \in K_4\}$ has Lebesgue measure zero. It follows by the monotone convergence theorem that for λ_x-a.e. X, $S_n^{(4)}(t, X) \to S^{(4)}(t, X)$ and hence $\exp[-S_n^{(4)}(t, X)] \to \exp[-S^{(4)}(t, X)]$.

We end the proof by noting that in the above arguments all the steps except II(ii) can also be shown only under the assumption (13).

Acknowledgements

The author wishes to thank Professor W. Schulze and Professor M. Demuth for their invitation to the Lambrecht Symposium. He is also grateful to Professor M. Demuth and Professor R. Seiler for their kind hospitality at the Karl-Weierstrass Institut für Mathematik and the Technische Universität Berlin in December 1991. Thanks are expressed to Mr. T. Tsuchida and Professor M. Tsuchiya for useful discussions.

References

Bilingsley P.(1968), Convergence of Probability Measures. Wiley, New York.

Ichinose T.(1987), The nonrelativistic limit problem for a relativistic spinless particle in an electromagnetic field. J.Functional Analysis **73**, 233-257.

Ichinose T.(1989), Essential selfadjointness of the Weyl quantized relativistic Hamiltonian. Ann. Inst. H. Poincaré, Phys. Théor. **51**, 265-298.

Ichinose T.(1991), Remarks on the Weyl quantized relativistic Hamiltonian. Note di Matematica (to appear).

Ichinose T. and Tamura H.(1986), Imaginary-time path integral for a relativistic spinless particle in an electromagnetic field. Commun. Math. Phys. **105**, 239-257.

Ichinose T. and Tsuchida T.(1991a), On Kato's inequality for the Weyl quantized relativistic Hamiltonian, Preprint.

Ichinose T. and Tsuchida T.(1991b), On essential self-adjointness of the Weyl quantized relativistic Hamiltonian, Preprint.

Ikeda N. and Watanabe S.(1981), Stochastic Differential Equations and Diffusion Processes. North-Holland/Kodansha, Amsterdam, Tokyo.

Kato T.(1976), Perturbation Theory for Linear Operators. 2nd ed., Springer, Berlin-Heidelberg-New York.

Landau L.D. and Lifschitz E.M.(1975), Course of Theoretical Physics, Vol.2, The Classical Theory of Fields. 4th revised English ed., Pergamon Press, Oxford.

Leinfelder H. and Simader C.G.(1981), Schrödinger operators with singular magnetic vector potentials. Math. Z. **176**, 1-19.

Nagase M. and Umeda T.(1987), On the essential self-adjointness of quantum Hamiltonians of relativistic particles in magnetic fields. Sci. Rep., Col. Gen. Educ. Osaka Univ., **36**, 1-6.

Nagase M. and Umeda T.(1990), Weyl quantized Hamiltonians of relativistic spinless particles in magnetic fields. J. Functional Analysis, **92**, 136-154.

Reed M. and Simon B.(1979), Methods of Modern Mathematical Physics, IV : Analysis of Operators. Academic Press, New York.

Shubin M.A.(1987), Pseudodifferential Operators and Spectral Theory. Springer, Berlin-Heidelberg.

Simon B.(1979), Functional Integration and Quantum Mechanics. Academic Press, New York.

Author's address:

Takashi Ichinose
Department of Mathematics
Kanazawa University
Kanazawa 920
Japan

Operator Theory:
Advances and Applications, Vol. 57
© 1992 Birkhäuser Verlag Basel

SPECTRAL ASYMPTOTICS FOR THE FAMILY
OF COMMUTING OPERATORS

Victor Ivrii

Abstract. Accurate spectral asymptotics for the family of com mut-
ing pseudo-differential operators are obtained. Some applica tions for
highly accurate spectral asymptotics for operators with periodic Haml-
tonian flow is given.

1. General theory

Let $A_j = A_j(x, hD, h)$ $(j = 1, ..., l)$ be h-pseudo-differential operators
on manifold X. We assume that

(1.1) Either $X = \mathbb{R}^d$ or X is a compact closed C^K-manifold or X is a
compact C^K-manifold with a boundary $\partial X \in C^K$ with large enough
$K = K(d)$, $d = \dim X$. In the third case A_j are differential operators and their original domains are given by boundary value conditions
$B_i u|_{\partial X} = 0$ for $i = 1, ..., \mu$ (originally A_j are given on smooth functions
and then we take their closures).

Moreover, let us assume that

(1.2) Operators A_j are self-adjoint operators in $L^2(X, \mathsf{H})$ where H
is a Hermitian space[1] and commute (i.e. their spectral projectors

---

[1] Surely, one can consider operators acting in the Hermitian bundles.

Typeset by $\mathcal{A}_{\mathcal{M}}\mathcal{S}$-TₑX

commute).

Then

$$(1.3) \qquad A_j = \int_{\mathbb{R}^l} \tau_j \, d_\tau E(\tau)$$

where $\tau = (\tau_1, ..., \tau_l)$, $E(\tau) = E_1(\tau_1)...E_l(\tau_l)$, $E_j(\tau_j)$ is the spectral projector of A_j. We are interested in the semi-classical asymptotics of $\mathrm{Tr}\, QE(\tau'; \tau)$ where $E(\tau'; \tau) = (E_1(\tau_1) - E_1(\tau_1'))...(E_l(\tau_l) - E_l(\tau_l'))$ and Q is an appropriate h-pseudo-differential operator.

To give more precise description of operators in question we assume that

(1.4) Either A_j are h-differential operators of orders m_j or $A_j \in \tilde{\Psi}^{m_j}_{\rho,\gamma,h}$ where classes of h-pseudo-differential operators $\Psi_{\rho,\gamma,h}$ and $\tilde{\Psi}^m_{\rho,\gamma,h}$ are introduced in [1] for $\rho, \gamma \in (0, 1]$, $\rho\gamma \geq h^{1-\delta}$, $\delta > 0$ arbitrarily small.

The main result of this section is

THEOREM 1.1.   *Let operators $A_1, ..., A_l$ satisfy* (1.1), (1.2), (1.4). *Let*

$$(1.5) \qquad \Omega \subset \{(x, \xi) \in T^*X, |x| \leq c, |\xi| \leq c\}, \mathrm{dist}(\Omega, \partial X) \geq \epsilon_0$$

and

$$(1.6) \quad |D^\alpha_{x,\xi,h} A_j(x, \xi, h)| \leq c$$
$$\forall (x, \xi) \in \Omega \quad \forall h \in (0, h_0) \quad \forall \alpha : |\alpha| \leq K$$

with $K = K(d, l, s)$ *where s is arbitrary and $\epsilon_0 > 0$, $h_0 > 0$ are arbitrarily small constants.*

Moreover let us assume that the following microhyperbolicity condition is fulfilled:

(1.7) *For every $\zeta \in \mathbb{S}^l$ and every $(x, \xi) \in \Omega$ there exists $T \in T_{(x,\xi)}T^*X$ with $|T| \leq 1$ such that*

$$\langle T a_\zeta(x, \xi)v, v\rangle \geq \epsilon_0 |v|^2 - c \sum_{1 \leq j \leq l} |a_j(x, \xi)v|^2 \quad \forall v$$

where $a_\zeta = \sum_{1 \leq j \leq l} \zeta_j a_j(x, \xi)$ and $a_j = A_j(x, \xi, 0)$ are principal symbols of A_j. Let $Q = Q(x, hD, h)$ be h-pseudo-differential operator with the *symbol satisfying (1.6) and supported in* $\Omega_{\epsilon_0} = \{(x,\xi), \mathrm{dist}((x,\xi),\complement\Omega) \geq \epsilon_0$. *Let* $\phi_j \in C^K([-1,1])$ *and*

$$|D^p \phi_j| \leq c \quad \forall p \leq K \quad \forall j = 1, ..., l.$$

Then estimates

$$(1.8) \quad \mathcal{R}_1 = |\mathrm{Tr}\ E(\tau';\tau)Q - h^{-d}\int \varkappa_0(\tau';\tau)| \leq$$

$$Ch^{-d+l+1} \sum_{1 \leq i \leq l} \prod_{1 \leq j \leq l, j \neq i} (1 + |\tau_j - \tau_j'|/h)$$

$$\forall \tau_1', ..., \tau_l \in [-\varepsilon, \varepsilon],$$

$$(1.9) \quad \mathcal{R}_1 = |\int_{\mathbb{R}^l} \phi_1(\tau_1'/L_1)...\phi_l(\tau_l'/L_l)d_{\tau'}E(\tau'',\tau') -$$

$$\sum_{0 \leq n \leq N} h^{-d+n} \int_{\mathbb{R}^l} \phi_1(\tau_1'/L_1)...\phi_l(\tau_l'/L_l)\varkappa_n'(\tau')d\tau'| \leq$$

$$Ch^{-d}(h/\min_{1 \leq j \leq l} L_j)^s$$

$$\forall L_j \in (h, \varepsilon) \quad j = 1, ..., l$$

hold for small enough constant $\varepsilon > 0$.

REMARK 1.2.   (i) Even for $l = 1$ estimate (1.9) of this theorem is stronger than theorem 4.3.9 from [2].

(ii) We leave to the reader to formulate the similar assertion in the case when Ω isn't disjoint from the boundary. In this case all the conditions of §5.3 [3] are assumed to be fulfilled and we assume that locally $X = \{x_1 \geq 0\}$, $Q = Q(x, hD', h)$, $x' = (x_2, ..., x_d)$; then $\Omega \cap \{x_1 = 0\}$ is a cylindrical domain with respect to ξ_1). In this case we assume that the modified standard microhyperbolicity (in an appropriate multidirection condition)of §5.3 [3] is fulfilled for boundary value problems (A_ζ, B) where $A_\zeta = \sum_j \zeta_j A_j$ and the modification is the same as above: in the right-hand expression we replace $|a(x,\xi)_\zeta v|^2$ or $\|a_\zeta(x', D_1, \xi')v\|_+^2$ by $\sum_j |a_j(x,\xi)v|^2$ or $\sum_j \|a_j(x', D_1, \xi')v\|_+^2$ respectively.

(iii) Moreover, if

$$(1.10) \qquad \sum_{1 \leq j \leq l} |a_j(x,\xi)v|^2 \geq \epsilon_0 |v|^2 \quad \forall v \quad \forall(x,\xi) \in \Omega$$

then $\varkappa_0 = 0$ and the right-hand expressions of estimates (1.8),(1.9) contain additional factor h.

(iv) Moreover, let us assume in frames of theorem or (ii),(iii) that $L_j \leq h^{1-\delta}$ with $\delta > 0$ for $j = k, ..., l$ (with $k = 1, ..., l$). Then for estimate (1.9) the microhyperbolicity condition should be checked only for ζ with $\zeta_k = ... = \zeta_l = 0$. In this case an estimate similar to (1.8) with mollification on $(\tau_k, ..., \tau_l)$ and with $k - 1$ instead of l in the right-hand expression is true.

(v) Coefficients $\varkappa'_n(\tau_1, ..., \tau_l)$ are the formal calculus coefficients of

$$\mathrm{Tr} \left(\prod_{1 \leq j \leq l} \mathrm{Res}_{\mathbf{R}}(\tau_j - A_j)^{-1} \times Q \right)$$

where $\mathrm{Res}_{\mathbf{R}} f(\tau) = \frac{1}{2\pi i}(f(\tau - i0) - f(\tau + i0))$. In particular

$$(1.11) \qquad \varkappa_0(\tau'; \tau) = (2\pi)^{-d} \int \mathrm{tr}\, \mathcal{E}(x, \xi; \tau', \tau) q(x, \xi) dx d\xi$$

where $\mathcal{E}(x, \xi; \tau', \tau)$ is the corresponding projector for commuting symbols $a_1(x, \xi), ..., a_l(x, \xi)$.

(vi) Under conditions to Hamiltonian trajectories the above remainder estimates can be improved.

(vii) Certain results can be obtained for the Riesz means. $\square$

SKETCH OF THE PROOF. The proof of theorem 1.1 (as well as proofs of the similar assertions near the boundary) uses the same arguments as in [2,3]. We consider multiparametrical unitary group

$$U(t) = \exp ih^{-1}(t_1 A_1 + ... + t_l A_l) = U_1(t_1)...U_l(t_l),$$
$$U_j(t_j) = \exp ih^{-1}t_j A_j.$$

If operator A_ζ is microhyperbolic at supp Q in an appropriate direction then results of [1] yield immediately that $\mathrm{Tr}\, U(t)Q$ is negligible uniformly with respect to $t = \zeta t'$, $h^{1-\delta} \leq t' \leq T_0$ where $T_0 > 0$ is an appropriate constant. Therefore, if A_ζ is microhyperbolic at supp Q for every $\zeta \in \mathbf{S}^{l-1}$ then $\mathrm{Tr}\, U(t)Q$ is negligible uniformly with respect to $h^{1-\delta} \leq |t| \leq T_0$. On the other hand for $|t| \leq h^{1/2+\delta}$ the Schwartz kernels of $U_j(t_j)$ are constructed by the successive approximation method in [2] and therefore under the microhyperbolicity condition we have $\mathrm{Tr}\, U(t)Q$

in $B(0, T_0) \subset \mathbf{R}^l$. Then using Tauberian arguments modified for "multi-time" case we obtain estimates

$$\mathrm{Tr}\ E(\tau', \tau)Q \le Ch^{-d+l} \quad \forall \tau, \tau' : |\tau -' \tau| \le c$$

and

$$(1.12) \qquad \mathrm{Tr}\ E(\tau', \tau)Q \le Ch^{-d}(|\tau_1 - \tau_1'| + h)...(|t_l - \tau_l'| + h).$$

Let us apply operator $\phi_1(hD_{t_1}/L_1)...\phi_k(hD_{t_l}/L_l)$ to $\bar\chi_1(t_1)...\bar\chi_l(t_l)U(t)$ and set $t_1 = ... = t_l = 0$ where $\bar\chi_j \in C_0^K([-1,1])$ equal 1 in $[-1/2, 1/2]$. We obtain (1.9). Moreover, one can replace here $\phi_j(\tau_j'/L_j)$ by $\phi_j((\tau_j' - \tau_j)/L_j)$ with $|\tau_j| \le \varepsilon$ and integrate by parts. Replacing τ' by τ'' and taking appropriate partition of unity in the box $\{\tau_j' + h \le \tau_j'' \le \tau_j - h \quad j = 1, ..., l\}$ (without loss of generality one can assume that $\tau_j' \le \tau_j\ \forall j$) we obtain estimate (1.8) modified by the following way: we replace τ', τ by λ', λ, multiply by

$$h^{-2l}\phi_1((\lambda_1' - \tau_1')/h)...\phi_l((\lambda_l' - \tau_l')/h)\phi_1((\lambda_1 - \tau_1)/h)...\phi_l((\lambda_l - \tau_l)/h)$$

and integrate on λ', λ (inside of $|.|$). However (1.11) yields that the same estimate for difference of this modified expression and the original one provided $\int \phi_j(\tau_j)d\tau_j = 1$ for $\forall j$.

On the other hand if condition (1.10) is fulfilled on supp Q then standard elliptic arguments yield that

$$T^{-l}|F_{t \to h^{-1}\tau}\chi_T(t)\mathrm{Tr}\ U(t)Q|$$

is negligible uniformly with respect to $|\tau| \le 2\varepsilon$ and $T \ge h^{1-\delta}$ where $\varepsilon > 0$ is a sufficiently small constant, $\chi \in C_0^K(B(0,1))$ is an arbitrary fixed function. Then standard Tauberian arguments yield estimate

$$\|E(\tau'; \tau)Q\| \le Ch^s \quad \forall \tau', \tau : |\tau'| \le \varepsilon, |\tau| \le \varepsilon.$$

Taking small partition of unity we obtain estimates (1.8),(1.9) under microhyperbolicity condition (1.7). $\square$

2. Corollaries. Degenerate case.

In this section we discuss different corollaries of theorem 1.1. First of all let us assume that symbols $a_1, ..., a_r$ are scalar with $r = 1, ..., l$. Then condition (1.7) is equivalent to the pair of conditions

$$(2.1) \qquad |a_1| + ... + |a_r| + |da_1| + ... + |da_r| \geq \epsilon_1 \quad \forall(x, \xi) \in \Omega$$

for some constant $\epsilon_1 > 0$ and

(2.2) At $\Sigma = \{(x, \xi) \in \Omega, a_1 = ... = a_r = 0\}$ condition (1.7) is fulfilled for symbols $a_{r+1}, ..., a_l$ with T tangent to Σ.

The second condition is empty for $r = l$.

Let us assume now that

(2.3) All the symbols $a_1, ..., a_l$ are scalar.

Then instead of (1.7) we have (2.1) with $r = l$. Let us reject this condition assuming however that (2.1) with $r = l - 1$ is fulfilled. In this case without loss of generality one can assume microlocally that $\sigma(A_j) = \xi_j$ for $j = 1, ..., l - 1$ where $\sigma(A_j)$ are complete symbols of A_j. Moreover, then $\sigma(A_l)$ doesn't depend on $x_1, ..., x_r$. Let us consider symbol $b = \sigma(A_l)|_\Lambda$ with $\Lambda = \{x_1 = ... = x_r = \xi_1 = ... = \xi_r = 0\}$. Let us introduce scaling function γ at Λ linked with this symbol:

$$(2.4) \qquad\qquad \gamma = \epsilon(|b| + |db|^2)^{1/2} + h^{1/2}$$

(see §4.3 of [2]). Applying method partition-dilatation (see §4.3 [2]) with scaling function 1 on $x_1, ..., x_r$ (and γ^2 with respect to $\xi_1, ..., \xi_r$) and γ with respect to $x_{r+1}, ..., x_d$ (and γ with respect to $\xi_{r+1}, ..., \xi_d$) we obtain

THEOREM 2.1.   *Let all the conditions of theorem 1.1 excluding condition (1.7) be fulfilled. Moreover let conditions (2.3) and (2.1) with $r = l - 1$ be fulfilled. Let symbol $a' = a_l|_\Sigma$ satisfy condition*

(2.5) $|a'| + |d_\Sigma a'| \leq \epsilon \Rightarrow \mathrm{Hess}_\Sigma a'$ has two eigenvalues f_1 and f_2 with $|f_1| \geq \epsilon, |f_2| \geq \epsilon$ with constant $\epsilon > 0$.

Then estimates (1.8) with additional factor $(|\log h| + 1)$ in the right-hand expression holds.

Moreover under condition

$(2.5)_+$ $|a'| + |d_\Sigma a'| \le \epsilon \Rightarrow \mathrm{Hess}_\Sigma a'$ has two eigenvalues of the same sign f_1 and f_2 with $|f_1| \ge \epsilon, |f_2| \ge \epsilon$ with constant $\epsilon > 0$

estimate (1.8) *holds.*

REMARK 2.2. (i) Applying more delicate arguments of [4] one can under condition

$(2.5)'$ $|a'| + |d_\Sigma a'| + |\mathrm{Hess}_\Sigma a'| \ge \epsilon$ with some constant $\epsilon > 0$

obtain estimate (1.8) with additional factor $h^{-\delta}$ in the right-hand expression with arbitrary small $\delta > 0$.

(ii) I think that applying propagation singularities arguments (see theorem 4.3.14 [2]) under condition (2.5) one can prove estimate (1.8).

(iii) Starting from theorem 2.1 and applying the same arguments one can prove estimate (1.8) with additional factor $(|\log h|+1)^{l-r}$ under conditions (2.3) and (2.1) (with $r \le l - 1$) and under certain restriction to symbols $a'_k = a_{r+k}|_\Sigma$, $k = 1, ..., l - k = p$. In the non-uniform form (i.e. for fixed $a_1, ..., a_l$) this condition is given in terms of fundamental matrices (skew-Hessians) of $a'_1, ..., a'_l$ at point where $a'_1, ..., a'_p, da'_1, ..., da'_p$ vanish (provided Q is supported in the small neighbourhood of this point); these matrices F_i commute (due to equalities $\{a_i, a_j\} = 0$).

3.Operators with the periodic Hamiltonian flow

Now we discuss how to apply the results of sections 1,2 to operators with the periodic Hamiltonian flows of the principal symbols. So let us consider now one scalar operator A with the principal symbol a and let us assume that

(3.1) Either X is a compact closed manifold and $A \in \tilde{\Psi}^{(1)}$ or $X = \mathbb{R}^d$ and symbol A satisfies inequalities

$$|D^\alpha_{(x,\xi)}a| \le C(1 + |x| + |\xi|)^{2-|\alpha|} \quad \forall \alpha : |\alpha| \le K.$$

In this case one says that operator A' is negligible if $h^{-s}\Lambda_s A'\Lambda_s$ is uniformly bounded for large enough s where $\Lambda_s = (1 + |x|^2 + h^2|D|^2)^{s/2}$.

Let us assume that

(3.2) $$\Phi_1(x, \xi) = (x, \xi)$$

where Φ_t is the Hamiltonian flow generated by a. Then

$$e^{ih^{-1}A} \equiv e^{iB}$$

for appropriate h-pseudo-differential operator operator B. Under certain condition to A operator B is really "smaller" than I and can be replaced by ηB with $\eta = h^r$. On the other hand, perturbing A by $\mu A'$ with $h^{r+1} \leq \mu \leq h^\delta$ we obtain equality

$$(3.3) \qquad e^{ih^{-1}A} \equiv e^{i\eta B}$$

with $\eta = \mu h^{-1}$. So let us assume *that (3.3) is fulfilled with $\eta \in [h^n, h^{\delta-1}]$ with arbitrarily small exponent $\delta > 0$.* More precisely this initial job is discussed in §4.5 of [2]. It is easy to see that (3.3) is also fulfilled with B replaced by $B' + (h + \eta)B''$ where

$$B' = \int_0^1 e^{-ih^{-1}tA} B e^{ih^{-1}tA} dt$$

and B' is an uniformly bounded h-pdo. Continuing this process one can replace B in (3.3) by operator which commutes with A modullo $O(h^s)$; moreover, at every step one can replace B by $(B + B^*)/2$; so one can assume without loss of generality that

(3.4) B is symmetric operator and $\|[B, A]\| \leq Ch^s$.

We need first to treat the case when

$$(3.5) \qquad e^{ih^{-1}A} \equiv I.$$

In this case Spec $A \subset \cup_{n \in \mathbb{Z}} I_n$ where $I_n = [(2\pi n - C_0\eta)h, (2\pi n + C_0\eta)h]$ and $\eta \leq Ch^s$.

Treating $\mathrm{Tr}\ \Phi(D_t)U(t)|_{t=0}$ one can prove easily

THEOREM 3.1. *Let conditions* $(3.1), (3.2), (3.5)$ *be fulfilled and moreover let us assume that*

$$(3.6) \qquad \mathrm{dist}((x, \xi), \Phi_t(x, \xi)) \geq \epsilon t(1 - t) \quad \forall t \in (0, 1) \quad \forall (x, \xi) \in \Sigma_0$$

where $\Sigma_\tau = \{a = \tau\}$ *is assumed to be compact; this condition means that there is no subperiodic trajectory. Then* $\forall \tau = 2\pi nh$ *with* $|\tau| \leq \varepsilon$

the Schwartz kernel of $E(\tau - \zeta, \tau + \zeta)$ is the Lagrangian distribution with the Lagrangian manifold $(x, \xi, x', \xi') \in \Sigma_\tau, \exists t(x, \xi) = \Phi_t(x', -\xi')$; $\zeta = C_0 h^s$ here[2].

Let B be self-adjoint operator commuting with A. Then one can treat distribution of eigenvalues of B on $\mathrm{Ran}\,(\tau - \zeta, \tau + \zeta)$. The usual analysis of [2] modified by rather obvious way yeilds

THEOREM 3.2. *Let A satisfy conditions of theorem 3.2 and let B be self-adjoint operator commuting with A. Let us assume that condition (2.1) is fulfilled for $a_1 = a$ and $a_2 = b$ at $\Sigma_{\tau,0}$; $|\tau| \leq \varepsilon$. Then $\forall \tau = 2\pi n h$ with $|\tau| \leq \varepsilon$*

$$(3.7) \quad |\mathrm{Tr}\, E_A(\tau - \zeta, \tau + \zeta) E_B(\lambda', \lambda) Q - \varkappa_0 h^{1-d}(\tau; \lambda', \lambda)| \leq C h^{2-d}$$

$$\forall \lambda', \lambda : |\lambda'| \leq \varepsilon, |\lambda| \leq \varepsilon$$

and moreover estimate similar to (1.9) also has place. Here

$$(3.8) \quad \varkappa_0 = (2\pi)^{-d} \int_{\{\Sigma_\tau, \lambda' \leq b \leq \lambda\}} q(x, \xi) dx d\xi : da.$$

REMARK 3.3. (i) Moreover, similar results hold for a family of commuting operators B_j (commuting also with A). Moreover, A can replaced be by a family of commuting operators A_i also.

(ii) In frames of theorem 3.2 method of partition and dilatation can be applied. Therefore one can replace condition (1.7) by condition (2.5) or $(2.5)_+$.

(iii) Let us consider A satisfying (3.1)-(3.3). Without loss of generality one can assume that (3.4) is fulfilled also. Then operator $A^0 = A - B$ satisfies conditions of theorem 3.1. This yields that there exists $B' \equiv B$ commuting with A^0 and then under condition (3.6) theorem 3.2 provides all the results of §4.5[2] and even slightly better (under (2.5)-type conditions) for operator $A' = A^0 + B'$; these results obviously remains true for A.

References

[1] V.Ivrii. (1990), Semiclassical microlocal analysis and precise spectral asymptotics. Preprint 1. Ecole Polytechnique, Preprint M964.1190, November 1990.

---

[2] However $E(\tau - \zeta; \tau + \zeta)$ is not Fourier integral operator because $\dim \mathrm{Ker}\, d\pi_X = \dim \mathrm{Ker}\, d\pi_Y = 1 (\neq 0)$ for projectors $\pi_X : \Lambda \to T^* X$, $\pi_Y : \Lambda \to T^* Y$.

[2] V.Ivrii. (1991), Semiclassical microlocal analysis and precise spectral asymptotics. Preprint 2. Ecole Polytechnique, Preprint M969.0191, January 1991

[3] V.Ivrii. (1991), Semiclassical microlocal analysis and precise spectral asymptotics. Preprint 3. Ecole Polytechnique, Preprint M971.0291, February 1991.

[4] V.Ivrii. (1991), Semiclassical microlocal analysis and precise spectral asymptotics. Preprint 6. Ecole Polytechnique, Preprint M1018.1091, October 1991.

[5] V.Ivrii. (1991), Semiclassical spectral asymptotics. Summer School on Semiclassical Analysis, Nantes, June, 1991. To appear in Asterisque.

[6] Y.Colin de Verdiere. (1979), Sur les spectres des opérateurs elliptiques á bicaractéristiques toutes périodiques, Comment. math. helv., 54, no 3, p.508–522 and references here.

[7] Y.Colin de Verdiere. (1980), Spectre conjoint d'opérateurs qui commutent. Math. Z., 171, p.51–73 and references here.

[8] A.-M.Charbonnel. (1988) Comportement semi-classique du spectre conjoint d'opérateurs pseudodifferentielsqui commutent. Asympt. Anal. 1, p.227–261 and references here.

[8] V.V.Guillemin, S.Sternberg. (1985) On the spectra of commuting pseudo-differential operators. Lect. Notes Pure Appl. Math., 48, p.149–165.

[9] A.Weinstein. (1977), Asymptotics of the eigenvalues; clusters for Laplacian plus a potential. Duke Math. J., 44, p.883–892.

Author's address

Victor Ivrii
Centre de Mathematiques
Ecole Polytechnique
91128 Palaiseau, FRANCE
ivrii@orphee.polytechnique.fr

Pseudo differential operators with negative definite functions as symbol : Applications in probability theory and mathematical physics

Niels Jacob

Abstract. We discuss pseudo differential operators $a(x,D)$ with a symbol $a(x,\xi)$ which is with respect to ξ a continuous negative definite function. Such operators do occur in probability theory and mathematical physics.

Introduction

In this paper a pseudo differential operator is any extension of an operator $a(x,D)$ defined on $C_0^\infty(\mathbb{R}^n)$ by

$$(0.1) \qquad a(x,D)u(x) = (2\pi)^{-n/2} \int_{\mathbb{R}^n} e^{ix \cdot \xi} \, a(x,\xi) \, \hat{u}(\xi) \, d\xi \, ,$$

where $a : \mathbb{R}^n \times \mathbb{R}^n \to \mathbb{R}$ is a function which is continuous and for any fixed $x \in \mathbb{R}^n$ the function $\xi \to a(x,\xi)$ is negative definite in the sense of Beurling and Deny. Such operators do occur naturally in the theory of Markov processes and recently they had also been considered in mathematical physics. In general these symbols do not belong to a classical symbol class. In particular a continuous negative definite function need not be differentiable. For this reason one can not use standard techniques to handle these operators. But it is still possible to get results for some non–trivial classes of operators. In this paper we want to present recent investigations on these operators.

1. Continuous negative definite functions

We start with

Definition 1.1. A function $a : \mathbb{R}^n \to \mathbb{C}$ is said to be **negative definite** if for all $m \in \mathbb{N}$ and $x^1,...,x^m \in \mathbb{R}^n$ the matrix $(a(x^i)+\overline{a(x^j)}-a(x^i-x^j))_{i,j=1,...,m}$ is positive Hermitian.

A standard reference for negative definite functions is the book [1] .

Example 1.1. A. For $0 \le s \le 1$ the function $\xi \mapsto |\xi|^{2s}$ is a continuous negative definite function on $\mathbb{R}^n$. **B.** The function $\xi \mapsto (|\xi|^2 + m^2)^{1/2} - m$, $m \ge 0$, is a continuous negative definite function on $\mathbb{R}^n$. **C.** Let $a_j : \mathbb{R}^{n_j} \to \mathbb{R}$, $j = 1,2$, be two continuous negative definite functions. Then the function $a(\xi,\eta) := a_1(\xi) + a_2(\eta)$ is con–

tinuous and negative definite on $\mathbb{R}^{n_1+n_2}$. D. Any continuous function $a : \mathbb{R} \to \mathbb{R}$ which is even, non-negative and whose restriction to $[0,\infty)$ is increasing and concave is negative definite.

An important representation theorem for continuous negative definite functions is the Lévy– Khinchine formula which reads for real–valued functions as follows (see [11], p.5–09) :

A real–valued continuous negative definite function $a : \mathbb{R}^n \to \mathbb{R}$ has the representation

$$(1.1) \qquad a(\xi) = c + Q(\xi) + \int_{\mathbb{R}^n} (1-\cos(\xi\cdot\eta)) \frac{1+|\eta|^2}{|\eta|^2} \, d\sigma(\eta) ,$$

where $c \geq 0$ is a constant, Q is a non–negative quadratic form on $\mathbb{R}^n$ and σ is a positive measure on $\mathbb{R}^n$ which does not charge the origin and has finite total mass.

Lemma 1.1. Let $a : \mathbb{R}^n \to \mathbb{R}$ be a continuous negative definite function. Then we have

$$(1.2) \qquad 0 \leq a(0) \leq a(\xi) \leq c_a(1+|\xi|^2)$$

for all $\xi \in \mathbb{R}^n$. Further $a^{1/2}$ is also a continuous negative definite function.

In order to get "good" estimates for pseudo differential operators having a symbol $a(x,\xi)$ which is with respect to ξ a continuous negative definite function we need

Lemma 1.2. ([26], p.157) Let $a : \mathbb{R}^n \to \mathbb{R}$ be a continuous negative definite function. Then we have for all $\xi,\eta \in \mathbb{R}^n$

$$(1.3) \qquad |a(\xi)-a(\eta)| \leq 4a^{1/2}(\xi)a^{1/2}(\xi-\eta) + a(\xi-\eta) .$$

Moreover, W.Hoh proved, see [22], Lemma 2.2, :

Lemma 1.3. Let $a^2 : \mathbb{R}^n \to \mathbb{R}$ be a continuous negative definite function and let a be its square root. Then for all $\xi, \eta \in \mathbb{R}^n$ we have

$$(1.4) \qquad |a(\xi) - a(\eta)| \leq a(\xi-\eta) .$$

Note that (1.3) and (1.4) were used as substitutes for the mean value theorem when proving commutator estimates for the operators $a^2(D)$ and $a(D)$, respectively.

2. Why to consider pseudo differential operators with a symbol being continuous and negative definite in the second variable?

We will show that pseudo differential operators with a symbol which is independent of x and a continuous negative definite function with respect to ξ arise naturally as generators of translation invariant Feller semigroups, i.e. certain convolution semigroups, and they do arise also as generators of translation invariant Dirichlet forms. Thus in both cases a stochastic process could be associated with such an operator. These processes are well known, they are the Lévy processes or processes with stationary and independent increments.

Definition 2.1. Let $(\mu_t)_{t\geq0}$ be a family of probability measures on $\mathbb{R}^n$. It is called a **convolution semigroup** of probability measures if $\mu_t \to \epsilon_0$ vaguely for $t \to 0$, ϵ_0 being the Dirac measure, and $\mu_{s+t} = \mu_s * \mu_t$ for all $s,t \geq 0$.

The next result is well known and essentially due to S.Bochner, see [1], Theorem 8.3.

Theorem 2.1. A family $(\mu_t)_{t\geq0}$ of probability measures on $\mathbb{R}^n$ is a convolution semi-group if and only if there exists a complex–valued negative definite function a on $\mathbb{R}^n$ such that for $t \geq 0$

$$(2.1) \qquad \hat{\mu}_t(\xi) = e^{-ta(\xi)}.$$

Now let $(\mu_t)_{t\geq0}$ be a convolution semigroup of probability measures on $\mathbb{R}^n$ and consider the semigroup $(T_t)_{t\geq0}$ of operators T_t defined on $C_0^\infty(\mathbb{R}^n)$ by

$$(2.2) \qquad (T_t u)(x) = \int_{\mathbb{R}^n} u(x+y)\, \mu_t(y).$$

A formal calculation gives for the generator of the semigroup

$$(2.3) \qquad Au(x) = \lim_{t\to 0}\left[\frac{T_t u - T_0 u}{t}\right](x) = F^{-1}_{\xi\mapsto x}(-\overline{a(\xi)}\hat{u}(\xi))(x).$$

Assuming (as in the following) a to be real–valued we finally get

$$(2.4) \qquad Au(x) = -(2\pi)^{-n/2}\int_{\mathbb{R}^n} e^{ix\cdot\xi}\, a(\xi)\, \hat{u}(\xi)\, d\xi.$$

Remember that the set of all convolution semigroups on $\mathbb{R}^n$ is in one–to–one correspondence to the set of all Lévy processes with state space $(\mathbb{R}^n,B^n)$. The correspondence of a Lévy process $(\Omega,A,P,(X_t)_{t\geq0})$ to a convolution semigroup $(\mu_t)_{t\geq0}$ is given by

$$(2.5) \qquad \mu_t = P_{X_t-X_0},$$

where P_X denotes the distribution of the random variable $X : \Omega \to \mathbb{R}^n$.

It is clear that all information about the process is contained in the function a, i.e. in the operator (2.4). However, starting with (2.4) one can ask whether certain perturbations of it do also generate a stochastic process and one can try to obtain results for this process by studying the operator. Thus we look at an operator

$$(2.6) \qquad -a(x,D)u(x) = -(2\pi)^{-n/2}\int_{\mathbb{R}^n} e^{ix\cdot\xi}\, a(x,\xi)\, \hat{u}(\xi)\, d\xi,$$

and try to find conditions in order that it generates a stochastic process. In the next section we will discuss conditions in order that the operator (2.6) generates a Feller semigroup and therefore a Feller process.

Let us again consider the operator (2.4) and define the bilinear form E^a on $C_0^\infty(\mathbb{R}^n)$ by

$$(2.7) \qquad\qquad E^a(u,v) \;=\; \int_{\mathbb{R}^n} a(\xi)\,\hat{u}(\xi)\,\overline{\hat{v}(\xi)}\,d\xi\;.$$

It turns out that this form is a closed form on $L^2(\mathbb{R}^n)$ with domain
$$D(E^a) \;=\; \{\,u \in L^2(\mathbb{R}^n)\,,\, E^a(u,u) < \infty\,\}\;.$$

Since we are only interested in real-valued symbols let us change our notation. Instead of (2.4) consider

$$(2.8) \qquad\qquad a^2(D)u(x) \;=\; (2\pi)^{-\text{L}/2}\int_{\mathbb{R}^n} e^{ix\cdot\xi}\,a^2(\xi)\,\hat{u}(\xi)\,d\xi\;,$$

where now $a^2 : \mathbb{R}^n \to \mathbb{R}$ is a continuous negative definite function. Introducing the scale of Hilbert spaces

$$(2.9)\quad H^{a^2,s}(\mathbb{R}^n) \;=\; \{\,u \in L^2(\mathbb{R}^n)\,,\, \|u\|_{s,a^2}^2 = \int_{\mathbb{R}^n}(1+a^2(\xi))^{2s}\,|\hat{u}(\xi)|^2\,d\xi < \infty\,\}\;,$$

we find $D(E^{a^2}) = H^{a^2,1/2}(\mathbb{R}^n)$.

The bilinear form E^{a^2} is not only a closed, symmetric, non-negative form, but it has also the following remarkable Markovian property :

$$(2.10) \qquad\quad u \in H^{a^2,1/2}(\mathbb{R}^n)\;\text{ implies }\;(0\vee u)\wedge 1 \in H^{a^2,1/2}(\mathbb{R}^n)\;\text{ and }$$

$$(2.11) \qquad\quad E^{a^2}((0\vee u)\wedge 1,(0\vee u)\wedge 1) \;\leq\; E^{a^2}(u,u)\;.$$

In general we give

Definition 2.2. Let E be a symmetric, closed, non-negative bilinear form with dense domain $D(E) \subset L^2(\mathbb{R}^n)$. It is called a **Dirichlet form** if (2.10) and (2.11) holds.

The notion of a Dirichlet form was introduced by A.Beurling and J.Deny, see [2], but also [11] and [37] and as the standard reference the monograph [17]. An important result in [2] says that all translation invariant Dirichlet forms are given by (2.7). Using the well known correspondence between closed symmetric non-negative forms and non-positive self-adjoint operators on $L^2(\mathbb{R}^n)$ we can ask whether the operator (2.6) generates a Dirichlet form. In section 4 we will construct certain Dirichlet forms by starting with an operator given by (2.6).
From the remarks made above, it is clear that the class of pseudo differential operators under consideration is suitable to handle certain problems in the theory of stochastic processes. However it turns out that some of these operators do also play an important role in mathematical physics. Moreover, the fact that a process corresponds to the operator is often used in these studies. The operator which is the most interesting for mathematical physicists is given by

$$(2.12) \qquad\qquad H_1 u(x) \;=\; (2\pi)^{-n/2}\int_{\mathbb{R}^n} e^{ix\cdot\xi}\,((|\xi|^2 + m^2)^{1/2} - m)\,\hat{u}(\xi)\,d\xi\;,$$

which is a relativistic Hamiltonian. The spectral analysis for the corresponding

Schrödinger operator $H_1 + V$, where V is a potential, had been investigated be-sides others by I. Daubechies [7], [8], I.Daubechies and E.Lieb [9], I.Herbst [20] and more recently by R. Carmona and coauthors [4] – [5]. These authors considered also as an approximation of H_1 the operator

$$(2.13) \qquad H_2 u(x) = (2\pi)^{-n/2} \int_{\mathbb{R}^n} e^{ix \cdot \xi} \, |\xi| \, \hat{u}(\xi) \, d\xi$$

and in [34] – [35] F.Nardini discussed Schrödinger operators related to

$$(2.14) \qquad H_3 u(x) = (2\pi)^{-n/2} \int_{\mathbb{R}^n} e^{ix \cdot \xi} \, (|\xi|^2 + 1)^{1/2} \, \hat{u}(\xi) \, d\xi \, .$$

The operators H_1, H_2 and H_3 do all belong to the class under consideration. But note that the operator

$$(2.15) \qquad H_4 u(x) = ((-i\nabla - A(x))^2 + m^2)^{1/2} u(x)$$

defined as a Weyl pseudo differential operator and considered by T.Ichinose [24] could not considered within our class. Besides these concrete operators let us mention the paper of I.Herbst and A.Sloan [21] and that of M.Demuth and J.van Casteren [10] who considered more general operators.

3. On a class of Feller semigroups generated by pseudo differential operators

In this section we mainly follow our paper [27] and discuss examples from [23] and [31]. First we recall some basic facts about Feller semigroups and their generators.

Definition 3.1. A family of bounded linear operators $T_t : C_\infty(\mathbb{R}^n) \to C_\infty(\mathbb{R}^n)$ is called a **Feller semigroup** if

(F.1) $\qquad\qquad T_{s+t} = T_s T_t$ and $T_0 = \mathrm{id}$ (semigroup property);

(F.2) $\qquad\qquad \lim_{t \to 0} \|T_t u - u\|_\infty = 0$ (strongly continuous);

(F.3) $\qquad\qquad 0 \leq T_t u \leq 1$ if $0 \leq u \leq 1$ (contraction and positivity preserving).

The generators of Feller semigroups are characterized by

Theorem 3.1. ([14],p.165) Let $D(A)$ be a linear subspace of $C_\infty(\mathbb{R}^n)$ and let $A : D(A) \to C_\infty(\mathbb{R}^n)$ be a linear operator. Suppose further

(A.1) $D(A)$ is dense in $C_\infty(\mathbb{R}^n)$;

(A.2) A satisfies the positive maximum principle on $D(A)$, i.e.if $u \in D(A)$ and $x_0 \in \mathbb{R}^n$ such that $\sup_{x \in \mathbb{R}^n} u(x) = u(x_0) \geq 0$ then it follows that $Au(x_0) \leq 0$;

(A.3) for some $\lambda \geq 0$ the operator $\lambda - A$ maps $D(A)$ onto a dense subspace of $C_\infty(\mathbb{R}^n)$.

Then A has a closed extension which is the generator of a Feller semigroup.

The next theorem is due to Ph.Courrège :

Theorem 3.2. ([8],p.2–37) Let $a : \mathbb{R}^n \times \mathbb{R}^n \to \mathbb{R}$ be a continuous function such that for each $x \in \mathbb{R}^n$ the function $\xi \mapsto a(x,\xi)$ is negative definite. Then the operator $-a(x,D)$ defined on $C_0^\infty(\mathbb{R}^n)$ by

$$(3.1) \qquad -a(x,D)u(x) \;=\; -\,(2\pi)^{-n/2} \int_{\mathbb{R}^n} e^{ix\cdot\xi}\, a(x,\xi)\, \hat{u}(\xi)\, d\xi$$

satisfies the positive maximum principle on $C_0^\infty(\mathbb{R}^n)$.

Unfortunately it is not possible to prove (A.3) for the operator (3.1) when its domain is $C_0^\infty(\mathbb{R}^n)$. In order to overcome these difficulties let us consider an operator with symbol as in (3.1) which satisfies in addition

$$(3.2) \qquad c_1(1+a^2(\xi)) \;\leq\; a(x,\xi) \;\leq\; c_2(1+a^2(\xi))$$

for $|\xi|$ large and a fixed continuous negative definite function $a^2 : \mathbb{R}^n \to \mathbb{R}$. Moreover assume

$$(3.3) \qquad c_0(1+|\xi|^2)^t \;\leq\; (1+a^2(\xi))$$

for some $t \in (0,1]$. It follows that the space $H^{a^2,s}(\mathbb{R}^n)$ is now continuously embedded into the Sobolev space $H^{st}(\mathbb{R}^n)$. Now we propose the following procedure in order to apply Theorem 3.1 :

Since by the Sobolev embedding theorem we have $H^r(\mathbb{R}^n) \subset C_\infty(\mathbb{R}^n)$ for $r > \frac{n}{2}$, see

[33],p.121, we try to define the operator $a(x,D)$ on $H^{a^2,m+1}(\mathbb{R}^n)$ for some large $m \in \mathbb{N}$ and try to prove

i) $a(x,D)(H^{a^2,m+1}(\mathbb{R}^n)) \subset H^{a^2,m}(\mathbb{R}^n) \subset C_\infty(\mathbb{R}^n)$;

ii) $(a(x,D)+\lambda)(H^{a^2,m+1}(\mathbb{R}^n)) \subset C_\infty(\mathbb{R}^n)$ is dense for some $\lambda \geq 0$;

iii) $-a(x,D)$ satisfies the positive maximum principle on $H^{a^2,m+1}(\mathbb{R}^n)$.

It turns out that for some classes of operators we get indeed generators of Feller semigroups. First we have

Theorem 3.3. ([28], Theorem 6.3) Let $0 < r \leq 1$ and set $\lambda^{2r}(\xi_j) = |\xi_j|^{2r}$, $1 \leq j \leq n$.

Further suppose that arbitrarily often differentiable functions $b_j : \mathbb{R}^n \to \mathbb{R}$ are given.

Assume in addition that all partial derivatives of b_j are bounded and $\dfrac{\partial b_j}{\partial x_j} = 0$,

$b_j(x) \geq d_1 > 0$ for all $x \in \mathbb{R}^n$, and that for some $x_0 \in \mathbb{R}^n$

$$(3.4) \qquad \max_{1 \leq j \leq n} \; \sup_{x \in \mathbb{R}^n} \; |b_j(x) - b_j(x_0)| \;\leq\; c(n,r)\cdot d_1$$

holds ($c(n,r)$ explicitly known). Then the operator $\displaystyle\sum_{j=1}^{n} b_j(x)\,\lambda_j^{2r}(D_j)$ defined on

$H^{\infty}(\mathbb{R}^n)$ extends to a generator of a Feller semigroup.

The next result is taken from [31]:

Let $p : \mathbb{R}^n \times \mathbb{R}^n \to \mathbb{R}$ be a continuous function such that for fixed $x \in \mathbb{R}^n$ the function $p(x,.) : \mathbb{R}^n \to \mathbb{R}$ is negative definite. Further suppose that for some $x_0 \in \mathbb{R}^n$ we have

$$p(x,\xi) = p(x_0,\xi) + (p(x,\xi) - p(x_0,\xi)) = p_1(\xi) + p_2(x,\xi) \,,$$

where p_1 and p_2 satisfy the following assumptions :

P.1.   There exists a continuous negative definite function $a^2 : \mathbb{R}^n \to \mathbb{R}$ such that

$$(3.5) \qquad\qquad |p_1(\xi)| \le \gamma_1(1+a^2(\xi))$$

holds for all $\xi \in \mathbb{R}^n$.

P.2.q Let a^2 be as in P.1. Assume that for some $q \in \mathbb{N}$ the function $p_2(.,\xi) : \mathbb{R}^n \to \mathbb{R}$ is q–times differentiable and that for any $\alpha \in \mathbb{N}_0^n$, $|\alpha| \le q$, there exists a function $\varphi_\alpha \in L^1(\mathbb{R}^n)$ such that (3.6) holds:

$$(3.6) \qquad\qquad |\partial_x^\alpha p_2(x,\xi)| \le \varphi_\alpha(x)\,(1+a^2(\xi)).$$

P.3. $p_1(\xi) \ge \gamma_0 a^2(\xi)$ for all $\xi \in \mathbb{R}^n$, $|\xi| \ge \rho \ge 0$,

P.4 $\gamma_b \le (1-\epsilon)\gamma_0$ for some $0 < \epsilon < 1$, where γ_b is given by

$$(3.7) \qquad\qquad \gamma_b = c_a\,\tilde{\gamma}_q \sum_{|\alpha|\le q} \|\varphi_\alpha\|_{L^1} \int_{\mathbb{R}^n} (1+|\tau|^2)^{(1-q)/2}\,d\tau\,,\ q > n+1 .$$

P.5. $\gamma_c \le \gamma_0/\sqrt{12}$, where γ_0 is given in P.3 and γ_c by

$$(3.8) \qquad\qquad \gamma_c = \tilde{\gamma}_q \sum_{|\alpha|\le q} \|\varphi_\alpha\|_{L^1} \int_{\mathbb{R}^n} (1+|\tau|^2)^{-q/2}\,d\tau\,,\ q > n .$$

Here c_a is given by (1.2) and $\tilde{\gamma}_q$ is a constant such that

$$(1+|\xi|^2)^{q/2} \le \tilde{\gamma}_q \sum_{|\alpha|\le q} |\xi^\alpha|$$

holds.

Theorem 3.4. ([31], Theorem 5.2) suppose that $p(x,D)$ satisfies P.1 − P.5 with q sufficiently large and that a^2 fulfills (3.3). Then $-p(x,D)$ has a closed extension which is the generator of a Feller semigroup on $\mathbb{R}^n$.

The proof of Theorem 3.3 and that of Theorem 3.4 goes as follows :

1. Define the bilinear form B on $C_0^\infty(\mathbb{R}^n)$ by $B(u,v) = (p(x,D)u,v)_0$. Prove $|B(u,v)| \leq c\|u\|_{1/2,a^2}\|v\|_{1/2,a^2}$ and $B(u,u) \geq c_1\|u\|_{1/2,a^2}^2 - c_2\|u\|_0^2$. This implies that for λ sufficiently large for each $f \in L^2(\mathbb{R}^n)$ there exists a weak solution $u \in H^{a^2,1/2}(\mathbb{R}^n)$ of $(p(x,D)+\lambda)u = f$, i.e. $B_\lambda(u,\varphi) = (f,\varphi)_0$ for all $\varphi \in C_0^\infty(\mathbb{R}^n)$.

2. Prove that $f \in H^{a^2,m_0}(\mathbb{R}^n)$ implies always $u \in H^{a^2,m_0+1}(\mathbb{R}^n)$ for the weak solution.

3. Use an approximation argument to show that $p(x,D)$ with domain $H^{a^2,m_0+1}(\mathbb{R}^n)$ satisfies the positive maximum principle.

4. On a class of Dirichlet forms generated by pseudo differential operators

We will discuss a result of a joint paper with W.Hoh, [23], where we proved that certain pseudo differential operators $a(x,D)$ do generate a Dirichlet form on $L^2(\mathbb{R}^n)$. In this case we could reduce the regularity assumptions on the coefficients b_j.

Theorem 4.1. ([23]) For $1 \leq j \leq n$ let $a_j^2 : \mathbb{R}^n \to \mathbb{R}$ be a continuous negative definite functions. Further let $b_j \in L^\infty(\mathbb{R}^n)$ be independent of x_j and assume $b_j(x) \geq d_0 > 0$ for all $j = 1,...,n$. Then $a(x,D) = \sum_{j=1}^n b_j(x)\, a_j^2(D_j)$ generates a Dirichlet form

$$B(u,v) = \sum_{j=1}^n \int_{\mathbb{R}^n} b_j(x)\, a_j(D_j)u(x)\, a_j(D_j)v(x)\, dx$$

with domain $H^{a^2,1/2}(\mathbb{R}^n)$, where $a^2(\xi) = \sum_{j=1}^n a_j^2(\xi_j)$.

The proof of this theorem uses certain estimates for B, for example a Gårding inequality

$$(4.1) \qquad B(u,u) \geq d_0\|u\|_{1/2,a^2}^2 - d_0\|u\|_0^2,$$

and the Lévy–Khinchine formula which enables us to show

$$B(u,u) = \int_{\mathbb{R}^n} \int_{\mathbb{R}^n} (u(x+y) - u(x))^2\, J(x,dy)\, dx$$

with a certain measure $J(x,dy)$.

5. Some remarks to the techniques

In principle the proofs split into two parts. We have to show certain a priori estimates and secondly we have to take care on the compability with the order structure, i.e we have to prove the positive maximum principle or the Markovian property. The second part is always proved by using the Lévy–Khinchine formula and this fact is responsible that we have to use symbols which are continuous negative definite functions with respect to the second variable. In order to get the required a priori estimates we try to follow the classical Hilbert space methods as they are used for uniformly elliptic differential operators. However, one important tool is not avaiable: Leibniz' role. But the fact which is really needed is that commutators of differential operators with functions are expressed as lower order differential operators. Thus a substitute for Leibniz' role should be obtained by proving commutator estimates. Let us quote a typical example for such a commutator estimate :

Theorem 5.1. Let p_2 satisfy P.2.q. Moreover let $t \in \mathbb{R}^n$, $s \geq 0$ and $N \in \mathbb{N}$ such that

$0 < s-N \leq 1$. Then we have for all $u \in H^{a^2,s+t+1/2}(\mathbb{R}^n)$

$$(5.2) \qquad \| [(1+a^2(D))^s, p_2(x,D)] u \|_{t,a^2} \leq c \| u \|_{s+t+1/2,a^2} .$$

provided that $q > n+2s+2|t|+4N+4$.

The proof of Theorem 5.1 follows the lines of the standard proof for commutator estimates for pseudo differential operators, see [36]. But instead of Peetre's inequality we have to use certain properties of continuous negative definite functions, in particular we need Lemma 1.2 and Lemma 1.3. A first version of Theorem 5.1 is given in [26] and [29]. A great improvement was Theorem 3.1 in [22] given by W.Hoh. The final form of the theorem was proved in [31]. Let us note that Theorem 5.1 is also a considerable improvement of Corollary 2.8 in [21].
It is the fact that we need a commutator estimate as (5.2) which determines our regularity assumptions on the coefficients when constructing a Feller semigroup.

6. What can analysis do for probability theory?

In order to construct a stochastic process from a Feller semigroup $(T_t)_{t\geq0}$ one has

to define transition probabilities $p_t(.,.)$ on $\mathbb{R}^n \times B^n$, which is in principle done by

$$(6.1) \qquad p_t(x,A) = T_t \chi_A(x) ,$$

where χ_A is the characteristic function of the Borel set A . Using these transition probabilities the Kolmogorov theorem gives the existence of a process ([13] or [14]).

On the other hand to any regular Dirichlet form E with domain $D(E) \subset L^2(\mathbb{R}^n)$ one can construct a Hunt process $(\Omega,A,(X_t)_{t\geq0},(P_x)_{x\in\mathbb{R}^n\cup\{\infty\}})$ which is uniquely determined up to a set of capacity zero. The proof of the existence of the process involves much about the potential theory of Dirichlet forms in order to construct certain transition probabilities, but in a final step one also has to use the Kolmogorov theorem to arrive at a process. These results for Dirichlet forms are due to M.Fukushima [15], see also his monograph [17].
In both cases one is interested in further properties of the process, for example :
1. Do there exist modifications of the process with nice paths?
2. Do the transition probabilities have densities?

3. Is it possible to describe the time asymptotics of the transition probabilities?
4. Is the process transient or recurrent?
5. Does the process have certain ergodicity properties?
6. When is the process uniquely determined, i.e. when is there no exceptional set?

Since we constructed the process by analytical means, our task is now to transform analytical results into probabilistic one. It turns out that certain a priori estimates already proved in order to get the Feller semigroup or the Dirichlet form are sufficient to arrive at such results. First let us recall some general results of this type.

Theorem 6.1. ([16]) Suppose for a symmetric regular Dirichlet form E on $L^2(\mathbb{R}^n)$

$$\|u\|_{L^q}^2 \le c(E(u,u) + c_0\|u\|_0^2)$$

for some $q > 2$ and $c_0 \ge 0$. Then there exists a Borel set N of zero capacity such that

$\mathbb{R}^n \backslash N$ is invariant and the transition probability $p_t(x,.)$ is absolutely continuous with

respect to the Lebesgue measure for each $t > 0$ and $x \in \mathbb{R}^n \backslash N$.

Moreover a result of N.Varopoulos [40] (see also [3]) implies that if
$$(6.1) \qquad\qquad \|u\|_{L^q}^2 \le c_1(B_\lambda(u,u) + c_2\|u\|_0^2)$$

is valid for all $u \in H^{a^2,1/2}(\mathbb{R}^n)$, $\nu \in (2,\infty)$, $q = 2\nu/(\nu-2) > 0$, then it holds

Theorem 6.2. Let (6.1) hold for all $u \in H^{a^2,1/2}(\mathbb{R}^n)$. Then there exists a constant $d > 0$ such that for the semigroup $(T_t^\lambda)_{t>0}$ generated by B_λ on $L^2(\mathbb{R}^n)$ the estimate

$$||T_t^\lambda||_{L^1-L^\infty} \le d\frac{e^{c_2't}}{t^{\nu/2}}$$

holds for $t > 0$.

Now consider the Dirichlet form constructed in Theorem 4.1 and assume in addition that a^2 fulfills the estimate
$$(6.2) \qquad\qquad c_s(1+|\xi|^2)^{s/2} \le 1 + a^2(\xi)\,,\, 0 < s \le 1\,.$$
Then we find for $\lambda \ge 0$ by the Sobolev embedding theorem that
$$\|u\|_{L^q}^2 \le c\|u\|_{s/2}^2 \le c'\|u\|_{1/2,a^2}^2 \le c''(B(u,u) + (\lambda+d_0)\|u\|_0^2)\,,$$
where we used (4.1), i.e. Gårding's inequality. Thus we can apply these results to the constructed Dirichlet form and the corresponding Markov process.
Let $L(x,D)$ be the operator considered in Theorem 3.3. For this operator we have a further analytical result :

Theorem 6.3. ([18], Theorem 3.1) Let L be the Friedrichs extension of $L(x,D)$. Then for any $k \in \mathbb{N}$ we have $D(L^k) = H^{2ks}(\mathbb{R}^n)$.

This result is proved by purely analytical means. But combining it with the theory of (r,p)–capacities developped by M.Fukushima and H.Kaneko [19] and H.Kaneko [34] we arrive at the following ergodicity result :

Theorem 6.4. ([18], Theorem 4.2) Let $(\Omega, A, (X_t)_{t \geq 0}, (P_x)_{x \in \mathbb{R}^n \cup \{\infty\}})$ be the Markov process associated with the operator $L(x,D)$ from Theorem 3.3. Then for each $f \in L^2(\mathbb{R}^n)$, there exists $h \in L^2(\mathbb{R}^n) \cap C_\infty(\mathbb{R}^n)$ such that

$$(6.3) \qquad \lim_{\substack{t \to \infty \\ x \in \mathbb{R}^n}} \sup | E_x(f(X_t)) - h(x) | = 0$$

and h satisfies $E_x(h(X_t)) = h(x)$, $x \in \mathbb{R}^n$, $t > 0$, where E_x denotes the expectation with respect to P_x .

7. Further results and remarks

In the last section we pointed out that we can use analytical statements to get probabilistic results. However, once it is known that one can associate a (Markov) process with the operator under consideration it is possible to use properties of the process to get analytical results. Let us mention some of these results.

We have seen that certain relativistic Hamiltonians do belong to our class. In order to study the corresponding Schrödinger operator $H + V$, it is often helpful to use a Feynman–Kac formula, i.e. the formula

$$(7.1) \qquad (e^{-tH}f)(x) = E_x\{f(X_t)e^{-\int_0^t V(X_s)ds}\} ,$$

where $(X_t)_{t \geq 0}$ is the Markov process generated by H . In [38] $-$ [39] a Feynman $-$ Kac formula was used to get a lot of results for Schrödinger operators associated with the generator of a Feller semigroup, in particular the potential theory of these operators have been investigated. Further interesting results have been obtained in [4] and [5]. Besides other things these authors do study the decay of the eigenfunctions of the operator $H_1 + V$, where H_1 is given by (2.12). They found that not only exponential decay can occur but also polynomial decay, depending on the potential and the fact whether or whether not the Lévy process generated by H_1 is transient or recurrent. In particular they found that if this Lévy process is recurrent, then the Schrödinger operator $H_1 + V$ has at least one negative bounded state whenever V is non–positive, non–identically zero, and bounded potential with compact support (The last statement is quoted from [7], Theorem V.1.)

Reference

[1]   Berg, C., and Forst, G. : Potential theory on locally compact Abelian groups. Ergebnisse der Mathematik und ihrer Grenzgebiete , II.Ser. Bd.87, Springer Verlag, Berlin $-$ Heidelberg $-$ New York, (1975).

[2]   Beurling, A., and Deny, J. : Dirichlet spaces. Proc. Natl. Acad. Sci. U.S.A. 45 (1959) 208 $-$ 215.

[3]   Carlen, E.A., Kusuoka, S., and Stroock, D.W. : Upper bounds for symmetric Markov transition functions. Ann. Inst. Henri Poincaré, Probabilités et Statistiques, Sup. no 2 Vol. 23 (1987) 245 $-$ 287.

[4]   Carmona, R. : Path integrals for relativistic Schrödinger operators. In : Proc.

Northern Summer School in Mathematical Physics. Aarhus 1988. Lect. Notes in Physics Vol. 345, 65 − 92. Springer Verlag, Berlin − Heidelberg − New York, (1989).

[5]   Carmona, R., Masters, W.C., and Simon, B. : Relativistic Schrödinger operators : Asymptotic behavior of the eigenfunctions. J. Funct. Anal. 91 (1990) 117 − 142.

[6]   Courrège, Ph. : Sur la forme intégro–différentielle des opérateurs de C_K^∞ dans C satisfaisant du principe du maximum. Sém. Théorie du Potentiel (1965/66) 38 p.

[7]   Daubechies, I. : An uncertainty principle for fermions with generalized kinetic energy. Comm. Math. Phys. 90 (1983) 311 − 320.

[8]   Daubechies, I. : One electron molecules with relativistic kinetic energy : Properties of the discrete spectrum. Comm. Math. Phys. 94 (1984) 523 − 535.

[9]   Daubechies, I., and Lieb, E. : One electron relativistic molecules with Coulomb interaction. Comm. Math. Phys. 90 (1984) 497 − 510.

[10]   Demuth, M., van Casteren, J. A. : On spectral theory of selfadjoint Feller generators. Rev. Math. Phys. 1 (1989) 325 − 414.

[11]   Deny, J. : Sur les espaces de Dirichlet. Sém. Théorie du Potentiel (1957) 12 p.

[12]   Deny, J. : Méthodes Hilbertiennes et théorie du potentiel. In : Potential Theory, C.I.M.E., Roma (1970), 123 − 201.

[13]   Dynkin, E.B. : Markov processes. Vol. 1. Die Grundlehren der mathematischen Wissenschaften Bd. 121, Springer Verlag, Berlin − Göttingen New York, (1965).

[14]   Ethier, S.N., and Kurtz, Th.G. : Markov processes − characterization and convergence. Wiley Series in Probability and Mathematical Statistics, John Wiley & Sons , New York (1985).

[15]   Fukushima, M. : On the generation of Markov processes by symmetric forms. Proc. 2nd Japan − USSR Symposium on Probability Theory. Lect. Notes Math. 330 , p. 46 − 79, Springer Verlag, Berlin − Heidelberg − New York, 1973.

[16]   Fukushima, M. : On an L^p−estimate of resolvents of Markov processes. Publ. R.I.M.S. 13 (1977) 277 − 284.

[17]   Fukushima, M. : Dirichlet forms and Markov processes. North Holland Math. Library Vol.23, North Holland Publ. Comp., Amsterdam − Oxford − New York, (1980).

[18]   Fukushima, M., Jacob, N., and Kaneko, H. : On (r,2)−capacities for a class of elliptic pseudo differential operators.(submitted)

[19]   Fukushima, M., and Kaneko, H. : On (r,p)−capacities for general Markov semigroups. In : "Infinite–dimensional analysis and stochastic processes". Proc. USP−meeting at Bielefeld 1983, Research Notes in Math. 124 , Pitman, Boston, Massachuetts, − London, (1985), p.41–47.

[20]   Herbst, I. : Spectral theory of the operator $(p^2 + m^2)^{1/2} - Ze^2/r$. Comm. Math. Phys. 53 (1977) 285 − 294.

[21]   Herbst, I.W., and Sloan, A.D. : Perturbation of translation invariant positivity preserving semigroups on $L^2(\mathbb{R}^n)$. Trans. Amer. Math. Soc. 236 (1978) 325 − 360.

[22]   Hoh, W. : Some commutator estimates for pseudo differential operators with negative definite functions as symbol. (Preprint 1991)

[23]   Hoh, W., and Jacob, N. : Some Dirichlet forms generated by pseudo differential operators. Bull. Sc. Math.(in press)

[24]   Ichinose, T. : Essential selfadjointness of the Weyl quantized relativistic

Hamiltonian. Ann. Inst. Henri Poincaré − Physique théorique 51 (1989) 265 − 298.

[25]   Jacob, N. : A Gårding inequality for certain anisotropic pseudo differential operators with non − smooth symbols. Osaka J. Math. 26 (1989) 857 − 879.

[26]   Jacob, N. : Commutator estimates for pseudo differential operators with negative definite functions as symbol. Forum Math. 2 (1990) 155 − 162.

[27]   Jacob, N. : Feller semigroups, Dirichlet forms, and pseudo differential operators.Forum Math.(in press)

[28]   Jacob, N. : A class of elliptic pseudo differential operators generating symmetric Dirichlet forms.(submitted)

[29]   Jacob, N. : Further pseudo differential operators generating Feller semigroups and Dirichlet forms.(submitted)

[30]   Jacob, N. : Pseudo differential equations and harmonic functions in Dirichlet spaces. (Preprint 1991)

[31]   Jacob, N. : A class of Feller semigroups generated by pseudo differential operators.(Preprint 1991).

[32]   Kaneko, H. : On (r,p)−capacities for Markov processes. Osaka J. Math. 23 (1986) 325 − 336.

[33]   Kumano−go, H. : Pseudo−differential operators. M.I.T. Press, Cambridge, Massachusetts, − London, (1981)

[34]   Nardini, F. : Spectral analysis of a pseudodifferential operator related to the relativistic Stark effect. B. U. M. I. (6) 3 B (1984) 53 − 73.

[35]   Nardini, F. : Exponential decay for eigenfunctions of the two body relativistic Hamiltonian. J. Analyse Math. 47 (1986) 87 − 109.

[36]   Oleinik, O.A., and Radkevic, E.V. : Second order equations with non−negative characteristic form. Amer. Math. Soc., Providence R.I. − Plenum Press, New York − London, (1973).

[37]   Silverstein, M.L. : Symmetric Markov processes. Lecture Notes in Mathematics, vol. 426, Springer Verlag, Berlin − Heidelberg − New York, (1974).

[38]   Sturm, K.−T. : Störung von Hunt−Prozessen durch signierte additive Funktionale. Dissertation, Friedrich−Alexander−Universität Erlangen−Nürnberg (1989).

[39]   Sturm, K.−T. : Gauge theorems for resolvents with application to Markov processes. Probab. Th. Rel. Fields 89 (1991) 387 − 406.

[40]   Varopoulos, N. : Hardy−Littlewood theory for semigroups. J. Funct. Anal.63 (1985) 240 − 260.

Niels Jacob
Mathematisches Institut
Universität Erlangen − Nürnberg
Bismarckstraße 1a
D − 8520 Erlangen
Deutschland

Operator Theory:
Advances and Applications, Vol. 57
© 1992 Birkhäuser Verlag Basel

One dimensional Schrödinger operators with high potential barriers.

W.Kirsch, S.A.Molchanov, L.A. Pastur

1   INTRODUCTION

A quantum mechanical particle moving in one-dimensional half-space under the influence of a potential V is described by the Hamiltonian

$$H_\alpha = -\frac{d^2}{dx^2} + V(x) \tag{1}$$

acting on $L^2[0, \infty)$ with boundary conditions :

$$\psi(0)\cos\alpha + \psi'(0)\sin\alpha = 0 \qquad \alpha \in [0, 2\pi) \tag{2}$$

It is well known that H has purely discrete spectrum if $V(x) \to \infty$ as $x \to \infty$. In fact,

A. Molchanov's theorem [3] tells us that purely discrete spectrum of H is equivalent with

$$\int_x^{x+1} |V(x)|\, dx \to \infty \text{ as } x \to \infty \tag{3}$$

Spencer and Simon [10] considered (non-negative) potentials V with $\overline{\lim}_{x\to\infty} V(x) = \infty$ in a qualified way. They proved that H has no *absolutely continuous spectrum* if there is a sequence $x_n \nearrow \infty$ such that

$$V(x) \geq V_n \text{ on } [x_n - h_n, x_n + h_n] \tag{4a}$$

where

$$V_n \to \infty \tag{4b}$$

and

$$h_n\sqrt{V_n} \to \infty \tag{4c}$$

The results of Spencer and Simon were extended to a wider class of potentials by Hislop and Nakamura [4].

In this note we investigate the question of *pure point spectrum* for operators of the above type. We will have to impose stronger conditions then (4) on the behavior of V at infinity. Even then we will be able to prove pure point spectrum (i.e. $\sigma_c(H_\alpha) = \phi$) only for typical values of the boundary condition α. This restriction is common in this type of questions. In fact, we proved in [7] that no growth condition like (4) can force H to have pure point spectrum for *all* α. The method employed here is based on the Simon-Spencer-paper and on Kotani's trick [8],[12],[2],[6]. In [7] the present authors proved analogous results for discrete Schrödinger operators.

2   STATEMENT OF THE RESULT

In the following we will assume for simplicity that $V(x) \geq 0$, but our method allows us to treat V that are unbounded below. We suppose furthermore that V is continuous, but $V \in L^2_{loc}$, not too negative, would be enough.

Let V satisfy conditions 4, i.e.

$$V(x) \geq V_n \quad \text{on} \quad [x_n - h_n, x_n + h_n] \tag{4a}$$

where

$$V_n \to \infty \tag{4b}$$

and

$$h_n \sqrt{V_n} \to \infty \tag{4c}$$

We also call $l_n := x_{n+1} - x_n > 0$. Again for simplicity we suppose $l_n \geq \beta > 0$.

Theorem 1 :   If

$$\sum_n (l_n^{3/2} + l_{n+1}^{3/2}) e^{\gamma \sqrt{V_n} h_n} < \infty \tag{5}$$

for some $\gamma < 1$, then

$$\sigma_c(H_\alpha) = \phi \quad \text{for Lebesgue-almost all} \quad \alpha \in [0, 2\pi)$$

(i.e. H_α has pure point spectrum).

Remark : The technique gives also results for energies $E < \lim V_n$ even if V_n does not converge to infinity.

3   A CRITERION FOR PURE POINT SPECTRUM

To prove theorem 1 we use Kotani's method in a version suitable for our problem. Let us denote by $R(x,y) = R_E^\alpha(x,y) = (H_\alpha - E)^{-1}(x,y)$ the Green's function of H_α.

Theorem 2 :   If

$$\int_0^\infty \left| R_{E+i0}^{\alpha_0}(0,y) \right|^2 dy < \infty \tag{6}$$

for some $\alpha_0 \neq 0$ and almost all $E \in (a,b)$ then $\sigma_c(H_\alpha) \cap (a,b) = \phi$ for Lebesgue-almost all α.

Remark : Note that $\alpha_0 = 0$ leads to $R_E^0(0,y) \equiv 0$. This is the reason to exclude $\alpha_0 = 0$ in the theorem. As the proof shows we could instead require

$$\int_0^\infty \left| D_1 R_{E+i0}^0(0,y) \right|^2 dy < \infty$$

where D_1 is the derivative with respect to the first variable.

Proof : We sketch a proof of theorem 2.

Let us call N_0 the set of energies E for which (6) does not hold. By assumption the Lebesgue measure $|N_0|$ of N_0 is zero. Standard ODE techniques imply that

$$\int \left| R_{E+i0}^\alpha(0,y) \right|^2 dy < \infty \tag{7}$$

for all α, if E is neither in N_0 nor an eigenvalue of H_α, i.e. (7) holds for all α and $E \neq N_0 \cup \epsilon(H_\alpha)$ ($\epsilon(H\alpha)$ is the set of eigenvalues of H_α).

Again by ODE techniques, it follows that the spectral measure μ_f^α of H_α for a function $f \in C_0^\infty(\mathbb{R}_+)$ fulfills :

$$\int \frac{1}{(E+\lambda)^2} \mu_f(d\lambda) < \infty \tag{8}$$

for $E \notin N_0 \cup \epsilon(H_\alpha)$.

By the De-Valle-Poussin-theorem the spectral measure of H_α is concentrated on $N_0 \cup \epsilon(H_\alpha)$. Thus $\sigma_{ac}(H_\alpha) = \phi$ for each α, since $|N_0 \cup \epsilon(H_\alpha)| = 0$ and $\sigma_{sc}(H_\alpha) \subset N_0$ since $\epsilon(H_\alpha)$ is countable. Consequently the singular continuous spectrum of H_α is concentrated on a set N_0 of Lebesgue measure zero, which is *independent* of α.

If we denote by σ^α the spectral measure corresponding to H_α we have

$$\int_0^{2\pi} \sigma^\alpha(A)\, d\alpha = c|A| \tag{9}$$

(see [9],[1],[8])

It follows that

$$\int_0^{2\pi} \sigma_{sc}^\alpha(\mathbb{R})\, d\alpha = \int_0^{2\pi} \sigma_{sc}^\alpha(N_0)\, d\alpha \leq$$

$$\int_0^{2\pi} \sigma^\alpha(N_0)\, d\alpha = c|N_0| = 0$$

so $\sigma_{sc}^\alpha(\mathbb{R}) = 0$ for almost all $\alpha \in [0, 2\pi]$. $\blacksquare$

4   A RESOLVENT EXPANSIONS

In the following we drop α in notations like H_α, R^α etc. . So R denote the Green's function of $H = H_\alpha$. By $\overline{H}$ we denote the operator H with Dirichlet boundary conditions at the points $\{x_n\}_{n\in\mathbb{N}}$. By $\tilde{H}$ we mean the operator H with Dirichlet boundary conditions at all the points $x_n^\pm = x_n \pm \epsilon h_n$ with a constant $\epsilon \leq 1$ to be fixed later. We write an expansion of R in terms of the resolvent $\overline{R}$ and $\tilde{R}$ of $\overline{H}$ and $\tilde{H}$. We note that from the Green's formula (see e.g. [5]):

$$R(0,x) = \overline{R}(0,x) + \sum_{n=1}^{\infty} R(0,x_n)D_1\overline{R}(x_n,x)$$

$$= \overline{R}(0,x) + \sum_{n=1}^{\infty} \tilde{R}(0,x_n)D_1\overline{R}(x_n,x)$$

$$+ \sum_{n,m=1}^{\infty} R(0,x_m^-)D_1\tilde{R}(x_m^-,x_n)D_1\overline{R}(x_n,x)$$

$$+ \sum_{n,m=1}^{\infty} R(0,x_m^+)D_1\tilde{R}(x_m^+,x_n)D_1\overline{R}(x_n,x) \tag{10}$$

For each fixed $x \in [0,\infty)$ only finitely many terms in (10) are nonzero since $\overline{R}(x,y) = 0$ (resp. $\tilde{R}(x,y) = 0$) if there is a Dirichlet boundary condition between x and y. More precisely, suppose $x \in (x_n,x_{n+1})$ then the double sums in (10) read

$$R(0,x_n^-)\,D_1\tilde{R}(x_n^-,x_n)\,D_1\overline{R}(x_n,x) + R(0,x_n^+)\,D_1\tilde{R}(x_n^+,x_n)\,D_1\overline{R}(x_n,x) +$$
$$R(0,x_{n+1}^-)\,D_1\tilde{R}(x_{n+1}^-,x_{n+1})\,D_1\overline{R}(x_{n+1},x) + R(0,x_{n+1}^+)\,D_1\tilde{R}(x_{n+1}^+,x_{n+1})\,D_1\overline{R}(x_{n+1},x)$$
$$\tag{10}$$

and similar for the other terms.

The first two sums in (10), i.e.

$$\overline{R}(0,x) + R(0,x_1)D_1\overline{R}(x_1,x) \tag{11}$$

are finite at $E + i0$ for almost all E.

Thus we have

$$\int_0^\infty |R(0,x)|^2\, dx \leq A(E) + C|R(0,x_n^-)|^2|D_1\tilde{R}(x_n^-,x_n)|^2 \int_{x_n}^{x_{n+1}} |D_1\overline{R}(x_n,x)|^2\, dx$$

$$+ \text{ similar terms} \tag{12}$$

In defining $x_n^\pm = x_n \pm \epsilon h_n$ we left open the choice of $0 < \epsilon < 1$. We now *integrate* the right hand side of (12) with respect to ϵ form $1 - \epsilon_0$ to 1, obtaining

$$\int_0^\infty |R(0,x)|^2\, dx \leq A(E) + \tilde{C}\sum_{R=1}^{\infty} \sup_{1-\epsilon_0 < \epsilon < 1} |D_1\tilde{R}(x_n^-,x_n)|^2$$

$$\int_{x_n}^{x_{n+1}} |D_1\tilde{R}(x_n,x)|^2\, dx \cdot \int_{x_n-h_n}^{x_n-(1-\epsilon_0)h_n} |R(0,y)|^2\, dy \tag{13}$$

setting

$$b_n = b_n(E) = \tilde{C} \sup_{1-\epsilon_0 < \epsilon < 1} \left| D_1 \tilde{R}(x_n^-, x_n) \right|^2 \int_{x_n}^{x_{n+1}} \left| D_1 \overline{R}(x_n, x) \right|^2 dx \qquad (14)$$

and $B(E) = \sup_n b_n(E)$ we conclude

$$\int_0^\infty |R(0, x)|^2 \, dx \leq A(E) + B(E) \int_0^\infty |R(0, x)|^2 \, dx \qquad (15)$$

Consequently, $\int_0^\infty |R(0, x)|^2 \, dx < \infty$ if $B(E) < 1$. For this consequence it is enough to have $b_n(E) \to 0$ as $n \to \infty$ (for almost all E), since we may drop finitely many x_n, by redoing the whole estimate with the sequence $\{x_n\}_{n=N_0}^\infty$ replacing $\{x_n\}_{n=1}^\infty$.

5 RESOLVENT ESTIMATE

To show that $b_n(E) \to 0$ for almost all E is the objective of the present section.

First we estimate $D_1 \tilde{R}(x_n^-, x_n)$. In this expression both x_n^- and x_n belong to the region $I_n = [x_n - h_n, x_n + h_n]$ where the potential $V(x) \geq V_n$.

By taking n large enough we have $V_n > E$. $\tilde{R}(x_n^-, x)$ is the resolvent kernel of H on $L^2(I_n)$ with Dirichlet boundary conditions at the boundary of I_n with E *below* the spectrum. Consequently $\tilde{R}(x, y)$ as well as $D_1 \tilde{R}(x, y)$ decay exponentially in $|x - y|$. More precisely, one shows that

$$\left| D_1 \tilde{R}(x_n^-, x_n) \right| \leq C \, e^{-\sqrt{V_n - E} \, |x_n - x_n^-|} \qquad (16)$$

Next, we estimate $D_1 \overline{R}(x_n, x)$. Call $\Delta_n = \Delta_n(E)$ the distance of $\sigma(H_n)$ and E, where H_n denotes the operator H restricted to $[x_n, x_{n+1}]$ with Dirichlet boundary conditions. It turns out, that

$$\int_{x_n}^{x_{n+1}} \left| D_1 \overline{R}(x_n, x) \right|^2 dx \leq C_1 + C_2 \frac{l_n}{\Delta_n^2} \qquad (17)$$

where $l_n = x_{n+1} - x_n$. (The constant C_2 may depend on E in a continuous way). At the first glance, the right hand side of (17) looks pretty ugly, since Δ_n will be very small for certain E ! However, "generically" it is not too small, as the following Lemma shows :

Lemma : If $\sum_n l_n \alpha_n < \infty$ then for almost all E :

$$\Delta_n(E) \geq \alpha_n$$

for n large enough.

Proof :

Consider the set

$$A_n = \{E \in [K, K+1] \mid \Delta_n(E) < \alpha_n\}$$

Then

$$|A_n| \le 2\alpha_n \sharp(\sigma(H_n) \cap [K + K + 1]) \le C\alpha_n l_n$$

So

$$\sum |A_n| < \infty$$

by assumption. By the Borel - Cantelli - Lemma we conclude that

$$|\{E \mid \Delta_n(E) < \alpha_n \text{ for infinitely many } n\}| = 0$$

$$\blacksquare$$

Collecting our estimates we obtain :
For almost all E there exists an $N_0 = N_0(E)$ such that

$$b_n(E) \le C e^{-2(1-\epsilon_o)\sqrt{V_n - E}\,h_n} l_n \cdot \frac{1}{\alpha_n^2} \tag{18}$$

for $n \ge N_0$ provided

$$\sum l_n \alpha_n < \infty \tag{19}$$

By assumption of our theorem we know (19) for $\alpha_n = l_n^{1/2} e^{-\gamma\sqrt{V_n}\,h_n}$ for some $\gamma < 1$ so (almost surely in E, for $E \ge 0$)

$$b_n(E) \le C\, e^{2((1-\epsilon_0)-\gamma)\sqrt{V_n - E}\,h_n} \;\to\; 0 \tag{20}$$

by an appropriate choice for ϵ_0.

This finishes the proof ot Theorem 1.

6   EXAMPLES

In this final section we briefly discuss two examples.

a)   Let q_n be independent identically distributed random variables, $f \ge 0$ a continuous function with support in $[-1/2, 1/2]$, $f(x) \ge 1$ for $|x| \le \delta$. We set

$$V(x) = \sum_{i=o}^{+\infty} q_i f(x - i) \tag{21}$$

Call P_0 the distribution of q_n. We assume that supp $P_0 \subset [0, \infty)$ is not bounded and set $x_n = \inf\{i \mid q_i \ge n\}$. Then condition (4a) is satisfied with $h_n = \delta$ and $V_n = n$. We estimate :

$$E(l_n^{3/2}) \le E(x_n^{3/2}) \le \sum_{m=0}^{\infty} m^{3/2}(1 - p_n)^{m-1} p_n \tag{22}$$

where $p_n = P(q_0 \geq n)$. (22) can be further estimated by :

$$\leq C_0 p_n (1 - p_n) \sum_{m=0}^{\infty} m(m-1)(1-p_n)^{m-2} \leq C_0 + C_1 \frac{1}{p_n^2} \qquad (23)$$

Consequently :

$$E \left(\sum_{n=1}^{\infty} l_n^{3/2} \, e^{\gamma \sqrt{V_n} h_n} \right) \leq C + \sum_{n=1}^{\infty} \frac{e^{-\gamma \sqrt{n} \delta}}{p_n^2}$$

So the above sum is P-almost surely finite, e.g. if $P(q_0 \geq n) \geq e^{-Cn^\beta}$ for some $\beta < \frac{1}{2}$. Note that this includes cases where the Lyapunov-exponent for V is not defined.

b)   Our next example is a bounded potential, namely

$$V(x) \; = \; \cos(2\pi x^{1-\epsilon}) \quad \text{for} \quad 0 < \epsilon < 1 \qquad\qquad 24$$

This slowly oscillating potential was considered by Stolz [11] who proved that $H = -\frac{d^2}{dx^2} + V$ (on $L^2(\mathbb{R})$) has purely absolutely continuous spectrum in $(1, \infty)$. He also posed the question about the nature of $\sigma(H)$ in $(-1, 1)$. The present considerations are an unpublished result by Kirsch and Stolz. Theorem 1 is not directly applicable. However, our proof above shows that the following holds :

Proposition :   If for some $x_n \to \infty$ and some h_n, $l_n = x_{n+1} - x_n$, $V(x) \geq V_n$ on $[x_n - h_n, x_n + h_n]$, $V_n \to V$ and

$$\sum_n l_n^{3/2} e^{-\sqrt{V_n} h_n} < \infty \qquad (25)$$

the $H_\alpha \; = \; -\frac{d^2}{dx^2} + V$ with boundary conditions (24) has pure point spectrum in $(-\infty, V_\infty)$ for almost all α.

In our example, and in fact in a whole class of examples, condition (25) can be verified easily. Take $x_n = n^{1/(1-\epsilon)}$. Then $V(x_n) = 1$. By expanding the cos-function one shows that $V(x) \geq 1 - \delta$ in a neighborhood $[x_n - h_n, x_n + h_n]$ of x_n with h_n of the order of n^ϵ. Thus

$$\sum_n l_n^{3/2} \, e^{-\sqrt{V_n} h_n} \leq \sum n^{\frac{1}{1-\epsilon} \frac{3}{2}} e^{-(1-\delta) n^\epsilon} < \infty \qquad (25)$$

This proves that H_α has pure point spectrum in $[-1, 1]$ for almost all boundary conditions α. For the potential (24) it follows that the spectrum below 1 is at least singular. Moreover, similar reasoning can be used to prove that $H_\alpha + \lambda \chi_{(-1/2, 1/2)}$ has pure point spectrum in $(-\infty, 1)$ for Lebesgue-almost all coupling constants λ.

7   REFERENCES

[1]   Carmona, R.; J. Funct. Anal. **51**, 229-258 (1983)

[2]   Delyon, F.; Levy, Y.; Souillard, B.; J. Stat. Phys. **41** 375-388 (1985)

[3]   Glazman, I. M. ; Direct methods of qualitive spectral analysis, Israel Program of Scientific Translation, 1965

[4]   Hislop, P.; Nakamura, S.; Rev. Math. Phys. **2**, 479-494 (1990)

[5]   Holden, H.; Martinelli, F.; Commun. Math. Phys. **93**, 197-217 (1984)

[6]   Howland, J.; J. Funct. Anal. **74**, 52-80 (1987)

[7]   Kirsch, W.; Molchanov, S.A.; Pastur, L.A.; Funct. Anal. Appl. **24**, 176-186 (1990)

[8]   Kotani, S.; Contemp. Math **50**, 277-286 (1986)

[9]   Levitan, Sargsjan; Introduction to Spectral Theory, Amer. Math. Soc.

[10] Simon, B.; Spencer, T.; Commun. Math. Phys. **125**, 113-125 (1989)

[11] Stolz, G.; Bounded solutions and absolute continuity of Sturm-Liouville operators, Preprint

[12] Simon, B.; Wolff, T.; Commun. Pure. Appl. Math. **39**, 75-90 (1986)

W.Kirsch
Institut für Mathematik und SFB 237
Ruhr-Universität Bochum
D 4630 Bochum, Germany

S.A.Molchanov
Department of Mathematics
Moscow University
Moscow,Russia

L.A.Pastur
Division of Mathematics
Institute for Low-Temperature Physics
Kharkov, Ukraine

Operator Theory:
Advances and Applications, Vol. 57
© 1992 Birkhäuser Verlag Basel

General boundary value problems in region with corners

A.I.Komech, A.E.Merzon

1 Introduction

We consider the following boundary value problems:

$$\begin{aligned}
Au(x) &= \sum_{|\alpha|\leq 2} a_\alpha(x)\partial_x^\alpha u(x) = f(x), \ x \in M,\\
Bu(x) &= \sum_{|\alpha|\leq m} b_\alpha(x)\partial_x^\alpha u(x) = g(x), \ x \in \partial M.
\end{aligned} \tag{1.1}$$

Here M is a bounded region in $I\!\!R^n$ with smooth corner Γ on the boundary ∂M. This means that each point $x \in \Gamma$ has a neighbourhood in M, which is diffeomorphic to $Q \times I\!\!R^{n-2}$, where Q is a plane angle of some magnitude $\varphi(x) < \pi$ or $\varphi(x) > \pi$. We assume that A is an elliptic operator, $a_\alpha(x)$ and $b_\alpha(x)$ are complex valued functions, and Shapiro–Lopatinsky condition holds for (1.1) on $\partial M\backslash\Gamma$. We suppose $u \in H^s(M)$, $f \in H^{s-2}(M)$, and $g \in H^{s-m-\frac{1}{2}}(\partial M\backslash\Gamma)$, where $s > m + \frac{1}{2}$.

Our main question is: when the problem (1.1) is a Fredholm (i.e. $dimKer(A,B) < \infty$ and $dimCoker(A,B) < \infty$)? The answer was found
1) in [6] – for $Bu(x) = \sum_{j=1}^n b_j(x)\frac{\partial u}{\partial x_j}$ and real $b_j(x)$, then
2) in [7] – for general problems (1.1) in the case $\varphi(x) < \pi$,
3) in [10] – for general problems (1.1) for arbitrary $\varphi(x) \in (0, 2\pi)$ and real coefficients $a_\alpha(x), b_\alpha(x)$.

Here we extend the method of [7] for the case $\varphi(x) > \pi$ too. Our method is valid for the case of complex coefficients a_α, b_α, which is important in diffraction theory. At the present time we have extended the method [7] also for the systems of elasticity [12] and of electrodynamics [13].

2   Boundary problems in an angle

As is shown in [7], to investigate the problem (1.1), it suffices to find the exact solution of corresponding similar model problem with a parameter in the plane angle Q:

$$A_0 u(x) \;=\; \sum_{|\alpha|\leq 2} a_\alpha^0 \partial_x^\alpha u(x) = f(x), \; x \in Q,$$

$$B_1 u(x)\mid_{\gamma_1} \;=\; g_1(x), \; B_2 u(x)\mid_{\gamma_2} = g_2(x). \tag{2.1}$$

Here γ_1, γ_2 – two sides of Q, γ_l is diffeomorphic to $I\!\!R_+ = \{x \in I\!\!R : x > 0\}$, $\overline{\gamma_1} \cup \overline{\gamma_2} = \partial Q$; $B_l u(x) = \sum_{|\alpha|\leq m_l} b_{\alpha l}^0 \partial_x^\alpha u(x)$, $l = 1,2$; $a_\alpha^0, b_\alpha^0 \in \mathbb{C}$; A_0 is "strong elliptic", i.e. $\sum_{|\alpha|\leq 2} a_\alpha^0 (-i\xi)^\alpha \neq 0$, $\forall \xi \in I\!\!R^2$. The problem to find an exact solution for (2.1) was confronting us for a long time: Firstly, for $B_l u = u$ or $\frac{\partial u}{\partial n}$ and $A_0 = \Delta + \omega^2$ (not strong elliptic!) the problem (2.1) was solved in [1, 2] (see also [3]), in connection with diffraction problem on the wedges.

Secondly, for $B_l u = \frac{\partial u}{\partial n} + \sigma_l u$ in [4].

Then, for $B_l u = b_{1l}\frac{\partial u}{\partial x_1} + b_{2l}\frac{\partial u}{\partial x_2} + \sigma_l u$ in [6].

At last in [7] – for arbitrary A and B_l with complex coefficients, when the magnitude φ of angle Q is $< \pi$. Later in [10] – for arbitrary A_0 and B_l with real coefficients and arbitrary $\varphi \in (0, 2\pi)$. The method [10] is applicable to $A_0 = \Delta - 1$, but it seems, it is not applicable to $A_0 = \Delta + \omega^2$, $\omega \in I\!\!R$. For $A_0 = \Delta + \omega^2$ and arbitrary B_l with complex coefficients the problem (2.1) was solved in [11] by the method [7]. Here we expose, for the first time, the extention of method [7] for case $\varphi > \pi$. For convenience we expose briefly the case $\varphi < \pi$ too (following [7]).

3   Reduction to the equation in the whole plane

Without loss of generality we may assume that $\varphi = \frac{\pi}{2}$ when $\varphi < \pi$ and $\varphi = \frac{3\pi}{2}$ when $\varphi > \pi$. So we consider only $Q = K_+ \equiv \{x \in \mathbb{R}^2 : x_1 > 0, x_2 > 0\}$, or $Q = K_- \equiv \mathbb{R}^2 \backslash \overline{K_+}$. For simplicity of exposition we describe the method in the particular case of the Dirichlet problem

$$(\Delta - k^2)u(x) = 0, \; x \in Q = K_\pm,$$
$$u(x_1, 0) = g_1(x_1), \; x_1 > 0; \quad u(0, x_2) = g_2(x_2), \; x_2 > 0. \tag{3.1}$$

Here $k \in \mathbb{C}$, $Re\ k \neq 0$ (The case $Re\ k = 0$ was considered in [11] for $\varphi < \pi$). For simplicity of exposition we seek only solutions of (3.1) $u \in C^\infty(\overline{Q}\backslash 0)$. (The case $u \in S'(Q) \equiv \{u(x) = v(x), x \in Q : v \in S'(\mathbb{R}^2)\}$ may be considered similarly). We suppose that for some $C, s, p \in \mathbb{R}$

$$| \partial_x^\alpha u(x) | \leq C(| x |^{-s} + | x |^p), \; x \in Q. \tag{3.2}$$

Similarly, we assume that $g_l \in C^\infty(\mathbb{R}_+)$ and for some C_l, s_l, p_l

$$| g_l(x_l) | \leq C_l(| x_l |^{-s_l} + | x_l |^{p_l}), \; x_l > 0, \; l = 1, 2 . \tag{3.3}$$

Let us introduce the Cauchy data of solution u on $\gamma_l = \{x \in \partial\overline{Q} : x_l > 0\}$, $l = 1, 2$:

$$u_1^0(x_1) = u(x_1, \pm 0) \quad , \quad u_1^1(x_1) = \frac{\partial u}{\partial x_2}(x_1, \pm 0), \; x_1 > 0,$$
$$u_2^0(x_2) = u(\pm 0, x_2) \quad ; \quad u_2^1(x_2) = \frac{\partial u}{\partial x_1}(\pm 0, x_2), \; x_2 > 0. \tag{3.4}$$

Here we choose "+" for $\varphi = \frac{\pi}{2}$ and "−" for $\varphi = \frac{3\pi}{2}$.

The following lemma shows that there exists a simple and natural connection between some extensions of $u(x)$ and of $u_l^\beta(x_l)$.

LEMMA: **3.1** [7, 11] Let u be a solution of (3.1). Then there exist $u_0 \in S'(\mathbb{R}^2)$ and $v_l^\beta \in S'(\mathbb{R})$, $\beta = 0, 1$, $l = 1, 2$, for which

$$u_0(x) = \begin{cases} u(x) & , x \in Q, \\ 0 & , x \in \mathbb{R}^2\backslash\overline{Q}; \end{cases} \qquad v_l^\beta(x_l) = \begin{cases} u_l^\beta(x_l) & , x_l > 0, \\ 0 & , x_l < 0. \end{cases} \tag{3.5}$$

and the following identity holds for $x \in \mathbb{R}^2$:

$$(\Delta - k^2)u_0 = \pm[\delta(x_2)v_1^1(x_1) + \delta'(x_2)v_1^0(x_1) + \delta(x_1)v_2^1(x_2) + \delta'(x_1)v_2^0(x_2)] \equiv d(x). \tag{3.6}$$

Here we also choose "+" for $\varphi = \frac{\pi}{2}$ and "−" for $\varphi = \frac{3\pi}{2}$.

In the remaining part of this paper we construct the "Cauchy data" v_l^β, $l = 1, 2$, $\beta = 0, 1$. If we know v_l^β, then we also know $d(x)$ from (3.6) . Then it is easy to express the solution u of boundary problem (3.1) . Namely, we apply Fourier transformation $\mathcal{F}_{x \to z}$ to (3.6):

$$(-z^2 - k^2)\tilde{u}_0(z) = \tilde{d}(z), \ z \in I\!\!R^2 (\text{here } z^2 \equiv z_1^2 + z_2^2 \ ). \tag{3.7}$$

But $-z^2 - k^2 \neq 0$ as $z \in I\!\!R^2$, because $Re \ k \neq 0$. Hence from (3.7) we obtain

$$\tilde{u}_0(z) = \frac{\tilde{d}(z)}{-z^2 - k^2} \Longrightarrow u(x) = u_0(x) = \mathcal{F}^{-1} \frac{\tilde{d}(z)}{-z^2 - k^2} = d * \varepsilon(x), \ x \in Q. \tag{3.8}$$

Here $\varepsilon(x) \in \mathcal{S}'(I\!\!R^2)$ is a fundamental solution of $\Delta - k^2$, which is unique.

REMARK: **3.2** By virtue of (3.6), the formula (3.8) is the representation of u as the sum of potentials of simple and double layers.

Plan of construction of v_l^β. Roughly speaking, the boundary conditions in (3.1) give us two equations involving four unknown functions v_l^β. Then we need two more equations, involving v_l^β. We construct one of such equations (the "connection equation") in the following section 4 for $\varphi = \frac{\pi}{2}$ and in section 5 for $\varphi = \frac{3\pi}{2}$, and the second equation (the "automorphy equation") in section 6. In this section 5 we shall also give a solution of obtained system of four equations involving v_l^β.

4   The "connection equation" on the Riemannian surface in the case $\varphi < \pi$

As $supp \ u_0 \subset \overline{K_+}$, then $\tilde{u}_0(z)$ is a holomorphic function in the tube $C\!\!\!\!\!\!\!\!K_+^* \equiv \{z \in C\!\!\!\!\!\!\!\!^2 : Im \ z_1 > 0, Im \ z_2 > 0\}$ by Paley–Wiener theorem [8]. Similarly, $\tilde{d}(z)$ is holomorphic in $C\!\!\!\!\!\!\!\!K_+^*$, and $\tilde{v}_l^\beta(z_l)$ is holomorphic for $Im \ z_l > 0$.

Let us introduce a Riemannian surface V of complex characteristics of the elliptic operator $\Delta - k^2$:

$$V = \{z \in C\!\!\!\!\!\!\!\!^2 : -z^2 - k^2 = 0\}; \ z^2 \equiv z_1^2 + z_2^2. \tag{4.1}$$

LEMMA: **4.1** *Let u be a solution of (3.1). Then for the $d(x)$ from (3.6) the following identity holds*

$$\tilde{d}(z) \equiv \tilde{v}_1^1(z_1) - iz_2 \tilde{v}_1^0(z_1) + \tilde{v}_2^1(z_2) - iz_1 \tilde{v}_2^0(z_2) = 0, \ z \in V^* \equiv V \cap C\!\!\!\!\!\!\!\!K_+^*. \tag{4.2}$$

PROOF: (3.7) holds for $z \in \mathbb{C}K_+^*$, hence for $z \in V^*$. But the left–hand side of (3.7) is zero for $z \in V^*$, hence the right–hand side also is zero. $\square$

REMARK: 4.2 The identity (4.2) is the relation between the Cauchy data v_l^β of solution u, arising on the complex characteristics of elliptic operator $\Delta - k^2$.

Let us also note that $\tilde{v}_l^\beta(z_l)$ are holomorphic functions in the regions $V_l^* \equiv \{z \in V : Im\ z_l > 0\}$.

Let us describe the topology structure of the surface V and the regions V^*, V_l^+. We introduce uniformization parameters θ and $\omega = i\theta$ on V by the relations

$$z_1 = ik \sin \theta = k\ sh\omega, \quad z_2 = ik \cos \theta = ik\ ch\omega, \quad \theta, \omega \in \mathbb{C}. \tag{4.3}$$

The maps $\theta \mapsto z$ and $\omega \mapsto z$ are universal coverings $\mathbb{C} \to V$. The projection of V on the plane $(Im\ z_1, Im\ z_2)$ is two–sheeted and its image lies out of the circle of the radius $|\ Re\ k\ |$. For $k \in \mathbb{R}$ we assume that $k > 0$ everywhere below. Then $Re\ \theta = Im\ \omega$ is the angle between $Im\ z$ and the axis $Im\ z_2$ (see fig.1). We introduce "the cuts" $\Gamma_1^\pm = \{z \in V : Im\ z_1 = 0, Im\ z_2 \gtrless 0\}$ and $\Gamma_2^\pm$ similarly, and also $\Gamma_{1,\epsilon}^- = \{z \in V : Im\ z_1 = \epsilon, Im\ z_2 < 0\}$, and $\Gamma_{2,\epsilon}^-$ similarly, $0 < \epsilon < k$ (see fig.1).

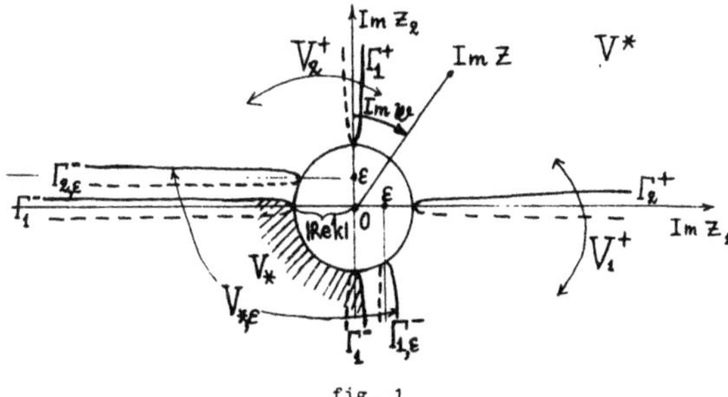

fig. 1.

Hence the regions V^*, V_l^+ for $k > 0$ are defined by the following inequalities respectively:

$$V^* : \quad 2\pi n < Im\ \omega < \tfrac{\pi}{2} + 2\pi n \quad , n \in \mathbb{Z}, \tag{4.4}$$

$$V_1^+ : \quad 2\pi n < Im\ \omega < \pi + 2\pi n \quad , n \in \mathbb{Z}, \tag{4.5}$$

$$V_2^+ : \quad -\tfrac{\pi}{2} + 2\pi n < Im\ \omega < \tfrac{\pi}{2} + 2\pi n \quad , n \in \mathbb{Z}. \tag{4.6}$$

Let us identify V^* and V_1^+, V_2^+ for $k > 0$ with the strips (4.4), (4.5), (4.6) respectively where $n = 0$, denoting them $\hat{V}^*$, $\hat{V}_1^+$, $\hat{V}_2^+$. Then the map $V_1^+ \cup V_2^+ \ni z \mapsto \omega \in \hat{V}_1^+ \cup \hat{V}_2^+$ is holomorphic. Let us denote by $\hat{v}_l^\beta(\omega)$ the liftings of v_I^β on V_I^+ by (4.3). So, for $k > 0$ the "connection equation" (4.2) for $\hat{v}_I^\beta$ takes the following form:

$$\hat{v}_1^1(\omega) + k\ ch\ \omega\ \hat{v}_1^0(\omega) + \hat{v}_2^1(\omega) - ik\ sh\ \omega\ \hat{v}_2^0(\omega) = 0 \text{ as } 0 < Im\ \omega < \frac{\pi}{2}. \qquad (4.7)$$

The solutions v_I^β of this equation we have constructed in [7].

5   The "connection equation" on the Riemannian surface in the case $\varphi > \pi$

In the case $\varphi > \pi$ there is no Paley–Wiener theorem. Then the reasons of section 3 are not applicable to derive the connection between the Cauchy data v_I^β, similar to (4.2).

Nevertheless we shall see below that in this case the same identity (4.2) also holds, but in some new meaning. Although, we must derive it by another method. This is the main result of the present paper.

Namely, as $supp\ u_0 \subset \overline{K_-}$, then $supp\ u_0 \cap K_+ = \emptyset$. Hence, taking into account, that $u_0 \in S'(\mathbb{R}^2)$, we have

$$\langle u_0(x), \psi(x) \rangle = 0,\ \forall \psi \in S(K_+) \equiv \{\psi \in S(\mathbb{R}^2) :\ supp\ \psi \subset K_+\}. \qquad (5.1)$$

By Parseval identity the latter is equivalent to

$$\langle \tilde{u}_0(z), \tilde{\psi}(-z) \rangle = 0. \qquad (5.2)$$

LEMMA: **5.1** *Let* $\psi \in S(K_+)$ *and for some* $\varepsilon_0 > 0$

$$\mid \psi(x) \mid \le C e^{-\varepsilon_0 x_1 - \varepsilon_0 x_2} \text{ as } x_1 > 0,\ x_2 > 0. \qquad (5.3)$$

Then for $0 < \varepsilon < \bar{\varepsilon} \equiv \min(\varepsilon_0, \frac{k}{\sqrt{2}})$

$$\langle \tilde{u}_0(z), \tilde{\psi}(-z) \rangle = -\int\limits_{\substack{Im\ z_1 = \varepsilon \\ Im\ z_2 = \varepsilon}} \frac{\tilde{d}(z)\tilde{\psi}(-z)}{z_1^2 + z_2^2 + k^2} dz. \qquad (5.4)$$

PROOF: By Paley–Wiener theorem $\tilde{d}(z)$ is holomorphic for $z \in \mathbb{C}K_+^*$, and $z^2 + k^2 \ne 0$ for $\mid Im\ z \mid < k$. Hence $\tilde{u}_0(z)$ is holomorpic in the region $Im\ z_1 > 0$, $Im\ z_2 > 0$, $\mid Im\ z \mid < k$ in virtue of (3.8), and

$$\mid \tilde{u}_0(z) \mid = \frac{\mid \tilde{d}(z) \mid}{\mid z^2 + k^2 \mid} \le C\frac{(1 + \mid z \mid)^\mu}{\mid Im\ z \mid^\nu},\ z \in \mathbb{C}K_+^*,\ \mid Im\ z \mid < k. \qquad (5.5)$$

Here μ and ν depend only on s and p from (3.2). Similarly, the function $\tilde{\psi}(z)$ is holomorphic for $Im\ z_1 > -\varepsilon_0$, $Im\ z_2 > -\varepsilon_0$ and is rapidly decreasing there. Hence (5.4) follows from the Cauchy integral formula. $\square$

Let us now transform identity (5.2), substituting $\tilde{u}_0(z)$ from (3.8) into (5.4), and taking $d(x)$ in the form (3.6) with the sign "$-$" as $\varphi = \frac{3\pi}{2}$. Then we obtain for $0 < \varepsilon < \bar{\varepsilon}$

$$\int\limits_{\substack{Im\ z_1 = \varepsilon \\ Im\ z_2 = \varepsilon}} \frac{\tilde{v}_1^1(z_1) - iz_2\tilde{v}_1^0(z_1) + \tilde{v}_2^1(z_2) - iz_1\tilde{v}_2^0(z_2)}{z_1^2 + z_2^2 + k^2}\tilde{\psi}(-z)dz = 0. \tag{5.6}$$

Now we are going to transform this integral identity to an algebraic identity, similar to (2.2), in two steps I and II.

I. We can write (5.6) in the form

$$\int\limits_{\substack{Im\ z_1 = \varepsilon \\ Im\ z_2 = \varepsilon}} \frac{\tilde{v}_1^1(z_1) - iz_2\tilde{v}_1^0(z_1)}{z_1^2 + z_2^2 + k^2}\tilde{\psi}(-z)dz + \int\limits_{\substack{Im\ z_1 = \varepsilon \\ Im\ z_2 = \varepsilon}} \frac{\tilde{v}_2^1(z_2) - iz_1\tilde{v}_2^0(z_2)}{z_1^2 + z_2^2 + k^2}\tilde{\psi}(-z)dz = 0. \tag{5.7}$$

The main idea is that we may "restrict" the integrals in (5.7) on Riemannian surface V by the Cauchy residues theorem. Indeed, the expression in the first integral is meromorphic for $Im\ z_2 < \varepsilon$, in the second – for $Im\ z_1 < \varepsilon$, and the poles lie on V in both cases. More precisely, let us factorize the symbol $z_1^2 + z_2^2 + k^2$ in the denominator in (5.7) as

$$z_1^2 + z_2^2 + k^2 = (z_2 - z_2^+(z_1))(z_2 - z_2^-(z_1)) = (z_1 - z_1^+(z_2))(z_1 - z_1^-(z_2)), \ z \in \mathbb{C}^2. \tag{5.8}$$

Here $Im\ z_1^\pm(z_2) \gtrless 0$ as $Im\ z_2 = \varepsilon$, and similarly, $Im\ z_2^\pm(z_1) \gtrless 0$ as $Im\ z_1 = \varepsilon$,

$$z_2^\pm(z_1) = \pm i\sqrt{z_1^2 + k^2}, \ z_1^\pm(z_2) = \pm i\sqrt{z_2^2 + k^2}. \tag{5.9}$$

Substituting the representations (5.8) in (5.9), we get by Cauchy residues theorem

$$\int\limits_{Im\ z_1 = \varepsilon} \frac{\tilde{v}_1^1(z_1) - iz_2^-(z_1)\tilde{v}_1^0(z_1)}{z_2^-(z_1) - z_2^+(z_1)}\tilde{\psi}(-z_1, -z_2^-(z_1))dz_1 +$$

$$+ \int\limits_{Im\ z_2 = \varepsilon} \frac{\tilde{v}_2^1(z_2) - iz_1^-(z_2)\tilde{v}_2^0(z_2)}{z_1^-(z_2) - z_1^+(z_2)}\tilde{\psi}(-z_1^-(z_2), -z_2)dz_2 = 0. \tag{5.10}$$

Note that $(z_1, z_2^-(z_1)) \in \Gamma_{1,\varepsilon}^-$ for $Im\ z_1 = \varepsilon$ and similarly $(z_1^-(z_2), z_2) \in \Gamma_{2,\varepsilon}^-$ for $Im\ z_2 = \varepsilon$. So (5.10) becomes

$$\int_{\Gamma_{1,\varepsilon}^-} \frac{\tilde{v}_1^1(z_1) - iz_2\tilde{v}_1^0(z_1)}{2z_2}\tilde{\psi}(-z)dz_1 + \int_{\Gamma_{2,\varepsilon}^-} \frac{\tilde{v}_2^1(z_2) - iz_1\tilde{v}_2^0(z_2)}{2z_1}\tilde{\psi}(-z)dz_2 = 0, \qquad (5.11)$$

because $z_2^+ = -z_2^-$ and $z_1^+ = -z_1^-$ in virtue of (5.9).

II. Now we are going to transform the integral identity (5.11) into an algebraic one, similar to (2.2). The main idea is that we may write the left hand side of (5.11) as the integral over boundary $\partial V_{*,\varepsilon} = \Gamma_{1,\varepsilon}^- \cup \Gamma_{2,\varepsilon}^-$ of the region $V_{*,\varepsilon} \equiv \{z \in V : Im\ z_1 < \varepsilon, Im\ z_2 < \varepsilon\}$ on V. Indeed, the function $\tilde{\psi}(-z)$ is holomorphic in $V_{*,\varepsilon}$. Then the annihilation of integral (5.11) suggest the idea that the integrated function is also holomorphic in $V_{*,\varepsilon}$. This gives us the algebraic relation between $\tilde{v}_l^\beta$ we need.

To adjust this idea we introduce the uniformization parameter from (4.3) as a new variable in the integral in (5.11). We choose the analytical branch of the function $\omega = \omega(z)$ over $V_\Sigma \equiv V_1^+ \cup V_{*,\varepsilon} \cup V_2^+$ to preserve holomorphy of $\tilde{\psi}(-z)$ in the $V_{*,\varepsilon}$, and holomorphy of $v_l^\beta(z_l)$ in V_l^+, too. In the variable ω the region $V_* \equiv \{z \in V : Im\ z_1 < 0, Im\ z_2 < 0\}$ is the union of the strips

$$-\pi + 2\pi n < Im\ \omega < -\frac{\pi}{2} + 2\pi n, \ n \in \mathbb{Z}. \qquad (5.12)$$

Let us choose an arbitrary analytical branch ω over V_*. For example let us choose $n = 0$ in (5.12). This branch maps analytically V_* on the strip (5.12) with $n = 0$ denoted by $\hat{V}_*$. Hence, its analytical continuation on V_Σ maps V_1^+ on the strip (4.5) with $n = -1$, denoted by $\hat{V}_1^+$, and V_2^+ on the strip (4.6) with $n = 0$, denoted by $\hat{V}_2^+$. Let us denote by $\hat{V}_{*,\varepsilon}$ the image of $V_{*,\varepsilon}$ by this map $z \mapsto \omega$, and by $\hat{\Gamma}_l^-$, $\hat{\Gamma}_{l,\varepsilon}^-$ the images of Γ_l^-, $\Gamma_{l,\varepsilon}^-$. Then $\partial\hat{V}_{*,\varepsilon} = \hat{\Gamma}_{1,\varepsilon}^- \cup \hat{\Gamma}_{2,\varepsilon}^-$ (see fig. 2).

So let us perform the change of the variables $z \mapsto \omega$ in (5.11) according to (4.3) and the branch selected. Then from (5.9) we get $z_1^\pm(z_2) = \mp k\ sh\omega$, $z_2^\pm(z_1) = \mp ik\ ch\omega$, and (5.11) becomes

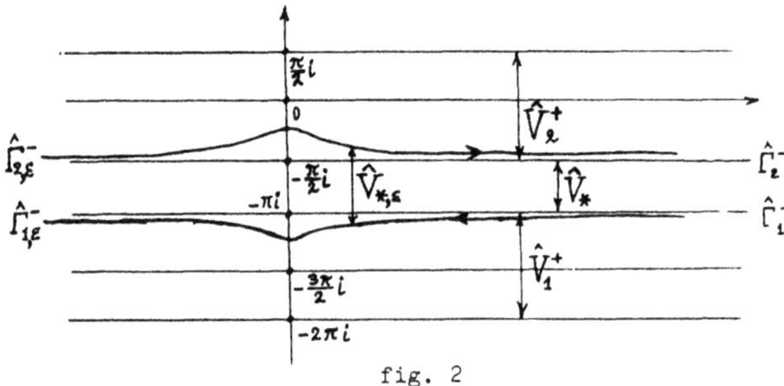

fig. 2

$$\int\limits_{\hat{\Gamma}_{1,\epsilon}^{-}} \frac{\hat{v}_1^1(\omega) + k\ ch\omega\ \hat{v}_1^0(\omega)}{2ik\ ch\omega}\check{\psi}(\omega)k\ ch\omega\ d\omega + \int\limits_{\hat{\Gamma}_{2,\epsilon}^{-}} \frac{\hat{v}_2^1(\omega) - ik\ sh\omega\ \hat{v}_2^0(\omega)}{2k\ sh\omega}\check{\psi}(\omega)ik\ sh\omega\ d\omega = 0.$$

$$(5.13)$$

Here $\hat{v}_l^\beta(\omega) \equiv \tilde{v}_l^\beta(z_l)$ and $\check{\psi}(\omega) \equiv \tilde{\psi}(-z)$, $\omega \in \overline{\hat{V}}_{*,\epsilon}$. We may rewrite (5.13) as

$$\int\limits_{\partial\hat{V}_{*,\epsilon}} \hat{v}(\omega)\check{\psi}(\omega)d\omega = 0. \qquad (5.14)$$

Here we denote

$$\hat{v}(\omega) = \begin{cases} \frac{1}{2i}(\hat{v}_1^1(\omega) + k\ ch\omega\ \hat{v}_1^0(\omega)) & ,\omega \in \hat{V}_1^+, \\ -\frac{1}{2i}(\hat{v}_2^1(\omega) - ik\ sh\omega\ \hat{v}_2^0(\omega)) & ,\omega \in \hat{V}_2^+. \end{cases} \qquad (5.15)$$

LEMMA: **5.2** *The function $\hat{v}(\omega)$, defined by (5.15) in the disjoint union $\hat{V}_1^+ \cup \hat{V}_2^+$ has an analytic continuation to the strip $\hat{V}_*$, lying between $\hat{V}_1^+$ and $\hat{V}_2^+$, and for some $C, q \in \mathbb{R}$*

$$| \hat{v}(\omega) | \leq C(1 + e^{|\omega|})^q, \ \omega \in \hat{V}_\Sigma \equiv \hat{V}_1^+ \cup \hat{V}_{*,\epsilon} \cup \hat{V}_2^+. \qquad (5.16)$$

To prove this lemma, we must, roughly speaking, substitute in(5.14) the Cauchy kernel, i.e. a function, like $\frac{1}{sh(\omega-\omega_0)}$, $\omega \in \hat{V}_{*,\epsilon}$, instead of $\check{\psi}(\omega)$. Then for $\omega_0 \in \hat{V}_{*,2\epsilon} \setminus \hat{V}_{*,\epsilon}$ and $2\epsilon < \bar{\epsilon}$ the identity

$$\mathcal{I}(\omega_0) \equiv \frac{1}{2\pi i} \int\limits_{\partial\hat{V}_{*,2\epsilon}} \hat{v}(\omega)\check{\psi}(\omega)d\omega = \hat{v}(\omega_0) \qquad (5.17)$$

would hold, by the Cauchy integral theorem, if (5.14) holds for such function $\check{\psi}(\omega) = \frac{1}{sh(\omega-\omega_0)}$ (see fig. 3).

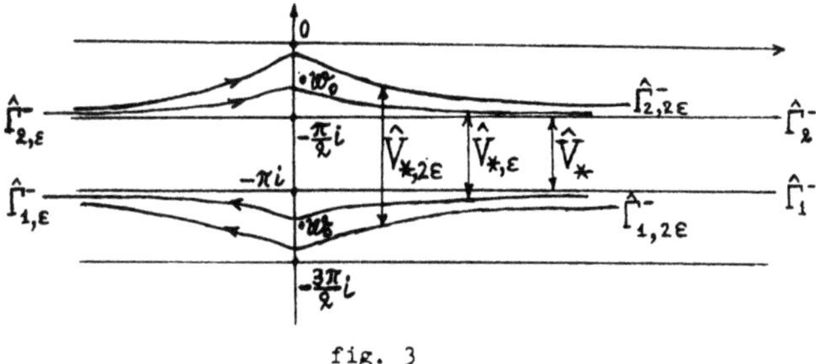

fig. 3

Hence, in the case the integral $\mathcal{I}(\omega_0)$ would define the analytic continuation of the function $\hat{v}$ we look for.

More precisely, let $\check{\psi}(\omega)$ be the Cauchy kernel

$$\check{\psi}(\omega) = \frac{\mathrm{const}(\beta,t,\tau)}{(k\ sh\omega - i)^N(ik\ ch\omega - i)^N(sh(\omega + i\beta) - sh(t + i\tau))} \qquad (5.18)$$

where $\beta, t \in \mathbb{R}$, $\tau \in (0,\pi)$, $N \gg s + p + 2$. For this $\check{\psi}$ the identity (5.14) holds $\forall \beta \in (0, \frac{\pi}{2})$ if $0 < \varepsilon \ll 1$, because the function (5.18) corresponds to

$$\psi(x) = C(\beta,t,\tau)(\delta(\bar{s}x_1 - \bar{c}x_2)(\theta(\bar{c}x_1 + \bar{s}x_2)e^{i\lambda(\bar{c}x_1 + \bar{s}x_2)}) * (x_1^N x_2^N \theta(x_1)\theta(x_2)e^{-x_1-x_2}), \qquad (5.19)$$

where $\bar{c} = \cos\beta$, $\bar{s} = \sin\beta$, $\lambda = sh(t + i\tau)$.

From lemma 4.2 it follows

COROLLARY: **5.3** *In the case $\varphi > \pi$ the following connection equation holds (instead of (4.2), (4.7)):*

$$[\hat{v}_1^1(\omega) + k\ ch\omega\ \hat{v}_1^0(\omega)]_1 + [\hat{v}_2^1(\omega) - ik\ sh\omega\ \hat{v}_2^0(\omega)]_2 = 0, \ \omega \in \hat{V}_\Sigma. \qquad (5.20)$$

Here $[f(\omega)]_l$ denotes the analytic continuation of a function $f(\omega)$ from $\hat{V}_l^+$ to $\hat{V}_\Sigma$.

REMARK: **5.4** The identity (5.20) formally coincides with (4.7). The main difference between (5.20) and (4.7) is the following: (4.7) holds for the functions $\hat{v}_l^\beta$, while (5.20) – for the analytic continuations of their specific combinations.

6   The automorphy equation and reduction to Riemann–Hilbert proplem

In the case $\varphi < \pi$ the equation (4.7) was solved in [7] by the automorphic function method due to Malyshev [5]. Similar procedure we apply to the case $\varphi > \pi$. We consider, for example, the Dirichlet problem (3.1). Then $u_1^0(x_1) = g_1(x_1)$, $x_1 > 0$ and $u_2^0(x_2) = g_2(x_2)$, $x_2 > 0$. Then by (3.5)

$$v_l^0(x_l) = g_l^0(x_l) + \sum_{n=0}^{N_l} C_{ln}\delta^{(n)}(x_l), \ x_l \in \mathbb{R}, \ l = 1,2, \qquad (6.1)$$

where g_l^0 is an extention of $g_l(x_l)$ on $\mathbb{R}$, $g_l^0 \in \mathcal{S}'(\mathbb{R})$, $g_l^0(x_l) = 0$ for $x_l < 0$; N_l depends only on s, p, s_l and p_l. Then (5.20) becomes

$$[\hat{v}_1^1(\omega) + k\ ch\omega(\hat{g}_1^0(\omega) + P_1(k\ ch\omega))]_1 + [\hat{v}_2^1(\omega) - ik\ sh\omega(\hat{g}_2^0(\omega) + P_2(ik\ sh\omega))]_2 = 0, \ \omega \in \hat{V}_*, \qquad (6.2)$$

where $P_l(z_l) \equiv \sum C_{ln}(-iz_l)^n$, $\hat{g}_l^0(\omega) \equiv \tilde{g}_l^0(z_l)$.

We are going now to eliminate the operator $[...]_l$ of analytic continuation out of (6.2). For this purpose let at first $g_1 = 0$ in (3.1). Then we may take $g_1^0 = 0$ in (6.1), (6.2). Hence in virtue of (6.2) the function $\hat{v}_1^1(\omega)$ has analytic continuation $[\hat{v}_1^1]_1$ from $\hat{V}_1^+$ to the region $\hat{V}_\Sigma$ and then

$$\hat{v}_1^1(\omega) + k\, ch\omega P_1(k\, sh\omega) + \hat{v}_2^1(\omega) - ik\, sh\omega(\hat{g}_2^0(\omega) + P_2(ik\, ch\omega)) = 0,\ \omega \in \hat{V}_2^+. \tag{6.3}$$

Here $\hat{v}_1^1(\omega)$ we write instead of $[\hat{v}_1^1(\omega)]_1$. Rewrite (6.3) as

$$\hat{v}_1^1(\omega) + \hat{v}_2^1(\omega) = \hat{G}_2(\omega),\ \omega \in \hat{V}_2^+. \tag{6.4}$$

Here $\hat{G}_2(\omega) \equiv -k\, ch\omega P_1(k\, sh\omega) + ik\, sh\omega(\hat{g}_2^0(\omega) + P_2(ik\, ch\omega))$ – the holomorphic function in $\hat{V}_2^+$.

Let us introduce the automorphisms $h_l : V \to V$, preserving z_l $(l = 1, 2)$:

$$h_1(z_1, z_2) = (z_1, -z_2),\ h_2(z_1, z_2) = (-z_1, z_2),\ (z_1, z_2) \in V. \tag{6.5}$$

Then h_l preseves V_l^+, $l = 1, 2$. Let us denote by $\hat{h}_l$ the lifting of h_l on $\hat{V}_l^+$. Then from (4.3) we get for corresponding ω

$$\hat{h}_1\omega = -\omega - 3\pi i,\ \hat{h}_2\omega = -\omega. \tag{6.6}$$

The main idea of the Malyshev method [4] is that the functions $\tilde{v}_l^*(z) \equiv \tilde{v}_l^1(z_l)$ are invariant with respect to $h_l : v_l^*(h_l z) = v_l^*(z)$, $z \in V_l^+$, $l = 1, 2$.

Then similarly

$$\hat{v}_l^1(\hat{h}_l\omega) = \hat{v}_l(\omega),\ \omega \in V_l^+,\ l = 1, 2. \tag{6.7}$$

These are the "automorphy equations" needed to derive v_l^1, $l = 1, 2$, involving in (6.4). Indeed, applying $\hat{h}_2$ to (6.4) we get in virtue of (6.7) for $l = 2$

$$\hat{v}_1^1(\hat{h}_2\omega) + \hat{v}_2^1(\omega) = \hat{G}_2(\hat{h}_2\omega),\ \omega \in \hat{V}_2^+. \tag{6.8}$$

Then, eliminating $\hat{v}_2^1$ from (6.8) and (6.4), we get

$$\hat{v}_1^1(\omega) - \hat{v}_1^1(\hat{h}_2\omega) = \hat{G}_2(\omega) - \hat{G}_2(\hat{h}_2\omega) \equiv \check{G}_2(\omega),\ \omega \in \hat{V}_2^+. \tag{6.9}$$

At last, we use (6.7) for $l = 1$, taking in mind that $\hat{v}_1^1$ is holomorphic in $\hat{V}_\Sigma$. Then from (6.7) we get the analytic continuation of $\hat{v}_1^1$ from $\hat{V}_\Sigma$ to $\hat{V}_\Sigma \cup \hat{h}_1\hat{V}_\Sigma = \{\omega \in \mathbb{C} : -\frac{3\pi}{2} < Im\, \omega < \frac{\pi}{2}\}$. Hence we may in (6.9) change $\hat{v}_1^1(\hat{h}_2\omega) = \hat{v}_1^1(\hat{h}_1\hat{h}_2\omega) = \hat{v}_1^1(\omega - 3\pi i)$ in virtue of (6.6) (Note that $\omega - 3\pi i \notin \hat{V}_\Sigma$ for $\omega \in \hat{V}_2^+$ - see fig. 2). At last, substituting $\hat{v}_1^1(\hat{h}_2\omega) = \hat{v}_1^1(\omega - 3\pi i)$ in (6.9), we get the Hasemann equation with the shift

$$\hat{v}_1^1(\omega) - \hat{v}_1^1(\omega - 3\pi i) = \check{G}_2(\omega),\ \omega \in \hat{V}_2^+. \tag{6.10}$$

It is easy to reduce (6.10) to Riemann–Hilbert problem and solve it as in [7].

So the case $g_1 = 0$ is studied now. Similarly we can consider the case $g_2 = 0$ and find the solution v_l^β of (5.20) in general case.

At last, the case of general operators B_l in (2.1) may be considered quite similarly. For this purpose one must use the expression of general boundary conditions in (2.1) with the Cauchy data v_l^β by the Cauchy–Kovalevskaya method, as in [7].

References

[1]   H.M.Macdonald, The electrical distribution on a cylinder bounded by two spherical surfaces cutting at any angle // Proc. Lond. Math. Soc.,**26**, 1895, p. 156–172.

[2]   A.Sommerfeld, Mathematisch Theorie der Diffraktion // Math. Ann., **47**, 1896, p. 317–341.

[3]   F.Oberhettinger, Diffraction of waves by a wedge // Comm. Pure Appl. Math., 1954, **7**, N 3, p. 551–563.

[4]   G.D. Malujinetz, Exitation, reflection and emission of the surface waves by a wedge with given impedances of the faces // Dokl. Acad. Nauk SSSR, 1958, **121**, N 3, p. 436–439 (in Russian).

[5]   V.A. Malyshev, Random walks, Wiener–Hopf equations in the quarter of a plane, Galois automorphisms. – Moscow State Univ. Publishing, Moscow, 1970 (in Russian).

[6]   V.G.Maz'ya, B.A.Plamenevskii, On the oblique derivative problem in a domain with piecewise smooth boundary // Func. Anal. 5 (1971), p. 102–103 (in Russian).

[7]   A.I.Komech, Elliptic boundary value problems on manifolds with piecesmooth boundary // Math. Sbornik, 1973, **92**(134), (in Russian).

[8]   A.I.Komech, Elliptic differential equation with constant coefficients in a cone // Vestnik of Moscow University, 1974, 2, p. 14–20 (in Russian).

[9]   A.E.Merzon, On the solvability of differential equations with constant coefficients in a cone // Soviet Math. Dokl., **14** (1973), N 4, p. 1012–1015.

[10] V.G.Maz'ya, B.A.Plamenevskii, On the boundary value problems for second order elliptic equations in the domain with edges // Vest. Leningrad. Univ., ser. math., 1975, N 1, p. 102–108 (in Russian).

[11] A.E.Merzon, The general boundary value problems for the Helmholtz equation in a plane angle // Uspekhi Matem. Nauk, 32:2, 1977, p. 219–220 (in Russian).

[12] A.I.Komech, A.E.Merzon, On exact asymptotics of the solutions of the elasticity system in the region with corners for general boundary value conditions (in preparation).

[13] A.I.Komech, A.E.Merzon, On the diffraction of electromagnetic waves on a wedge for general boundary value conditions (in preparation).

Alexander Komech
Dep. Math.–Mech.
Moscow State University
Moscow, 19899, Russia

Anatoly Merzon
Dep. Initial Education
Moscow State Pedagogical University
Moscow, 117571, Russia

Operator Theory:
Advances and Applications, Vol. 57
© 1992 Birkhäuser Verlag Basel

SOME RESULTS FOR NONLINEAR ELLIPTIC EQUATIONS IN CYLINDRICAL DOMAINS

By V.A.Kondratiev, O.A.Oleinik

In this lecture we describe some results about the behaviour of solutions of boundary–value problems in cylindrical domains for a class of second order nonlinear elliptic equations in a neighbourhoot of infinity. This problem arises in mathematical physics and it was considered in many papers (see, for example [?]–[?]).

We denote

$$\begin{aligned}
x &= (x_1, ..., x_n), \\
x' &= (x_1, ..., x_{n-1}), \\
S(a,b) &= \{x : x' \in \omega, a < x_n < b\}, \\
\sigma(a,b) &= \{x : x' \in \partial\omega, a < x_n < b\},
\end{aligned}$$

ω is a bounded domain in $I\!\!R^{n-1} = (x_1, ..., x_{n-1})$ with a smooth boundary $\partial\omega$.

In paper [?] the equation

$$\Delta u + g(x', u) = 0 \text{ in } S(0, \infty) \tag{1}$$

with boundary conditions

$$\frac{\partial u}{\partial \nu} = 0 \text{ on } \sigma(0, \infty) \tag{2}$$

or

$$u = 0 \text{ on } \sigma(0, \infty) \tag{3}$$

is considered, where ν is the exterior unit normal to $\sigma(0, \infty)$. Under some assumptions about the smoothness of the function $g(x', u)$ it is proved in [?] that if $u(x)$ is a positive solution of problem (??), (??) and $u(x)$ tends to zero as $x_n \to \infty$, μ_1 is the first eigenvalue of the problem

$$\begin{aligned}
-\Delta v - g_u(x', 0)v &= \mu v \text{ in } \omega, \\
\frac{\partial v}{\partial \nu} &= 0 \text{ on } \partial\omega,
\end{aligned}$$

$\mu_1 > 0$, and $\Phi(x')$ is a corresponding eigenfunction, then

$$
\begin{aligned}
u(x', x_n) &= \alpha\, e^{-\sqrt{\mu_1}x_n}\Phi(x') + O(e^{-\sqrt{\mu_1}x_n}), \\
\frac{\partial u}{\partial x_n}(x', x_n) &= -\alpha\sqrt{\mu_1}\, e^{-\sqrt{\mu_1}x_n}\Phi(x') + O(e^{-\sqrt{\mu_1}x_n}),
\end{aligned}
$$

where $\alpha = const > 0$.

A similar result is proved in [?] for positive solutions of problem (??), (??) with the condition $u(x) \to 0$ as $x_n \to \infty$.

In paper [?] we proved the following theorems.

THEOREM 1: *Suppose that $u \in C^2(S(-\infty, +\infty)) \cap C^0(\overline{S}(-\infty, +\infty))$ and $u = 0$ on*

$\sigma(-\infty, +\infty)$ *or $u \in C^2(S(-\infty, +\infty)) \cap C^1(\overline{S}(-\infty, +\infty))$ and $\frac{\partial u}{\partial \nu} = 0$ on $\sigma(-\infty, +\infty)$,*

$$
|\sum_{i,j=1}^{n} a_{ij}(x)\frac{\partial^2 u}{\partial x_i \partial x_j} - f(x, u)| \leq A\,|\nabla u|^{p_1} + B\,|\nabla u|, \tag{4}
$$

where $\nabla u = (\frac{\partial u}{\partial x_1}, ..., \frac{\partial u}{\partial x_n})$, $a_{ij}(x)$, p, p_1, A, B, $f(x, u)$ satisfy the following conditions:

$$
\alpha_1\,|\xi|^2 \leq \sum_{i,j=1}^{n} a_{ij}(x)\xi_i\xi_j \leq \alpha_2\,|\xi|^2\,, \alpha_1, \alpha_2 = const > 0, \tag{5}
$$

$\xi \in \mathbb{R}^n$, $a_{ij}(x)$ are bounded measurable functions, $f(x, u_1) - f(x, u_2) > 0$ for $u_1 > u_2 \geq 0$, $p_1 = 2p(1+p)^{-1}$, $f(x, u) \geq f_1(u)$, $f_1(y) \geq a_1 y^p$ for $y \geq 0$, $a_1 = const > 0$, $p > 1$, $A, B = const > 0$, $f_1(y)$ is a continuous function, the function $-f(x, -u)$ as a function of x and u satisfies the same condition as the function $f(x, u)$. Then $u \equiv 0$ in $S(-\infty, +\infty)$.

THEOREM 2: *If $u(x)$ is a solution of the inequality (??) in $S(0, \infty)$ with boundary conditions (??) or (??) on $\sigma(0, \infty)$, then $u(x', x_n) \to 0$ as $x_n \to \infty$.*

THEOREM 3: *Let $u(x)$ be a solution of the inequality*

$$
|\sum_{i,j=1}^{n} a_{ij}(x)\frac{\partial^2 u}{\partial x_i \partial x_j} - f(x, u)| \leq A\,|\nabla u|^{p_1} + B\,|\nabla_{x'} u| \tag{6}
$$

in $S(0, \infty)$ with the boundary conditions (??) or (??), where $\nabla_{x'} u = (\frac{\partial u}{\partial x_1}, ..., \frac{\partial u}{\partial x_{n-1}})$, $p_1 = 2p(p+1)^{-1}$, $p > 1$, $a_{ij}(x)$, A, B, $f(x, u)$, $-f(x, -u)$ satisfy the conditions of Theorem 1. Then

$$
|u(x)| \leq C\,|x_n|^{\frac{2}{1-p}}, \quad C = const > 0. \tag{7}
$$

It is easy to see that the equation

$$\Delta u - \mid u \mid^{p-1} u = 0 \ \ in \ \ S(o, \infty), \ p > 1, \tag{8}$$

has a solution

$$u(x) = C_p x_n^{\frac{2}{1-p}}, \ C_p = (\frac{2(1+p)}{(p-1)^2})^{\frac{1}{p-1}}, \tag{9}$$

which satisfies the boundary condition (??) on $\sigma(0, \infty)$. This means that the estimate (??) can not be improved for all class of inequalities (??) with the boundary condition (??).

THEOREM 4: *Let $u(x)$ be a solution of the equation*

$$\sum_{i,j=1}^{n} \frac{\partial}{\partial x_i}(a_{ij}(x)\frac{\partial u}{\partial x_j}) - f(x, u) = 0 \ \ in \ \ S(0, \infty) \tag{10}$$

with the boundary condition $u = 0$ on $\sigma(0, \infty)$, $uf(x, u) \geq 0$, a_{ij} satisfy the condition (??), $\mid u(x) \mid \leq C \mid x \mid^{\beta}$, $C = const > 0$, $\beta < \frac{1}{3}$. Then

$$\mid u(x) \mid \leq C_1 e^{-\alpha x_n}, \ C_1, \alpha = const > 0.$$

THEOREM 5: *Let $u(x)$ be a solution of the equation (??) in $S(0, \infty)$ with the boundary condition (??) on $\sigma(0, \infty)$, $f(x, u)u \geq 0$, $u(x) \to 0$ as $x_n \to \infty$, $n = 2$, $u(x)$ changes sign in $S(k, \infty)$ for any $k > 0$ (It means that $u(x)$ takes in $S(k, \infty)$ positive and negative values).*

Then there exists constants C and β such that

$$\mid u(x_1, x_2) \mid \leq C e^{-\beta x_2} \ \ in \ \ S(0, \infty).$$

The main result of the paper [?] is the following proposition.

THEOREM 6: *Let $u(x)$ be a solution in $S(0, \infty)$ of the equation*

$$\Delta u - Q(x) \mid u \mid^{p-1} u = 0, \tag{11}$$

with the boundary condition (??), $p > 1$, $c_1 \geq Q(x) \geq c_2 = const > 0$. Suppose that $u(x)$ changes sign in $S(k, \infty)$ for any $k > 0$. Then

$$| u(x', x_n) | \leq C e^{-\beta x_n}, \ C, \beta = const, \ \beta > 0. \tag{12}$$

This theorem is valid also for a more general class of equations than (??).

In this lecture we consider the equation (??) in $S(0, \infty)$ with the boundary condition $\frac{\partial u}{\partial \nu} = 0$ on $\sigma(0, \infty)$ and we suppose that $u(x)$ preserves sign in $S(0, \infty)$. It is proved in [?], using the maximum principle and the Hopf Lemma, that if $u(x) > 0$ in $S(0, \infty)$, then

$$C_p(x_n + \gamma)^{\frac{2}{1-p}} \leq u(x) \leq C_p x_n^{\frac{2}{1-p}}, \tag{13}$$

$$u(x) = C_p(x_n + \gamma_1)^{\frac{2}{1-p}} + v(x), \tag{14}$$

where C_p is defined by (??), $\gamma, \gamma_1 = const > 0$,

$$| v(x) | \leq C'(x_n^{\frac{1+p}{1-p}}), \ C' = const. \tag{15}$$

Here we prove that $v(x)$ has an exponential decay. Our main result for positive solutions of problem (??), (??) is the following theorem.

THEOREM 7: *Let $u(x)$ be a positive solution of the equation (??) in $S(0, \infty)$ with the boundary condition (??) on $\sigma(0, \infty)$, $p > 1$. Then*

$$u(x) = C_p(x_n + \gamma)^{\frac{2}{1-p}} + v(x), \tag{16}$$

where

$$| v(x) | \leq C \ e^{-\beta x_n},$$

$\gamma, C, \beta = const > 0$, C_p *is defined by (??), γ is a constant, depending of u.*

PROOF: In [?] the inequality (??) is proved. From the inequality (??) it follows that the set of nonnegative numbers δ such that

$$u(x) \leq C_p(x_n + \delta)^{\frac{2}{1-p}}$$

for $x_n \geq N_\delta$, is not empty. We denote

$$\delta_0 = \sup \delta.$$

Consider at first the case when $\delta_o < \infty$. From the definition of δ_0 it follows that

$$u(x) \leq C_p(x_n + \delta_0 - \varepsilon)^{\frac{2}{1-p}}$$

for any $\varepsilon = const > 0$ anf for $x_n > X_\varepsilon$ and in addition

$$u(x^\varepsilon) \geq C_p(x_n^\varepsilon + \delta_0 - \varepsilon)^{\frac{2}{1-p}}, \quad x_n^\varepsilon \to \infty. \tag{17}$$

Consider the function $v(x) = u(x) - C_p(x_n + \delta_0)^{\frac{2}{1-p}}$. It is evident that

$$v(x) \leq C_p(x_n + \delta_0 - \varepsilon)^{\frac{2}{1-p}} - C_p(x_n + \delta_0)^{\frac{2}{1-p}} \tag{18}$$

for $x_n > X_\varepsilon$. It follows from (??) that

$$v(x) \leq C_p(x_n + \delta_0)^{\frac{2}{1-p}}[(1 - \frac{\varepsilon}{x_n + \delta_0})^{\frac{2}{1-p}} - 1].$$

Therefore
$$v_+ = O(x_n^{\frac{1+p}{1-p}}) \quad \text{as} \quad x_n \to \infty, \tag{19}$$

where $v_+(x) = v(x)$, if $v(x) \geq 0$, and $v_+(x) = 0$, if $v(x) < 0$. It is easy to see that

$$v_-(x) = O(x_n^{\frac{2}{1-p}}) \quad \text{as} \quad x_n \to \infty, \tag{20}$$

where $v_-(x) = v(x)$, if $v(x) \leq 0$, and $v_-(x) = 0$, if $v(x) > 0$.

We define the function Z in the following way:

$$Z(x)(x_n + \delta_0)^{\frac{2}{1-p}} = u(x) - C_p(x_n + \delta_0)^{\frac{2}{1-p}}. \tag{21}$$

The function Z satisfies the equation

$$\Delta Z + 4[(1-p)(x_n + \delta_0)]^{-1}\frac{\partial Z}{\partial x_n} + \frac{s(1+p)}{(1-p)^2(x_n + \delta_0)^2}Z + A_p(x, Z)[Z(x_n + \delta_0)^{\frac{2}{1-p}}]^{-1}Z = 0, \tag{22}$$

where

$$A_p(x, Z) = 2C_p(1-p)^{-2}(1+p)(x_n + \delta_0)^{-2+\frac{2}{1-p}} - [C_p(x_n + \delta_0)^{\frac{2}{1-p}} + Z(x_n + \delta_0)^{\frac{2}{1-p}}]^p.$$

For the function Z we have an estimate from below

$$Z(x) = u(x)(x_n + \delta_0)^{\frac{-2}{1-p}} - C_p \geq C_p[(\frac{x_n + \gamma}{x_n + \delta_0})^{\frac{2}{1-p}} - 1] = O(\frac{1}{x_n})$$

and an estimate from above

$$Z(x) \leq C_p[(1 - \frac{\varepsilon}{x_n + \delta_0}) - 1] = o(\frac{1}{x_n}). \tag{23}$$

Therefore we get

$$| Z(x) | \le O(\frac{1}{x_n}), \ Z_+(x) \le o(\frac{1}{x_n}), \ x_n \to \infty. \tag{24}$$

Let us prove that $Z_-(x^\epsilon) = o(\frac{1}{x_n^\epsilon})$, where x^ϵ is defined by (??). According to (??) we have

$$Z(x^\epsilon) \ge C_p[1 - \frac{\epsilon}{x_n^\epsilon + \delta_0}]^{\frac{2}{1-p}} - C_p \ge \frac{2\epsilon C_p}{(1-p)(x_n^\epsilon + \delta_0)}. \tag{25}$$

It follows from (??) that $Z_-(x^\epsilon) = o(\frac{1}{x_n^\epsilon})$. Let us estimate the coefficients of the equation (??) for Z. We have

$$A_p(x, Z) = C_p \frac{2(1+p)}{(1-p)^2}(x_n + \delta_0)^{\frac{2p}{1-p}}(1 - (1 - \frac{Z}{C_p})^p).$$

Thus the function $Z(x)$ satisfies the equation

$$\Delta Z + a_1(x_n)\frac{\partial Z}{\partial x_n} + a_2(x_n)Z = 0,$$

where

$$a_1(x_n) \to 0, \ a_2(x_n) \to 0, \ a_1(x_n) = O(x_n^{-1}), \ a_2(x_n) = O(x_n^{-2}) \ \text{ as } \ x_n \to \infty. \tag{26}$$

We shall prove that $| Z(x_n) | = o(x_n^{-1})$ as $x_n \to \infty$. We have $Z_+(x) = o(x_n^{-1})$ according to (??). Consider the function

$$w(x) = -Z(x) + \frac{\epsilon}{x_n}.$$

It is easy to see that

$$w(x) \ge \frac{\epsilon}{2x_n} \ \text{ for } \ x_n > X_\epsilon, \tag{27}$$

since $Z_+(x) = o(x_n^{-1})$. For the sequence x_n^ϵ we have

$$w(x_n^\epsilon) \le 2\epsilon(x_n^\epsilon)^{-1}. \tag{28}$$

The function w satisfies the equation

$$\Delta w + a_1 w_{x_n} + a_2 w - 2\epsilon x_n^{-3} + a_1 \epsilon x_n^{-2} - a_2 \epsilon x_n^{-1} = 0 \tag{29}$$

in $S(0, \infty)$ with the boundary condition $\frac{\partial w}{\partial \nu} = 0$ on $\sigma(0, \infty)$. Let w_0 be a solution of the equation (??) in $S(T - 2, T + 2)$ with the boundary conditions

$$w_0(x', T - 2) = 0, \ w_0(x', T + 2) = 0, \ \frac{\partial w_0}{\partial \nu} = 0 \ \text{ on } \ \sigma(T - 2, T + 2). \tag{30}$$

The solution w_0 exists for T sufficiently large, since $a_2(x_n) \to 0$ as $x_n \to \infty$. Concider the function

$$W = w - w_0.$$

in $S(T - 2, T + 2)$. We have

$$\Delta W + a_1 W_{x_n} + a_2 W = 0 \quad \text{in} \quad S(T - 2, T + 2), \tag{31}$$

$$\frac{\partial W}{\partial \nu} = 0 \quad \text{on} \quad \sigma(T - 2, T + 2), \quad W = w \quad \text{for} \quad x_n = T + 2 \quad \text{and} \quad x_n = T - 2. \tag{32}$$

In order to prove that $W > 0$ in $S(T - 2, T + 2)$ we use the following well–known proposition:

Let $u(x)$ be a solution of the equation

$$\Delta u + \sum_{j=1}^{n} a^j u_{x_j} + au = f$$

in the domain $S(T - 2, T + 2)$ with the boundary condition $\frac{\partial u}{\partial \nu} = 0$ on $\sigma(T - 2, T + 2)$ and with $a(x)$ sufficiently small in $S(T - 2, T + 2)$. Then

$$|u(x)| \le c(\sup_{S(T-2,T+2)} |f| + \max_{x_n = T-2,\ x_n = T+2} |u|), \quad c = const. \tag{33}$$

From (??) and the estimate (??), applied to solution w_0 of the problem (??), (??), we get

$$W(x) \ge \frac{\varepsilon}{2x_n} - o(x_n^{-3}) \ge \frac{\varepsilon}{4x_n} \quad \text{in} \quad S(T - 2, T + 2).$$

For the function $W(x)$ which is a solution of the problem (??), (??) we apply the Harnack type theorem (see [?]): for any positive solution W of the problem (??), (??) in $S(T - 2, T + 2)$ the inequality

$$W(x) \le CW(x^0) \tag{34}$$

is valid, where $x \in S(T - 1, T + 1)$, $x^0 \in S(T - 1, T + 1)$. The constant C does not depend on W, x^0, x.

This proposition can be proved in the same way as the Harnack theorem is proved in [?].

It follows from (??) that

$$W(x) \le CW(x^\varepsilon) \le C_1 \varepsilon x_n^{-1}, \tag{35}$$

since $W > 0$ in $S(T - 2, T + 2)$ and $W(x^\varepsilon) = w(x^\varepsilon) - w_0(x^\varepsilon) \le \frac{2\varepsilon}{x_n^\varepsilon} + o((x_n^\varepsilon)^{-3}) \le \frac{3\varepsilon}{x_n^\varepsilon}$.

Therefore we get

$$w = W + w_0,$$

$$| w(x) |\leq C_1 \varepsilon x_n^{-1} + C_2(x_n^{-3}) \leq o(x_n^{-1}) \text{ for } x_n > X_\varepsilon \tag{36}$$

according to (??) and the estimate (??) for w_0. Since $Z(x) = \frac{\varepsilon}{x_n} - w(x)$, we obtain from (??) that

$$Z(x) \geq \frac{\varepsilon}{x_n} - o(x_n^{-1}) = o(xN^{-1}), \; Z_-(x) = o(x_n^{-1}). \tag{37}$$

From (??) and (??) we get

$$| Z(x) |= o(x_n^{-1}).$$

According to (??) we have

$$v(x) = u(x) - C_p(x_n + \delta_0)^{\frac{2}{1-p}} = Z(x)(x_n + \delta_0)^{\frac{2}{1-p}} = o(x_n^{\frac{p+1}{1-p}}). \tag{38}$$

Set

$$\theta(x_n) = C_p(x_n + \delta_0)^{\frac{2}{1-p}}.$$

For the function v we have the equation

$$\Delta v = g(x)v,$$

where

$$g(x) = \frac{| u |^{p-1} u - \theta^p}{u - \theta}. \tag{39}$$

It is evident that $g(x) = p | \tilde{u} |^{p-1}$, where

$$\tilde{u}(x) = \theta + o(x_n^\lambda), \; \lambda = \frac{p+1}{1-p},$$

as $x_n \to \infty$, according to (??). Therefore we have

$$g(x) = p(C_p)^{p-1} x_n^{-2}(1 + \frac{\delta_0}{x_n})^{-2}(1 + o(x_n^{-1})) = p(C_p)^{p-1} x_n^{-2} + o(x_n^{-3}).$$

Let us consider the function $V(x) = \mathcal{S}(x_n)v(x)$, where $\mathcal{S}(s) = 0$ for $s < \tau$, $\tau = const > 0$, $\mathcal{S}(s) = 1$ for $s > \tau + 1$, $\mathcal{S} \in C^\infty(\mathbb{R}^1)$. The function V satisfies the equation

$$\Delta V - g_1 V = f \text{ in } S(-\infty, +\infty), \tag{40}$$

$\frac{\partial V}{\partial \nu} = 0$ on $\sigma(-\infty, +\infty)$, f has a compact support, $g_1 = g$ for $x_n > \tau + 1$. In the paper [?] it is proved that the equation (??) has in $S(-\infty, +\infty)$ a solution V_0, which has such properties:

$$V_0(x) = O(e^{-\alpha x_n}), \; \alpha = const > 0, \text{ as } x_n \to \infty, \tag{41}$$

$$V_0(x) = ax_n + b + O(e^{\alpha x_n}), \; a, b, \alpha = const > 0, \; \alpha > 0, \; x_n \to -\infty \tag{42}$$

and for $V_0(x)$ the following estimate is valid

$$\int_{S(-\infty,+\infty)} \sum_{|\alpha|\leq 2} |\,\mathcal{D}^\beta V_0\,|^2\, e^{2\alpha x_n} dx \leq C_1 \int_{S(-\infty,+\infty)} |\,f\,|^2\, e^{2\alpha x_n} dx \leq C_2. \qquad (43)$$

It is wellknown that for the solution of the equation (??) with the boundary condition $\frac{\partial V_0}{\partial \nu} = 0$ on $\sigma(-\infty,+\infty)$ for $t > \tau + 1$ we have

$$\max_{x\in S(t-1,t+1)} |\,V_0(x)\,|^2 \leq C \int_{S(t-2,t+2)} |\,V_0\,|^2\, dx,$$

where $C = const > 0$ and C does not depend on V_0 (This is a De Giorgi type estimate (see [?], [?])).

For the solution $V_0(x)$ we get

$$\max_{x\in S(t-1,t+1)} |\,V_0(x)\,|^2 \leq C \int_{S(t-2,t+2)} |\,V_0\,|^2\, dx \leq C\, e^{-2\alpha(t-2)} \int_{S(t-2,t+2)} |\,V_0\,|^2\, e^{2\alpha x_n} dx \leq C_1\, e^{-2\alpha t}.$$

Therefore

$$|\,V_0(x)\,|^2 \leq C_2\, e^{-2\alpha x_n},\ \ C, C_1, C_2 = const > 0. \qquad (44)$$

Let us consider $V_1 = V - V_0$. It follows from (??) and (??) that $V_1(x) = o(x_n^\lambda)$ as $x_n \to \infty$ and according to (??) $V_1(x) = ax_n + b + O(e^{\alpha x_n})$ as $x_n \to -\infty$, $a, b, \alpha = const$, $\alpha > 0$. We shall prove that $a = 0$, $b = 0$. Suppose that $a < 0$. For V_1 we have the equation

$$\Delta V_1 - g_1 V_1 = 0\ \ \text{in}\ \ S(-\infty,+\infty).$$

Let us consider the equation

$$Y''(x_n) - \theta Y(x_n) = 0,\ \ -\infty < x_n < +\infty. \qquad (45)$$

It is easy to verify that $Y(x_n) = x_n^\lambda$, where $\lambda = \frac{p+1}{1-p}$ is a solution of equation (??) for $x_n > 0$. Since $a < 0$ and $V_1(x) \to 0$ as $x_n \to \infty$, we have $V_1 > 0$ in $S(-\infty,+\infty)$. We introduce the function

$$Y_1(x_n) = Y(x_n) + K x_n^{\lambda-\frac{1}{2}},\ \ K = const > 0.$$

It is easy to see that

$$\Delta Y_1 - g_1 Y_1 \geq 0\ \ \text{and}\ \ Y_1 > 0\ \ \text{in}\ \ S(0,\infty),$$

since $\Delta Y_1 - g_1 Y_1 = K C_0 x_n^{\lambda-\frac{5}{2}} + O(x_n^{-3+\lambda}) \geq 0$, if $x_n > 0$ and K is sufficiently large; $C_0 = const > 0$. Let us consider the function

$$E = CV_1 - Y_1.$$

We have

$$\Delta E - g_1 E \leq 0 \quad \text{in} \ \ S(1, \infty),$$

$E(x) \to 0$ as $x_n \to \infty$, $E > 0$, if C is sufficiently large, since $V_1 > 0$ in $S(-\infty, +\infty)$. The function $E(x)$ can not take negative values in $S(1, \infty)$, since $E(x)$ can not attain a negative minimum in $S(1, \infty)$, on $\sigma(1, \infty)$ and for $x_n = 1$. Therefore $E(x) \geq 0$ in $S(1, \infty)$,

$$CV_1 \geq x_n^\lambda + K x_n^{\lambda - \frac{1}{2}} \quad \text{in} \ \ S(1, \infty). \tag{46}$$

This inequality contradicts $V_1(x) = o(x^\lambda)$ as $x_n \to \infty$. In the same way we can get a contradiction if we suppose $a > 0$ or $b \neq 0$. Therefore, $V_1(x) \to 0$ as $x_n \to -\infty$. According to the maximum principle $V_1 \equiv 0$ in $S(-\infty, +\infty)$. We have $V = v$ for $x_n > \tau + 1$, $V_0(x) = O(e^{-\alpha x_n})$ as $x_n \to \infty$, $u(x) = v(x) + C_p(x_n + \delta_0)^{\frac{2}{1-p}}$. The theorem is proved in the case $\delta_0 < \infty$. Consider the case $\delta_0 = +\infty$. Since $u > 0$, according to (??)

$$u(x) \geq C_p(x_n + \gamma)^{\frac{2}{1-p}}.$$

Therefore, for any $\delta > 0$, which satisfies the inequality

$$u(x) \leq C_p(x_n + \delta)^{\frac{2}{1-p}},$$

we have $\delta < \gamma$. It means that $\delta_0 < \infty$. The theorem is proved. $\square$

References

[1] A.N.Kolomogorov, I.G.Petrovsky, N.S.Piskunov: The study of the equation, joint with a growth of quantity of substance. Bull. MGU, Math. Mech. (1937), v. 1, 1–26.

[2] A.I.Volpert: On propogation of waves, described by nonlinear parabolic equations. In: I.G.Petrovsky Selected papers. Differential equations, Theory of probability. Moscow, Nauka (1987), p. 333–358.

[3] H.Berestycki, L.Nirenberg: Some qualitative properties of solutions of semilinear elliptic equations in cylindrical domains. Analysis, ed. by P.Rabinovitz, Academic Press (1990), p. 114–164.

[4] V.A.Kondratiev, O.A.Oleinik: On asymptotic behaviour of solutions of some non-linear elliptic equations in unbounded domains. Proceedings of the conference, dedicated to L.Nirenberg. Longman ed. (to appear).

[5] E.M.Landis: Second order elliptic and parabolic equations. Moscow, Nauka (1971).

[6] E.De Giorgi: Sulla differenziabilita e l'analiticita delle estremali degli integrali. Mem. Acc. Sci. Torino (1957), 1–19.

[7] J.Moser: A new proof of De Giorgi theorem concerning the reqularity problem for elliptic differential equations. Comm. Pure and Appl. Math., **13**, Nr.3 (1960), 457–468.

Global Representation
of Lagrangian Distributions

by

A .Laptev, Yu.Safarov and D .Vassiliev

0. Let M be a C^∞-manifold without boundary, $\dim M = n$, and $T^*M\backslash 0$ be the cotangent bundle without the zero section. Let $A = A(x, D_x)$ be a first order elliptic pseudodifferential operator in the space of half-densities on M (as usual, $D_x = -i\partial_x$). Assume that the principal symbol a_1 of A is real and positive. We consider the fundamental solution $u(t, x, y)$ of the following initial value problem

$$D_t u \;+\; A(x, D_x)u \;=\; 0,$$

$$u(0, x, y) \;=\; \delta(x - y).$$

It is well known that u is a Lagrangian distribution of order zero associated with the Lagrangian manifold Λ, generated by the Hamiltonian flow with Hamiltonian a_1 (see, for example, [1],[2]). In this paper we propose a convenient method allowing a global parametrization of the Lagrangian manifold Λ, and, consequently, a global representation of the fundamental solution in the form

$$u(t, x, y) \;=\; (2\pi)^{-n} \int e^{i\varphi(t;x;y,\eta)} \, q(t; y, \eta) \, |d_\varphi(t; x; y, \eta)| \, d\eta.$$

For the sake of brevity we describe our construction for a fixed t, omitting it futher on. This paper can be regarded as a development of [3].

1. Let G be a smooth homogeneous canonical transformation in the cotangent bundle $T^*M\backslash 0$. For $(y, \eta) \in T^*M\backslash 0$ let us denote

$$G(y, \eta) = (x^*(y, \eta), \xi^*(y, \eta)).$$

Then $G(y, \lambda\eta) = (x^*(y, \eta), \lambda\xi^*(y, \eta))$ for $\lambda > 0$. We consider the Lagrangian manifold

$$\Lambda = \{(x, \xi), (y, -\eta) : (x, \xi) = G(y, \eta)\} \subset (T^*M\backslash 0) \times (T^*M\backslash 0).$$

It is clear that Λ is naturally parametrized by $(y, \eta) \in T^*M\backslash 0$, and this allows us to identify all objects (functions, half-densities, etc.) defined on Λ with those on $T^*M\backslash 0$.

A complex homogeneous function of degree 1

$$\varphi(x; y, \eta) \in C^\infty(M \times T^*M\backslash 0)$$

such that $\operatorname{Im}\varphi \geq 0$ is said to be the *phase function*. We shall always assume that $\operatorname{Im}\varphi(x;y,\eta) > 0$ for x lying outside a small neighbourhood of the point $x^*(y,\eta)$.

Denote by $\mathcal{F}$ the class of phase functions φ satisfying the following three conditions :

$$\varphi\big(x^*(y,\eta);y,\eta\big) \;=\; 0, \tag{1}$$

$$\varphi_x\big(x^*(y,\eta);y,\eta\big) \;=\; \xi^*(y,\eta), \tag{2}$$

$$\det \partial_{x\eta}\varphi\big(x^*(y,\eta);y,\eta\big) \;\neq\; 0. \tag{3}$$

Remark 1. The condition (3) is invariant. Indeed, when we change the coordinates $x \to \tilde{x}$ and $y \to \tilde{y}$ we obtain

$$\partial_{\tilde{\eta}\tilde{x}}\varphi \;=\; (\partial\tilde{y}/\partial y)\cdot\partial_{\eta x}\varphi\cdot(\partial\tilde{x}/\partial x)^{-1}$$

and therefore $\det \partial_{\tilde{\eta}\tilde{x}}\varphi$ is not equal to zero.

Lemma 2. *Any phase function φ satisfying the conditions (1) - (3) gives a global parametrization of the Lagrangian manifold Λ.*

Proof. Let us differentiate with respect to η the identity (1). In view of (2) we obtain

$$\varphi_{\eta_k}\big(x^*(y,\eta);y,\eta\big) \;+\; \xi^*(y,\eta)\cdot x^*_{\eta_k}(y,\eta) \;=\; 0.$$

Since the transformation G preserves the canonical 1-form $\xi\cdot dx$ we have

$$\xi^*\cdot x^*_{\eta_k} \;=\; 0, \quad \xi^*\cdot x^*_{y_k} \;=\; \eta_k. \tag{4}$$

Therefore

$$\varphi_\eta(x;y,\eta) \;=\; 0 \tag{5}$$

for $x = x^*(y, \eta)$. On the other hand, the Euler identity

$$\eta \cdot \varphi_\eta(x; y, \eta) = \varphi(x; y, \eta)$$

implies that $\varphi(x; y, \eta) = 0$ if $\varphi_\eta(x; y, \eta) = 0$. So $\varphi_\eta(x; y, \eta)$ can be equal to zero only if x is sufficiently close to $x^*(y, \eta)$. In view of (3) in a small neighbourhood of the point $x^*(y, \eta)$ the equation (5) may have only one solution with respect to x. Therefore the equation (5) has the only one global solution $x = x^*(y, \eta)$. By analogy, differentiating (1) with respect to y and taking into account (2), (4), we obtain

$$\varphi_y\big(x^*(y, \eta); y, \eta\big) = -\eta.$$

This completes the proof.

Denote

$$\Phi_{\eta\eta} = \Phi_{\eta\eta}(y, \eta) = \partial_{\eta\eta}\varphi\Big|_{x=x^*},$$

$$\Phi_{xx} = \Phi_{xx}(y, \eta) = \partial_{xx}\varphi\Big|_{x=x^*},$$

$$\Phi_{x\eta} = \Phi_{x\eta}(y, \eta) = \partial_{x\eta}\varphi\Big|_{x=x^*}, \tag{6}$$

$$\Phi_{\eta x} = \Phi_{\eta x}(y, \eta) = \Phi_{x\eta}^T(y, \eta).$$

The condition (3) is equivalent to the fact that the matrix $\Phi_{x\eta}(y, \eta)$ (or $\Phi_{\eta x}(y, \eta)$) is non-degenerate for all (y, η).

Remark 3. In view of (1) and (5) the symmetric matrix $\Phi_{\eta\eta}$ behaves as a tensor. Changing the coordinates $y \to \tilde{y}$ we obtain

$$\Phi_{\tilde{\eta}\tilde{\eta}} = (\partial\tilde{y}/\partial y) \cdot \Phi_{\eta\eta} \cdot (\partial\tilde{y}/\partial y)^T\Big|_{x=x^*}. \tag{7}$$

By analogy, since the function $\mathrm{Im}\,\varphi(x; y, \eta)$ of the variables x has a zero of the second order at the point $x = x^*(y, \eta)$, the imaginary part $\mathrm{Im}\,\Phi_{xx}$ of the matrix

Φ_{xx} is a tensor over the point x^*. This fact together with $\operatorname{Im}\varphi(x;y,\eta) \geq 0$ implies also

$$\operatorname{Im}\Phi_{xx} \geq 0. \tag{8}$$

On account of the conditions (1) and (2), in any coordinate system

$$\varphi(x;y,\eta) = (x - x^*)\cdot\xi^* + \frac{1}{2}\Phi_{xx}(x - x^*)\cdot(x - x^*) \tag{9}$$

$$+ O(|x - x^*|^3|\eta|), \qquad x \to x^*, \ |\eta| \to \infty$$

where $O(|x - x^*|^3|\eta|)$ is homogeneous with respect to η of degree one. Differentiating the identity (9) with respect to x and η, we obtain

$$\Phi_{x\eta} = \xi_\eta^* - \Phi_{xx}\cdot x_\eta^*, \quad \Phi_{\eta\eta} = -(x_\eta^*)^T\cdot\xi_\eta^* + (x_\eta^*)^T\cdot\Phi_{xx}\cdot x_\eta^*. \tag{10}$$

Now (8) and (10) yield

$$\operatorname{Im}\Phi_{\eta\eta} \geq 0.$$

The matrices x_η^* and ξ_η^* will be used very often later on. Note that when we change the coordinates $x \to \tilde{x}$ and $y \to \tilde{y}$ we have

$$\tilde{x}_{\tilde{\eta}}^* = (\partial\tilde{x}/\partial x)\cdot x_\eta^*\cdot(\partial\tilde{x}/\partial x)^T$$

i.e. x_η^* behaves as a tensor. The matrix ξ_η^* also behaves as a tensor with respect to y. However, this is not true with respect to x. Indeed, passing from coordinates x to $\tilde{x}$ we obtain

$$\tilde{\xi}^*(y,\eta) = (\partial x/\partial\tilde{x})^T\Big|_{\tilde{x}=\tilde{x}^*}\xi^*(y,\eta).$$

Differentiating this identity with respect to η we see that

$$(\partial x/\partial\tilde{x})^T\Big|_{\tilde{x}=\tilde{x}^*}\cdot\xi_\eta^*(y,\eta) = \tilde{\xi}_\eta^*(y,\eta) - C(y,\eta)\cdot\tilde{x}_\eta^*(y,\eta) \tag{11}$$

where $C = \{C_{ij}\}$ is the symmetric matrix-function with elements

$$C_{ij}(y,\eta) = \sum_k\left(\frac{\partial^2 x_k}{\partial\tilde{x}_i\,\partial\tilde{x}_j}\right)\Big|_{\tilde{x}=\tilde{x}^*}\xi_k^*(y,\eta).$$

Here the $\partial^2 x_k / \partial \tilde{x}_i \partial \tilde{x}_j$ are the second Taylor coefficients of $x_k(\tilde{x})$ at the point $\tilde{x} = \tilde{x}^*$ which for fixed (y, η) can be chosen arbitrarily. Thus given coordinates $\tilde{x}$, an arbitrary real symmetric matrix C_0 and a fixed point (y, η) we can find coordinates x such that in (11) $C(y, \eta) = C_0$.

2. To demonstrate the existence of phase functions satisfying the conditions (1) - (3) we shall prove the following lemma, which gives a natural example of the function φ.

Let us introduce on M a Riemannian metric g. For sufficiently close points $x \in M$, $y \in M$ let $\gamma_{y,x}(s)$ be the "shortest" geodesic connecting y and x, i.e. the geodesic defined in the normal system of coordinates with centre y. We shall choose the parameter s such that $\gamma_{y,x}(0) = y$, $\gamma_{y,x}(1) = x$. We denote $\nu(y, x) = \dot{\gamma}_{y,x}(0) \in T_y M$, and smoothly extend ν on $M \times M$. Let $b(x; y, \eta)$ be a smooth homogeneous function of degree one such that

$$b(x; y, \eta) \; = \; O(|x - y|^2 |\eta|), \quad x \to x^*,$$

$$\operatorname{Im} b_{xx}(x^*; y, \eta) \; > \; 0,$$

and

$$\operatorname{Im} b(x; y, \eta) \; > \; 0$$

for $x \neq x^*$.

Lemma 4. *The phase function*

$$\varphi(x; y, \eta) \; = \; \nu(x^*, x) \cdot \xi^* \; + \; b(x; y, \eta)$$

satisfies the conditions (1) - (3).

Proof. The conditions (1) and (2) are obviously fulfilled. Thus it remains to verify (3). Let us choose a local coordinate system. Then

$$\operatorname{Re} \Phi_{x\eta} \; = \; \xi_\eta^* \; - \; \operatorname{Re} \Phi_{xx} \cdot x_\eta^*,$$

$$\operatorname{Im} \Phi_{x\eta} \; = \; - \operatorname{Im} \Phi_{xx} \cdot x_\eta^*$$

where

$$\operatorname{Im} \Phi_{xx} \cdot x_\eta^* \; = \; \operatorname{Im} b_{xx}(x^*; y, \eta) \; > \; 0 \,.$$

Since the transformation G preserves the canonical 2-form $dx \wedge d\xi$ we have

$$(\xi_\eta^*)^T \cdot x_\eta^* \; - \; (x_\eta^*)^T \cdot \xi_\eta^* \; = \; 0 \tag{12}$$

and therefore

$$\operatorname{Re} \Phi_{x\eta}^T \cdot x_\eta^* \; - \; (x_\eta^*)^T \cdot \operatorname{Re} \Phi_{x\eta} \; = \; 0 \,.$$

It implies that

$$(\operatorname{Re} \Phi_{x\eta}^T \; - \; i \operatorname{Im} \Phi_{x\eta}^T) \cdot \operatorname{Im} \Phi_{xx}^{-1} \cdot (\operatorname{Re} \Phi_{x\eta} \; + \; i \operatorname{Im} \Phi_{x\eta})$$

$$= \; \operatorname{Re} \Phi_{x\eta}^T \cdot \operatorname{Im} \Phi_{xx}^{-1} \cdot \operatorname{Re} \Phi_{x\eta} + \; \operatorname{Im} \Phi_{x\eta}^T \cdot \operatorname{Im} \Phi_{xx}^{-1} \cdot \operatorname{Im} \Phi_{x\eta} \,. \tag{13}$$

The real matrix on the right hand side of (13) is non-negative, and we obtain
for any vector $\vec{c}$ from its kernel

$$x_\eta^* \, \vec{c} \; = \; 0 \,, \quad \xi_\eta^* \, \vec{c} \; = \; 0 \,.$$

But since the transformation G is non-degenerate then $\vec{c} = 0$. This completes
the proof.

Example 5. Given a Riemannian metric g on M we may take

$$\varphi(x; y, \eta) \; = \; \nu(x^*, x) \cdot \xi^* \; + \; \frac{i}{2} \, |\nu(x^*, x)|^2 \, |\eta| \,.$$

In particular, when $M = \mathbf{R}^n$ with Euclidian metric the conditions (1) - (3) are
fulfilled for the phase function

$$\varphi(x; y, \eta) \; = \; (x - x^*) \cdot \xi^* \; + \; \frac{i}{2} \, |x - x^*|^2 \, |\eta| \,.$$

We shall see below (subsection 2.4) that in general there is no real phase
function satisfying the conditions (1) - (3) globally. However, one can always
find the phase function which is real in a given small neighbourhood.

Proposition 6. *Let (y_0, η_0) be a fixed point from $T^*M\backslash 0$, and $x_0 = x^*(y_0, \eta_0)$. Then in a neighbourhood of the point x_0 there exists a coordinate system x such that*

$$\det \xi_\eta^*(y_0, \eta_0) \neq 0. \tag{14}$$

Proof. Let $\tilde{x}$ be arbitrary coordinates in a neighbourhood of the point x_0. In view of (12)

$$\tilde{x}_\eta^*(y_0, \eta_0) \ : \ \ker \tilde{\xi}_\eta^*(y_0, \eta_0) \ \longrightarrow \ \ker \left(\tilde{\xi}_\eta^*(y_0, \eta_0)\right)^T$$

and since the transformation G is non-degenerate, the rank of this map is maximal. Let C_0 be the orthogonal projection on $\ker \left(\tilde{\xi}_\eta^*(y_0, \eta_0)\right)^T$. Then $C_0 \cdot \tilde{\xi}_\eta^*(y_0, \eta_0) = 0$ and therefore the kernel of the matrix

$$\tilde{\xi}_\eta^*(y_0, \eta_0) \ - \ C_0 \cdot \tilde{x}_\eta^*(y_0, \eta_0)$$

contains only the zero vector. Choosing now coordinates x such that

$$C_{ij}(y_0, \eta_0) \ = \ \sum_k \left(\frac{\partial^2 x_k}{\partial \tilde{x}_i \partial \tilde{x}_j}\right)\Bigg|_{\tilde{x}=x_0} \xi_k^*(y_0, \eta_0)$$

(see (11)) we obtain (14). The proof is complete.

Obviously, if (14) is fulfilled then in a neighbourhood of the point $(x_0; y_0, \eta_0)$ the real phase function

$$(x - x^*) \cdot \xi^* \tag{15}$$

satisfies the conditions (1) - (3), and thus it gives a local parametrization of the Lagrangian manifold Λ.

Lemma 7. *Let U be a open subset of $T^*M\backslash 0$ and φ be a phase function satisfying (1) - (3) in U. Then any phase function $\varphi_0 \in \mathcal{F}$ can be continuously transformed in this class to a phase function $\varphi_1 \in \mathcal{F}$ such that $\varphi_1 = \varphi$ in U.*

Proof. Let us introduce on M a Riemannian metric. Choose a smooth non-negative function $\rho(y, \eta)$ such that $\rho = 1$ in U and $\rho \leq 1$ and consider

$$\varphi_s(x; y, \eta) \ = \ (1 - s)\varphi_0(x; y, \eta) \ + \ s\,\rho(y, \eta)\,\varphi_1(x; y, \eta)$$

$$+ \frac{i}{2}s\big(1 - s\rho(y,\eta)\big)\,|x - x^*|^2\,|\eta|, \quad 0 \le s \le 1. \tag{16}$$

By (8) and Lemma 4 φ_s satisfy (1) - (3) for all $s \in [0,1]$, and $\varphi_1 = \varphi$ in U. This completes the proof.

3. Let us introduce a smooth homogeneous "function" of order zero $d_\varphi(x; y, \eta)$ such that

$$|d_\varphi| = \sqrt{|\det \partial_{x\eta}\varphi|} \tag{17}$$

for x close to x^*. Under change coordinates $\sqrt{|\det \partial_{x\eta}\varphi|}$ behaves as a $\frac{1}{2}$-density with respect to x and as a $(-\frac{1}{2})$-density with respect to y, and we suppose that $|d_\varphi|$ has the same property.

Lemma 2 immediately implies the following theorem.

Theorem 8. *Let $u(x,y)$ be a Lagrangian distribution associated with the Lagrangian manifold Λ. Then for any phase function $\varphi \in \mathcal{F}$ there exists an amplitude $p(x; y, \eta)$ such that*

$$u(x,y) = (2\pi)^{-n} \int e^{i\varphi(x;y,\eta)}\, p(x; y, \eta)\, |d_\varphi(x; y, \eta)|\, d\eta \tag{18}$$

modulo a smooth function.

Remark 9. Usually Lagrangian distributions are supposed to be half-densities. When we consider p in (18) as a function on $M \times T^*M \backslash 0$ the properties of $|d_\varphi|$ yield that the integral $\int e^{i\varphi}p\, |d_\varphi|\, d\eta$ gives us exactly a half-density of x and y.

Proof of the Theorem 8. By the condition (3) the phase function φ is non-degenerate. Therefore according to the theorem 25.4.7. from [2,v.4], an oscillatory integral with an arbitrary phase function locally parametrizing Λ modulo a smooth function is equal to an oscillatory integral with the phase function φ. Since a Lagrangian distribution is a locally finite sum of such oscillatory integrals, the same is valid for u.

Definition 10. *The Lagrangian distribution u and the oscillatory integral in (18) are said to be of order m if the amplitude $p \in S^m$.*

Lemma 11. *The oscillatory integral (18) of order m can be written in the form*

$$(2\pi)^{-n} \int e^{i\varphi}\, p(x; y, \eta)\, |d_\varphi(x; y, \eta)|\, d\eta$$

$$= (2\pi)^{-n} \int e^{i\varphi}\, p(x^*; y, \eta)\, |d_\varphi(x; y, \eta)|\, d\eta$$

$$+ (2\pi)^{-n} \int e^{i\varphi}\, \tilde{p}(x; y, \eta)\, |d_\varphi(x; y, \eta)|\, d\eta \qquad (19)$$

with $\tilde{p} \in S^{m-1}$. If in a local coordinate system for x close to x^* we have

$$p(x; y, \eta) \; - \; p(x^*; y, \eta) \; = \; (x - x^*) \cdot r(x; y, \eta), \quad r \in S^m \qquad (20)$$

then in these coordinates

$$\tilde{p}(x; y, \eta) \; = \; L(x; y, \eta, D_\eta) r(x; y, \eta)$$

where L is a first order differential operator such that

$$L(x; y, \lambda\eta, \lambda^{-1} D_\eta) \; = \; \lambda^{-1} L(x; y, \eta, D_\eta), \quad \lambda > 0.$$

Proof. Without loss of generality we may assume that the amplitude p has a small support with respect to x. If p is equal to zero in a neighbourhood of the set $\{x = x^*\}$, then $\operatorname{Im}\varphi(x; y, \eta) > 0$ on the support of p, and (18) is a smooth function. Therefore we may assume also that x is sufficiently close to x^* and (20) holds.

Let us consider the oscillatory integral with the amplitude $(x - x^*) \cdot r$. We can replace $(x - x^*)e^{i\varphi}$ by $B^{-1}\nabla_\eta(e^{i\varphi})$, where $B = B(x; y, \eta)$ is a homogeneous non-degenerate matrix of degree zero, and $B(x^*; y, \eta) = -i\,\Phi_{x\eta}(y, \eta)$. Now, integrating by parts with respect to η we obtain an oscillatory integral with the same phase function and the amplitude

$$\tilde{p} \; = \; |d_\varphi|^{-1}\operatorname{div}_\eta\big(|d_\varphi|\,(B^{-1})^T r\big) \; \in \; S^{m-1}.$$

The proof is complete.

Remark 12. Lemma 11 implies that if $p(x^*; y, \eta) = 0$ then we can decrease the order of the amplitude by one. However, if the amplitude $p(x; y, \eta)$ has a zero of the order $2N - 1$ or $2N$ at $x = x^*$, generally speaking, the order of the amplitude is decreased by N. This happens because in the process of integrating by parts we differentiate the rest of the factors $(x - x^*)$ with respect to η. (In the special case $x^* = y$ these derivatives are equal to zero, and in this case the order of the amplitude can be decreased exactly by the order of the zero of p at $x = x^*$.)

Iterating the formula (19) we obtain the following

Corollary 13. *For any amplitude $p(x; y, \eta) \in S^m$ there exists an amplitude $q(y, \eta) \in S^m$ independent of x such that modulo a smooth function*

$$(2\pi)^{-n} \int e^{i\varphi} p(x; y, \eta) \, |d_\varphi(x; y, \eta)| \, d\eta \;=\; (2\pi)^{-n} \int e^{i\varphi} q(y, \eta) \, |d_\varphi(x; y, \eta)| \, d\eta .$$

By Theorem 8 and Corollary 13 any Lagrangian distribution $u(x, y)$ may be written modulo C^∞ in the form

$$u(x, y) \;=\; (2\pi)^{-n} \int e^{i\varphi(x; y, \eta)} \, q(y, \eta) \, |d_\varphi(x; y, \eta)| \, d\eta . \tag{21}$$

4. Obviously, there are many amplitudes $p(x; y, \eta)$ determining the same Lagrangian distribution (18). However, in (21) q is defined almost uniquely by u and φ. We deduce this fact from the next theorem, but first we formulate

Definition 14. *Let $C = C_1 + iC_2$ be a $n \times n$ complex symmetric matrix with $C_1 \geq 0$ (C_1 and C_2 are real symmetric matrices) and $\ker C = \ker C_1 \cap \ker C_2$. We denote by Π_C the orthogonal projection on $\ker C$ and introduce*

$$\det{}_+ C \;=\; \det(C + \Pi_C) .$$

We choose the branch of the argument of $\det_+ C$ *such that it is continuous with respect to* C *on the set of matrices* C *with a fixed kernel and is equal to zero when* $C_2 = 0$ *and* $C_1 > 0$.

If $C_1 = 0$ then we have

$$\operatorname{Arg} \det_+ C \;=\; \frac{\pi}{2} \operatorname{sgn} C_2 \tag{22}$$

where $\operatorname{sgn} C_2$ is the signature of C_2 (see [2,v.I]).

Theorem 15. *Let* q_m *be the leading term of the amplitude* q *in (21). Then the (non-smooth) function*

$$e^{-i\left(\operatorname{Arg}\det_+ (\Phi_{\eta\eta}/i)\right)/2} q_m \tag{23}$$

is uniquely determined by the Lagrangian distribution u *and is independent of the phase function* $\varphi \in \mathcal{F}$.

Remark 16. Since $\operatorname{Im} \Phi_{\eta\eta} \geq 0$ the determinant $\det_+ (\Phi_{\eta\eta}/i)$ is correctly defined, and by (7) its argument does not depend on the choice of coordinates.

Theorem 15 immediately implies

Corollary 17. *All the homogeneous terms of the amplitude* q *in (21) are uniquely determined by the Lagrangian distribution* u *and the phase function* $\varphi \in \mathcal{F}$.

Indeed, if u is a smooth function then by Theorem 15 all the homogeneous terms of the amplitude q are equal to zero. Thus, if u is represented by two different oscillatory integrals with the same phase function then all the homogeneous terms of the amplitudes must be equal.

The proof of Theorem 15 is based on two auxiliary lemmas. Let us fix a point (y_0, η_0) and denote $x_0 = x^*(y_0, \eta_0)$, $\xi = \xi^*(y_0, \eta_0)$. Choose a local coordinate system x in a neighbourhood of x_0 such that (14) is fulfilled, and define the matrix-function

$$\Psi \;=\; \Psi(y,\eta) \;=\; \begin{pmatrix} \Phi_{xx} & \Phi_{x\eta} \\ \Phi_{\eta x} & \Phi_{\eta\eta} \end{pmatrix}$$

Lemma 18. *Let (14) be fulfilled. Then at the point (y_0,η_0)*

$$\det{}_+ (\Psi/i) \;=\; \det{}^2 \xi_\eta^* \; \det{}_+^{-1}\!\left(i(x_\eta^*)^T \xi_\eta^*\right) \det{}_+ (\Phi_{\eta\eta}/i)$$

and the branches of arguments on the right and left hand sides determined by Definition 14 are the same.

Proof. In view of (12) we have for $\varepsilon > 0$

$$(x_\eta^* - \varepsilon\xi_\eta^*)^T \cdot (x_\eta^* + \varepsilon\xi_\eta^*) \;=\; (x_\eta^*)^T \cdot x_\eta^* \;-\; \varepsilon^2 (\xi_\eta^*)^T \cdot \xi_\eta^* . \qquad (24)$$

By (14) the matrix $(\xi_\eta^*)^T \cdot \xi_\eta^*$ is positive at the point (y_0,η_0). Multiplying (24) from both sides by $\left((\xi_\eta^*)^T \cdot \xi_\eta^*\right)^{-1/2}$ we see that for sufficiently small $\varepsilon > 0$ the matrix (24), and thus the matrix $x_\eta^* + \varepsilon\xi_\eta^*$, is non-degenerate.

The equalities (10) imply

$$\begin{pmatrix} (x_\eta^* + \varepsilon\xi_\eta^*)^T & I \\ 0 & I \end{pmatrix} \cdot \begin{pmatrix} \Phi_{xx}/i & \Phi_{x\eta}/i \\ \Phi_{\eta x}/i & \Phi_{\eta\eta}/i \end{pmatrix} \cdot \begin{pmatrix} x_\eta^* + \varepsilon\xi_\eta^* & 0 \\ I & I \end{pmatrix}$$

$$= \begin{pmatrix} (x_\eta^*)^T \cdot \xi_\eta^*/i & 0 \\ 0 & \Phi_{\eta\eta}/i \end{pmatrix} - i\varepsilon \begin{pmatrix} 2(\xi_\eta^*)^T \cdot \xi_\eta^* & (\xi_\eta^*)^T \cdot \Phi_{x\eta} \\ \Phi_{\eta x} \cdot \xi_\eta^* & 0 \end{pmatrix}$$

$$- i\varepsilon^2 \begin{pmatrix} (\xi_\eta^*)^T \cdot \Phi_{xx} \cdot \xi_\eta^* & 0 \\ 0 & 0 \end{pmatrix} . \qquad (25)$$

The deteminant $\det_+$ of the matrix on the left hand side of (25) is equal to

$$\det{}^2 (x_\eta^* + \varepsilon\xi_\eta^*) \, \det{}_+ (\Psi/i) \;=\; c_k \varepsilon^{2k} \, \det{}_+ (\Psi/i) \;+\; O(\varepsilon^{2k+1})$$

where $k = \dim \ker x_\eta^*$, and $c_k \neq 0$ is independent of φ.

Let $\tilde{C}$ be the restriction of the matrix

$$\begin{pmatrix} 2(\xi_\eta^*)^T \cdot \xi_\eta^* & (\xi_\eta^*)^T \cdot \Phi_{x\eta} \\ \Phi_{\eta x} \cdot \xi_\eta^* & 0 \end{pmatrix}$$

on the kernel of

$$\begin{pmatrix} (x_\eta^*)^T \cdot \xi_\eta^* & 0 \\ 0 & \Phi_{\eta\eta} \end{pmatrix} \ .$$

Then the determinant $\det_+$ of the matrix on the right hand side of (25) modulo $O(\varepsilon^{2k+1})$ is equal to

$$\det_+ \left[\begin{pmatrix} (x_\eta^*)^T \cdot \xi_\eta^*/i & 0 \\ 0 & \Phi_{\eta\eta}/i \end{pmatrix} - i\varepsilon \tilde{C} \right]$$

$$= \varepsilon^{2k} \det_+ \left((x_\eta^*)^T \cdot \xi_\eta^*/i\right) \det_+ \tilde{C} \det_+ (\Phi_{\eta\eta}/i)$$

(in this proof of this lemma the sign "=" means also equality of the branches of arguments). Since by (10), (12)

$$\ker \Phi_{\eta\eta} \;=\; (x_\eta^*)^T \cdot \xi_\eta^* \;=\; \ker x_\eta^*$$

and $\Phi_{x\eta} = \xi_\eta^*$ on $\ker x_\eta^*$, the matrix $\tilde{C}$ is independent of φ.

Thus at the point (y_0, η_0)

$$\det_+ (\Psi/i) \;=\; c \, \det_+ (\Phi_{\eta\eta}/i)$$

where $c \neq 0$ does not depend on φ. To compute the constant c we take the special phase function (15). For this phase function

$$\det_+ (\varphi_{\eta\eta}/i) \;=\; \det_+ \left(i(x_\eta^*)^T \cdot \xi_\eta^*\right)$$

$\operatorname{sgn} \Psi = 0$ and by (22) we have

$$\det_+ (\Psi/i) \;=\; \det{}^2 \xi_\eta^*$$

with zero branch of the argument. It leads to the formula

$$c = \det{}^2 \xi_\eta^* \, \det{}_+^{-1} \left(i(x_\eta^*)^T \xi_\eta^* \right) \Big|_{(y_0, \eta_0)}$$

which completes the proof.

Lemma 19. *Let u be the Lagrangian distribution (21) of order m, and q_m be the leading homogeneous term of the amplitude q. Let (14) be fulfilled. Then for any smooth function $\rho(x)$ with sufficiently small support which is equal to 1 at the point x_0 we have*

$$\int e^{-i\lambda x \cdot \xi_0} \rho(x) u(x, y_0) \, dx = \lambda^{n+m} e^{-i\lambda x_0 \cdot \xi_0}$$

$$\times \left(e^{i\pi \operatorname{sgn} \left((x_\eta^*)^T \cdot \xi_\eta^* \right)/4} \, e^{-i\left(\operatorname{Arg} \det_+ (\Phi_{\eta\eta}/i) \right)/2} \, |\det \xi_\eta^*|^{-1/2} \, q_m \right) \Big|_{(y_0, \eta_0)} \tag{26}$$

$$+ \, O(\lambda^{n+m-1}), \qquad \lambda \to +\infty.$$

Proof. Let us replace in the left hand side of (26) the distribution u by the oscillatory integral (21) and change the variables $\eta = \lambda\theta$. Then we obtain the integral

$$(2\pi)^{-n} \lambda^n \int e^{i\lambda(\varphi(x; y_0, \theta) - x \cdot \xi_0)} \rho(x) q(y_0, \lambda\theta) |d_\varphi(x; y_0, \theta)| \, dx d\theta.$$

Now we apply the stationary phase method. Remember that the equation

$$\varphi_\eta(x; y, \eta) = 0$$

has the unique solution $x = x^*$, and by (14) the equation

$$\varphi_x(x^*; y_0, \theta) - \xi_0 = \xi^*(y_0, \theta) - \xi_0 = 0$$

also has the unique solution $\theta = \eta_0$. Thus the function $\varphi(x; y_0, \theta) - x \cdot \xi_0$ has a unique stationary point $x = x_0, \theta = \eta_0$, and its Hessian coincides with the Hessian of φ with respect to x, η. In view of (10) we have

$$\begin{pmatrix} \Phi_{xx} & \Phi_{x\eta} \\ \Phi_{\eta x} & \Phi_{\eta\eta} \end{pmatrix} \cdot \begin{pmatrix} I & x_\eta^* \\ 0 & I \end{pmatrix} = \begin{pmatrix} \Phi_{xx} & \xi_\eta^* \\ \Phi_{\eta x} & 0 \end{pmatrix}.$$

It follows that

$$|\det_+ (\Psi/i)| = |\det \xi_\eta^*| \, |\det \Phi_{x\eta}| = |\det \xi_\eta^*| \, |d_\varphi|^2 \Big|_{x=x^*} \tag{27}$$

where $\Phi_{x\eta}$ and d_φ are defined in (6) and (17). Now by (3) and (14) the stationary point (x_0, η_0) is non-degenerate. Using the stationary phase formula we see that

$$\int e^{-i\lambda x \cdot \xi_0} \rho(x) u(x, y_0) \, dx =$$

$$\lambda^{n+m} e^{-i\lambda x_0 \cdot \xi_0} \left(\det_+ (\Psi/i)\right)^{-1/2} |d_\varphi| \, q_m \Big|_{(x_0; y_0, \eta_0)} + O(\lambda^{n+m-1}).$$

By (27) we obtain at the point $(x_0; y_0, \eta_0)$

$$\left(\det_+ (\Psi/i)\right)^{-1/2} |d_\varphi| = \left(\det_+ (\Psi/i)\right)^{-1/2} |\det \xi_\eta^*| \, |\det_+ (\Psi/i)|^{1/2}$$

$$= |\det \xi_\eta^*| \, e^{-i\left(\operatorname{Arg} \det_+ (\Psi/i)\right)/2}.$$

Now Lemma 18 yields

$$\operatorname{Arg} \det_+ (\Psi/i) = \operatorname{Arg} \det_+ (\Phi_{\eta\eta}/i) - \operatorname{Arg} \det_+ \left(i(x_\eta^*)^T \cdot \xi_\eta^*\right)$$

and by (22)

$$\operatorname{Arg} \det_+ \left(i(x_\eta^*)^T \cdot \xi_\eta^*\right) = \pi \operatorname{sgn}\left((x_\eta^*)^T \cdot \xi_\eta^*\right)/2. \tag{28}$$

It implies (26), and proves the lemma.

Proof of Theorem 15. The coefficient attached to λ^{n+m} in the asymptotic formula (26) for the Fourier transform of the distribution $\rho(x) u(x, y_0)$ depends on the choice of the coordinate system but not on the phase function φ. Therefore the function (23) also is independent on φ. Since this function can not depend on the coordinates x, it is uniquely determined by the Lagrangian distribution u. The theorem is proved.

References

1.   L.Hörmander, *The analysis of linear partial differential operators* I-IV, Springer Verlag, Berlin, Heidelberg, New York, 1984.

2. F.Treves, *Introduction to Pseudodifferential and Fourier Integral Operator* II, Plenum Press, New York and London, 1982.

3. A.Laptev and Yu.Safarov, *Global Parametrization of Lagrangian Manifolds and the Maslov Index*, Preprint, Linköping University, 1989.

A. Laptev
Department of Mathematics
Linköping University
S–581 83 Linköping, Sweden

Yu. Safarov
LOMI, Fontanka 27,
191 011 Leningrad, USSR

D. Vassiliev
Mathematics Division, MAPS,
University of Sussex, Falmer,
Brighton BN1 9QH, England

Operator Theory:
Advances and Applications, Vol. 57
© 1992 Birkhäuser Verlag Basel

Stable asymptotics of the solution to the Dirichlet problem for elliptic equations of second order in domains with angular points or edges

Vladimir G. Maz'ya and Jürgen Roßmann

Abstract. The Dirichlet problem for second order elliptic equations will be considered in plane domains with angular corners and in N-dimensional domains with smooth (N-2)-dimensional edges. The asymptotical decomposition of the solution will be found even in the case when the opening of the angle runs through a critical value.

Introduction

The present paper concerns the asymptotic behaviour of solutions to the Dirichlet problem for second order elliptic equations with real coefficients in plane domains with angular corners or n-dimensional domains with edges. In particular, we are interested in the case when the angle of the corner lies in a neighbourhood of some critical value or the angle on the edge runs through a critical value. The asymptotics of solutions of elliptic boundary value problems have been studied by many authors (e. g. M. Dauge (1988), V. A. Kondratjev (1977), V. G. Maz'ya and B. A. Plamenevskij (1978), V. G. Maz'ya and J. Rossmann (1988), B. W. Schulze (1991)). But in these papers (except that of B. W.Schulze who has given a very abstract description of the asymptotics near edges by means of analytic functionals) critical angles at the edges were either excluded or the authors considered only operators with constant coefficients

in a domain with constant angle. An explicit representation of the asymptotics of solutions of second order elliptic equations near so-called crossing points has been first announced by M. Costabel and M. Dauge (1990).
In order to illustrate the difficulties of this situation we consider the Dirichlet problem for the Laplacian in a plane domain G with an angular corner in the origin. For simplicity we assume that G coincides with a plane wedge K_α in a neighbourhood of the origin (see figure 1).

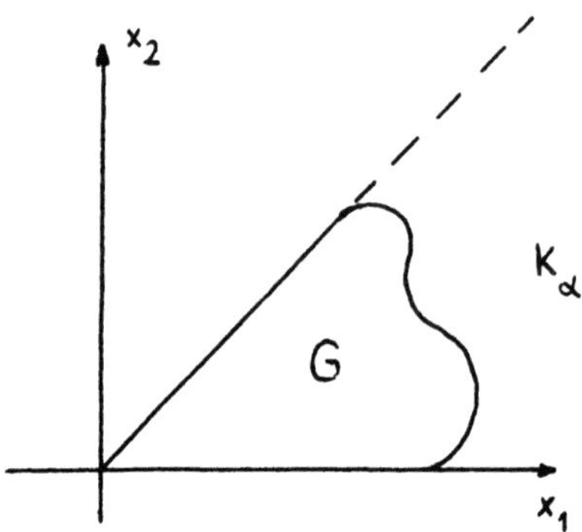

figure 1

Let $u \in \overset{\circ}{W}^1(G)$ be the unique variational solution of the Dirichlet problem

$$\Delta u = f \quad in \ G, \quad u = 0 \ on \ \partial G \tag{1}$$

where $f \in C^\infty(G)$. If $(D^\beta f)(0) = 0$ for $|\beta| \leq \ell-1$ then the solution of (1) admits the following decomposition (see e. g. Kondratjev (1967)):

$$u = \sum_{\nu:\nu\pi/\alpha<\ell+2} c_\nu \, r^{\nu\pi/\alpha} \, sin(\nu\pi\phi/\alpha) + u_1 \tag{2}$$

where $u_1 = O(r^{\ell+2-\varepsilon})$ with an arbitrary small positive ε. Here r, ϕ denote the polar coordinates, i.e. $r=(x_1^2+x_2^2)^{1/2}$, $\phi = arg(x_1+ix_2)$. For arbitrary $f \in C^\infty(G)$ one gets

$$u = \sum_\nu c_\nu \, r^{\nu\pi/\alpha} \, \sin(\nu\pi\phi/\alpha) + \sum_{\mu=0}^{1} \sum_{\nu=2}^{\ell+1} c_{\mu\nu} \, r^\nu \, \log^\mu r \; \psi_{\mu\nu}(\phi)$$

$$+ \, u_1 \qquad\qquad (3)$$

where $u_1 = O(r^{\ell+2-\varepsilon})$.

Here two different cases have to been considered. If the angle α is not equal to a "critical value" then the decomposition (3) does not contain logarithmic terms whereas in the so-called resonance case (i. e. α is equal to a critical value) logarithmic terms occur.

Let e. g. $\ell=1$, $\pi/3 < \alpha < 2\pi/3$. In order to obtain the asymptotic decomposition of the solution of (1) one has to solve the auxilliary problem

$$\Delta v = f(0) \; in \; K_\alpha \,, \qquad v = 0 \;\; on \; \partial K_\alpha \,. \qquad\qquad (4)$$

By means of an elementary ansatz one obtains the following solution v of (4):

$$v = \frac{f(0)}{4} \, r^2 \left\{ 1 - \cos(2\phi) + \frac{\cos(2\alpha)-1}{4 \sin(2\alpha)} \sin(2\phi) \right\} \qquad\qquad (5)$$

if $\alpha \neq \pi/2$ and

$$v = \frac{f(0)}{4} \, r^2 \, (1 - \cos(2\phi)) + \frac{f(0)}{\pi} \, r^2 \, \phi \, \cos(2\phi)$$

$$+ \frac{f(0)}{\pi} \, r^2 \, \log r \, \sin(2\phi) \qquad\qquad (6)$$

if $\alpha = \pi/2$. Hence, the solution of (1) has the representation

$$u = c_\alpha \, r^{\pi/\alpha} \, \sin(\pi\phi/\alpha) + \frac{f(0)}{4} \, r^2 \, (1 - \cos(2\phi))$$

$$+ \frac{f(0)}{4} \, \frac{\cos(2\alpha)-1}{\sin(2\alpha)} \, r^2 \, \sin(2\phi) + u_1 \qquad\qquad (7)$$

if $\alpha \neq \pi/2$ and

$$u = c'_{\pi/2} \, r^2 \, \sin(2\phi) + \frac{f(0)}{4} \, r^2 \, (1 - \cos(2\phi))$$

$$+ \frac{f(0)}{\pi} \, r^2 \phi \, \cos(2\phi) + \frac{f(0)}{\pi} \, r^2 \, \log r \, \sin(2\phi) + u_1 \qquad\qquad (8)$$

if $\alpha = \pi/2$, $u_1 = O(r^{3-\varepsilon})$.

The representation (7) has an essential disadvantage. It is nonstable relative to α near the critical angle $\alpha^* = \pi/2$. The coefficients c_α and $\dfrac{f(0)}{4} \dfrac{\cos(2\alpha)-1}{\sin(2\alpha)}$ tend to infinity if $\alpha \to \dfrac{\pi}{2}$.

However, if we rewrite (7) into the following form

$$u = c'_\alpha \, r^{\pi/\alpha} \, sin(\pi\phi/\alpha) + \frac{f(0)}{4} \, r^2 \, (1-cos(2\phi))$$

$$+ \frac{f(0) \, (\alpha-\pi/2) \, (cos(2\alpha)-1)}{4 \, sin(2\alpha)} \, \frac{r^2 sin(2\phi)-r^{\pi/2} sin(\pi\phi/\alpha)}{\alpha - \pi/2} + u_1 \tag{9}$$

with the "singular" functions $r^{\pi/\alpha} \, sin(\pi\phi/\alpha)$, $r^2 \, (1-cos(2\phi))$

and $\dfrac{r^2 sin(2\phi)-r^{\pi/\alpha} sin(\pi\phi/\alpha)}{\alpha - \pi/2}$ then the new coefficients $c'_\alpha =$

$c_\alpha + \dfrac{f(0) \, (cos(2\alpha)-1)}{4 \, sin(2\alpha)}$, $\dfrac{f(0)}{4}$ and $\dfrac{f(0) \, (\alpha-\pi/2) \, (cos(2\alpha)-1)}{4 \, sin(2\alpha)}$

are bounded relative to α. The purpose of this paper is to
find similar stable asymptotics for the solution to the
Dirichlet problem for arbitrary second order elliptic
equations with real coefficients in a neighbourhood of each
critical angle $\alpha^* = \frac{m}{k}\pi$ ($k=1,2, \ldots ,\ell+1; \; m=1, \ldots ,2k$). We
will see that the solution u can be written as a finite sum
of terms

$$A(\alpha) \, z^{t\pi/\alpha \, +\mu} \, \bar{z}^{-\nu} \, S_\chi(z,\alpha)$$

($z=x_1+ix_2$, $z^s=r^s e^{is\phi}$) and conjugate terms with bounded
coefficients $A(\cdot)$ and certain singular functions $\mathcal{S}_\chi(z,\alpha)$.
An analogous representation can be found for the solution of
the Dirichlet problem in a n-dimensional ($n\geq3$) domain with
smooth (n-2)-dimensional edges if the angle on the edge runs
through critical value.

Under the additional assumption that the coefficients of the
operators and the smooth components of the boundary are
analytic M. Costabel and M. Dauge (1991) have obtained
similar results for arbitrary second order differential
operators and for inhomogeneous boundary conditions of
order ≤ 1.

1. Critical angles, singular functions

Let ℓ be a given non-negative integer (which determines the
smoothness of the regular remainder in the asymptotics of the

solution). An angle $\alpha^* \in (0, 2\pi]$ is said to be critical (for the Laplacian) if

$$\alpha^* = \frac{m}{k}\pi \qquad (k=1,2, \ldots ,\ell+1; \; m=1,2, \ldots ,2k),$$

i. e. there exists an integer $k \in \{1, \ldots ,\ell+1\}$ such that $k\alpha^*/\pi$ is integer. For a fixed critical angle α^* we denote by $\Re_\ell(\alpha^*)$ the set of all tuples $x = (k_0, k_1, \ldots , k_n)$ of interger numbers $k_0, k_1, \ldots ,k_n$ such that $0=k_0<k_1< \ldots < k_n \leq \ell+1$ $(n\geq0)$ and $k_j\alpha^*/\pi$ are integer. Let $x = (k_0, \ldots ,k_n) \in \Re_\ell(\alpha^*)$. Then we define the function $S_x(z,\alpha)$ of the complex variable $z = x_1+ix_2 = r\,e^{i\phi}$ as follows

$$S_x(z,\alpha) = (\alpha-\alpha^*)^{-n}\, z^{k_n}\, \sum_{j=0}^{n} a_j\, z^{-k_j(1-\alpha^*/\alpha)} \tag{10}$$

where $a_j = -\displaystyle\prod_{\nu=1,\nu\neq j}^{n} \frac{k_\nu}{k_\nu - k_j}$ $(j=1, \ldots ,n)$, $a_0 = 1$.

Remark 1. *1) The functions $S_x(z,\alpha)$ are bounded with respect to α for each $z \in \mathbb{C}\backslash\{0\}$. The coefficients a_j in (10) are uniquely determined by the boundedness of S_x.*

2) The limit of $S_x(z,\alpha)$ for $\alpha\to\alpha^*$ is a polynomial of degree n relative to $\log z$.

Let α^* be a critical angle, $x = (k_0, k_1, \ldots , k_n) \in \Re_\ell(\alpha^*)$ and $U(\alpha^*)$ a neighbourhood of α^*. An expression of the form

$$A(z,\alpha) = (\alpha-\alpha^*)^{-m}\, z^{k_n}\, \sum_{j=0}^{n} c_j(\alpha)\, z^{-k_j(1-\alpha^*/\alpha)} \tag{11}$$

$(m \leq n$, integer) with coefficients $c_j \in C^\infty(U(\alpha^*))$ is said to be regular relative to α (for $\alpha\to\alpha^*$) if the limit

$$A(z,\alpha^*) = \lim_{\alpha\to\alpha^*} A(z,\alpha)$$

exists for every $z \in \mathbb{C}\backslash\{0\}$.

Lemma 1. *If the expression (11) is regular relative to α for $\alpha\to\alpha^*$ then there exist functions $A_s \in C^\infty(U(\alpha^*))$ such that*

$$A(z,\alpha) = \sum_{s=0}^{n} A_s(\alpha)\, z^{k_n-k_s}\, S_{x_s}(z,\alpha)$$

where $x_s = (k_0, k_1, \ldots ,k_s)$. The functions A_s are real if $c_0, c_1, \ldots ,c_n$ are real.

2. Stable asymptototics near corners with angles in a neighbourhood of a critical value

Let ℓ be a given non-negative integer. We consider the problem (1) with the right-hand side $f \in W^{\ell}_{-\varepsilon}(G)$ ($\varepsilon > 0$, small). Here $W^{\ell}_{-\varepsilon}(G)$ denotes the weighted Sobolev space with the norm

$$\|u\| = \left(\int_G \sum_{|\alpha| \leq \ell} r^{-2\varepsilon} |D^{\alpha}u|^2 \, dx \right)^{1/2} .$$

Note that every function from $W^{\ell}_{-\varepsilon}(G)$ can be written as a sum of a polynomial

$$P_{\ell}(u) = \sum_{\mu+\nu \leq \ell-1} \frac{1}{\mu! \, \nu!} \, (\partial^{\mu}_{x_1} \partial^{\nu}_{x_2})(0) \, x_1^{\mu} \, x_2^{\nu}$$

and a function $u' \in V^{\ell}_{-\varepsilon}(G)$ where $V^{\ell}_{-\varepsilon}(G)$ is defined as the closure of $C^{\infty}_0(\overline{G}\backslash\{0\})$ with respect to the norm

$$\|u\| = \left(\int_G \sum_{|\alpha| \leq \ell} r^{-2(\varepsilon+\ell-|\alpha|)} |D^{\alpha}u|^2 \, dx \right)^{1/2}$$

Lemma 2. Let α^* be a critical angle and $x=(k_0,k_1, \ldots,k_n) \in \mathcal{R}_{\ell}(\alpha^*)$. We consider the auxilliary problem

$$\frac{1}{4} \Delta v = \partial_z \partial_{\bar{z}} v = \partial^q_z \left(z^{\mu+t\pi/\alpha} \, \bar{z}^{\nu} \, S_x(z,\alpha) \right) \quad \text{in } K_{\alpha}$$

$$v = 0 \quad \text{on } \partial K_{\alpha} \tag{12}$$

$(\partial_z = \frac{1}{2}(\partial_{x_1} - i\partial_{x_2}), \; \partial_{\bar{z}} = \frac{1}{2}(\partial_{x_1} + \partial_{x_2}))$ where t, q,μ,ν are non-negative integers. Assume that $m-1 < k_{n+1}\alpha^*/\pi \leq m$ for $k_{n+1} = k_n+\mu+\nu+2-q$ (m integer) and that α lies in a sufficiently small neighbourhood $U(\alpha^*)$ of α. Then there exists a solution

$$v = \sum_{s=0}^n A_s(\alpha) \, z^{\mu+1-q+k_n-k_s+t\pi/\alpha} \, \bar{z}^{\nu} \, S_{x_s}(z,\alpha)$$

$$+ \sum_{s=0}^n B_s(\alpha) \, Re\{z^{k_{n+1}-k_s+t\pi/\alpha} \, S_{x_s}(z,\alpha)\}$$

$$+ \sum_{s=0}^{n+1} C_s(\alpha) \, Im\{z^{k_{n+1}-k_s+t\pi/\alpha} \, S_{x_s}(z,\alpha)\}$$

of (12) with coefficients A_s, B_s, $C_s \in C^{\alpha}(U(\alpha^*))$. Here $x_s = $

$(k_0, k_1, \ldots, k_s)$, $C_{n+1} \equiv 0$ if $k_{n+1}\alpha^*/\pi \neq m$.

Let now $L(x,D)$ be an arbitrary differential operator of second order with smooth real coefficients. Then without loss of generality we can assume that the principal part $L_0(0,D)$ of L with coefficients frozen in the origin coincides with the Laplacian. As before we further assume that G coincides with the plane wedge K_α in a neighbourhood of 0. By means of Lemma 2 and the result of Kondratjev (1967) one can prove the following theorem.

Theorem 1. *Let $u \in \overset{\circ}{W}{}^1(G)$ be a solution of the Dirichlet problem $Lu = f$ in G, $u = 0$ on ∂G where $f \in W^\ell_{-\varepsilon}(G)$. Suppose that the inner angle α of G in the corner 0 lies in a sufficiently small neighbourhood $U(\alpha^*)$ of the critical value α^*. Then u admits the decomposition*

$$u = \Sigma_\ell + u_\ell$$

where $u_\ell \in V^{\ell+2}_{-\varepsilon}(G)$ and Σ_ℓ is a finite sum of terms of the form $c(\alpha)\, z^{\mu+t\pi/\alpha}\, \bar{z}^{-\nu}\, S_\chi(z,\alpha)$ and $c(\alpha)\, \bar{z}^{\mu+t\pi/\alpha}\, z^\nu\, S_\chi(\bar{z},\alpha)$ (t, μ, ν non-negative, integer, $\chi=(k_0,k_1, \ldots ,k_n) \in \mathfrak{R}_\ell(\alpha^*)$, $\mu+\nu+k_n+t\pi/\alpha^* \leq \ell+1$, $c \in C^\infty(U(\alpha^*))$).

3. Applications to edge singularities

The singular functions $S_\chi(z,\alpha)$ defined in Section 1 can also be used for the description of edge asymptotics if the angle on the edge runs through a critical value. For simplicity we assume that the domain $G \subset \mathbb{R}^N$ coincides with the dihedral angle $\mathcal{D} = \{ (x_1,x_2,y) \in \mathbb{R}^N : 0 < \arg(x_1+ix_2) < \alpha(y) \}$ (see figure 2) in a neighbourhood of the edge point y_0.

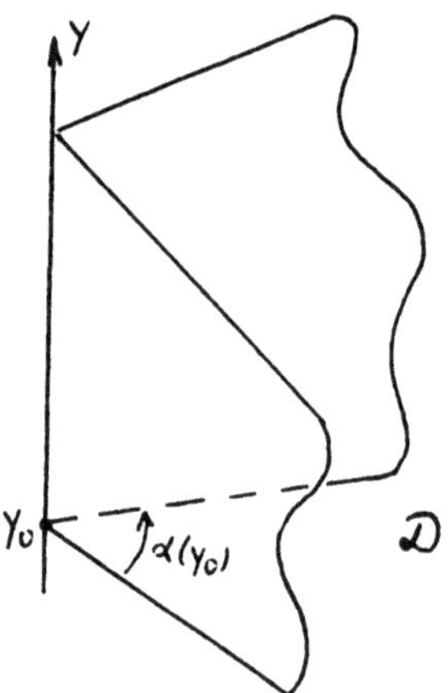

figure 2

We consider the problem

$$Lu = f \quad in \ G, \quad u = 0 \quad on \ \partial G \tag{13}$$

for the elliptic differential operator

$$L = \sum_{i+j+|\beta| \le 2} a_{ij\beta}(x,y)\, \partial_{x_1}^i\, \partial_{x_2}^j\, \partial_y^\beta$$

with smooth real coefficients $a_{ij\beta}$. Again without loss of generality we can suppose that $a_{2,0,0}(0,y) = a_{0,2,0}(0,y) \equiv 1$ and $a_{1,1,0}(0,y) \equiv 0$. Analogously to V. Maz'ya and J. Roßmann (1988) the following regularity assertion for the derivatives of the solution in the direction of the edge can be shown.

Lemma 3. *Let* $u \in \overset{\circ}{W}{}^1(G)$ *be a solution of (13) with* $f \in W^\ell_{-\varepsilon}(G)$. *Then* $\chi\, D_y^\beta u \in V^1_{-\varepsilon}(G)$ *for* $|\beta| \le \ell+1$. *Here* χ *is a smooth cut-off function equal to one in a neighbourhood of* y_0.

The following theorem has been shown in several papers (see e. g. V. Maz'ya and J. Roßmann (1988)) for the case that the angle $\alpha(y)$ of the dihedron $\mathfrak{D}$ is either not critical or constant. We now assume that α is variable and $\alpha(y_0)$ is critical, i. e. $\alpha(y_0) = \dfrac{m}{k}\,\pi$ $(k \in \{1, \ldots, \ell+1\},\ m \in \{1, \ldots, 2k\})$

Theorem 2. *Let* $u \in \overset{\circ}{W}{}^1(G)$ *be a solution of (13) and* $f \in W^\ell_{-\varepsilon}(G)$ $(\varepsilon > 0)$. *Then* u *admits the decomposition*

$$u = \Sigma_\ell + u_\ell$$

where $\chi u_\ell \in V^{\ell+2}_{-\varepsilon'}(G)$ $(0 < \varepsilon' < \varepsilon)$ and Σ_ℓ is a finite sum of expressions

$$\hat{c}(x,y)\,\partial_y^\beta\,\{\,z^{\mu+t\pi/\alpha}\,\bar{z}^{-\nu}\,S_x(z,\alpha(y))\,\}$$

and conjugate terms. Here t,μ,ν are integers, $t,\nu \geq 0$, $\mu+\nu \geq 0$, $x = (k_0,k_1,\ \ldots\ ,k_n) \in \mathfrak{R}_\ell(\alpha^*)$, and $\hat{c}$ is an extension of a function $c \in W^{\ell+1-\mu-\nu-|\beta|-k_n-\varepsilon'-t\pi/\alpha}(\mathbb{R}_y^{N-2})$.

Remark 2. *Theorem 2 includes the cases* $\alpha(y_0)=\pi$ *and* $\alpha(y_0)=2\pi$, *provided that the domain* G *is diffeomorphic to a dihedral angle* $\mathfrak{D}=\{\ (x_1,x_2,y) \in \mathbb{R}^N:\ 0 < \arg(x_1+ix_2) < \alpha(y)\ \}$ *with a function* α *only depending on the variable* y. *It can be shown that the sum* Σ_ℓ *consists only of terms of the form*

$$\hat{c}(x,y)\,\partial_y^\beta\,z^{\mu+t\pi/\alpha}\,\bar{z}^{-\nu}\qquad\text{and}\qquad \hat{c}(x,y)\,\partial_y^\beta\,\bar{z}^{-\mu+t\pi/\alpha}\,z^{\nu}$$

in these cases, i. e. the asymptotics of u *does not differ from that for non-critical angles.*

References:

Costabel, M. and Dauge, M. (1990), Edge asymptotics on a skew cylindre. Lecture given at the International Workshop "Analysis in domains and on manifolds with singularities", Breitenbrunn.

Costabel, M. and Dauge, M. (1991), General edge asymptotics of solutionsof second order elliptic boundary value problems.Preprints R91016 and R91017, Universite Pierre et Marie Curie, Paris.

Dauge, M. (1988), Elliptic boundary value problems in corner domains - smoothness and asymptotics of solutions. Lecture Notes in Mathematics, Vol. 1341, Springer-Verlag, Berlin.

Kondratjev, V.A. (1967), Boundary value problems for elliptic equations in domains with conical or angular points. Trudy Mosk. Mat. Obšč. 16, 209-292 (Russian).

Kondratjev, V. A. (1977), Singularities of the solution of

the Dirichlet problem for elliptic equations of second order
in a neighbourhood of an edge. Diff. Uravn. **13**, 11, 2026-2032
(Russian).

Maz'ya, V. G. and Plamenevskij, B. A. (1978), L_p-estimates
for solutions of elliptic boundary value problems in domains
with edges. Trudy Mosk. Mat. Obšč. **37**, 49-93 (Russian).

Maz'ya and J. Rossmann (1988), über die Asymptotik der
Lösungen elliptischer RWA in der Umgebung von Kanten.
Math. Nachrichten **138**, 27-53.

Maz'ya and J. Rossmann (1990), on a problem of Babushka
(Stable asymptotics of the solution of the Directed problem
for elliptic equations of second order in domains with
angular points) Preprint Lith-Mat-R-90-33, University
Linköping, Dep. of Math.

Maz'ya and J. Rossmann (1991), On the asymptotics of
solutions to the Dirichlet problem for second order elliptic
equations in domains with critical angles on the edges.
Preprint Lith-Mat-R-91-37, Univ. Linköping, Dep. of Math.

B.-W. Schulze, Regularity with continuous and branching
asymptotics for elliptic operators on manitolds with edges.
Integral equations and operator theory, **11**, 4, 557-602.

B.-W. Schulze (1991), Pseudo-differential operators on
manifolds with singularities.North-Holland/Elsevier, New York

Author's adress:

Dr. Jürgen Roßmann
Universität Rostock
Fachbereich Mathematik
Universitatsplatz 1
O-2500 Rostock
Bundesrepublik Deutschland

Operator Theory:
Advances and Applications, Vol. 57
© 1992 Birkhäuser Verlag Basel

MASLOV OPERATIONAL CALCULUS

AND NONCOMMUTATIVE ANALYSIS

DR. VLADIMIR NAZAIKINSKII, DR. BORIS STERNIN, DR. VICTOR SHATALOV

Abstract. Maslov's noncommutative operational calculus has proved to be a convenient tool to study various problems in algebra and differential equations. Here we give a brief description of its main ideas and notions together with several examples of applications. The detailed exposition may be found in "Introduction to Noncommutative Analysis (Maslov Operational Calculus)", a book by the present authors, to be published by Walter de Gruyter & Co.

1. General Noncommutative Analysis

The object of study in Maslov's operational method (Maslov V.P. (1973), Maslov V.P. (1987)) is the analysis of functions of several linear operators $A_1, \ldots, A_n$. First of all, let us consider how to define such functions.

This question has been completely studied for $n = 1$ (functions of a single operator). In this, case the definition goes as follows. We need to learn how to substitute an operator A into a function $f(x)$ instead of a numeric variable x, thus obtaining an operator, denoted by $f(A)$. For naturality reasons, the desired result of the substitution is clear for some particular functions $f(x)$. For example, if $f(x) = x^l$, then $f(A) = A^l$; if $f(x) = 1/(\lambda - x)$, then we should obtain the resolvent $R_\lambda(A)$; if $f(x) = e^{ikx}$, then $f(A)$ should be equal to the element $exp(ikA)$ of the one-parametric group, generated by the operator iA; if $f(x) = \theta(\lambda - x)$ and A is a self-adjoint operator, then

$f(A)$ should be equal to the projector $E_\lambda(A)$ an so forth.

Next, the correspondence $f \longmapsto f(A)$ should be linear and continuous. In particular, one may permute this mapping with integration or summation over parameters, say, $\int f(A, y)\, dy = F(A)$, where $F(x) = \int f(x, y)\, dy$.

Since series and integral representations (such as the Taylor series, the Cauchy integral, and the Fourier integral) allow us to build general functions from the elementary ones mentioned above, our considerations immediately lead to the following variants of the definition:

$$(1) \qquad f(A) = \int_{-\infty}^{\infty} f(\lambda)\, dE_\lambda(A);$$

$$(2) \qquad f(A) = \frac{1}{2\pi i} \oint_{\partial c} f(\lambda) R_\lambda(A)\, d\lambda;$$

$$(3) \qquad f(A) = \frac{1}{\sqrt{2\pi}} \int_{-\infty}^{\infty} \tilde{f}(k) e^{ikA}\, dk;$$

the list may be continued easily.

Of course, any of definitions (1) – (3) is valid for its own function and operator classes; e.g., in (2) $f(x)$ must be an analytic function in some domain C, containing the spectrum of the operator A. Note, however, that if more than one definition applies to a particular case, they do not contradict one another (except for pathological examples).

Definitions (1) – (3) satisfy various naturality properties; e.g., for fixed A, the mapping $f \longmapsto f(A)$ is a homomorphism from the ring of functions to the ring of operators.

Now let us try to make things clear for the case of several operators, using the same principles.

We face a difficulty immediately. Let

$$(4) \qquad\qquad f(x,y) = xy.$$

The attempt to define the function $f(A,B)$ of two operators A and B fails if the commutator

$$(5) \qquad\qquad [A,B] = AB - BA$$

is not equal to zero, since one cannot decide reasonably which of the two variants

$$(6) \qquad\qquad \begin{aligned} f(A,B) &= AB, \\ f(A,B) &= BA \end{aligned}$$

to prefer. (Moreover, why should the symmetric variant

$$f(A,B) = \frac{1}{2}(AB + BA),$$

suggested by H. Weyl, be rejected?)

A simple solution was invented by Feynman R.P. (1951). He proposed to supply operators with indices, or numbers, prescribing the order in which they act (Feynman ordering of operators). In formulas, these numbers are placed over the operators. Thus the operator A acts first on the vector u in the expression $f(\overset{1}{A},\overset{2}{B})u$, so that

$$(7) \qquad\qquad f(\overset{1}{A},\overset{2}{B}) = BA.$$

For the function $f(x,y)$ given by (4), we have, on the opposite,

$$(8) \qquad\qquad f(\overset{2}{A},\overset{1}{B}) = AB.$$

Note that in these notations

$$(9) \qquad\qquad \overset{1}{A}\overset{2}{B} = \overset{2}{B}\overset{1}{A} = BA.$$

Now the definitions (1) – (3) may be generalized for functions of n operators $A_1, \ldots, A_n$ with a chosen order of their action, say $\overset{1}{A_1}, \ldots, \overset{n}{A_n}$:

$$(10) \qquad f(\overset{1}{A_1}, \ldots, \overset{n}{A_n}) = \int f(\lambda_1, \ldots, \lambda_n) \, dE_{\lambda_n}(A_n) \ldots dE_{\lambda_1}(A_1),$$

or

$$(11) \quad f(\overset{1}{A_1}, \ldots, \overset{n}{A_n}) = \left(\frac{1}{2\pi}\right)^n \oint f(\lambda_1, \ldots, \lambda_n) R_{\lambda_n}(A_n) \ldots R_{\lambda_1}(A_1) \, d\lambda_1 \ldots d\lambda_n,$$

or

$$(12) \qquad f(\overset{1}{A_1}, \ldots, \overset{n}{A_n}) = \left(\frac{1}{2\pi}\right)^n \int \tilde{f}(k_1, \ldots, k_n) e^{ik_n A_n} \ldots e^{ik_1 A_1} \, dk_1 \ldots dk_n.$$

Just as in the case of functions of a single operator, these definitions do not contradict one another when more than one definition may be used. However, the mapping $f \longmapsto f(\overset{1}{A_1}, \ldots, \overset{n}{A_n})$ is *no more a homomorphism of rings* unless the operators $A_1, \ldots, A_n$ are pairwise commutative, and the study of its algebraic structure is a complicated stand-alone problem. (With commutativity requirement, the situation is quite similar to the one-dimensional case.)

We point out that definitions (10) – (12) are natural in the following sense:

a) (Correspondence principle.) If the function $f(x_1, \ldots, x_n)$ depends only on one of the arguments, say $f(x_1, \ldots, x_n) = \varphi(x_j)$, $f(\overset{1}{A_1}, \ldots, \overset{n}{A_n})$ is equal to the corresponding function of a single operator, $f(\overset{1}{A_1}, \ldots, \overset{n}{A_n}) = \varphi(A_j)$.

b) (Naturality with respect to morphisms.) Let $A_1, \ldots, A_n$ and $B_1, \ldots, B_n$ be linear operators in H_1 and H_2, respectively. Suppose that for some linear mapping $\tau : H_1 \to H_2$

$$\tau \circ A_j = B_j \circ \tau, \qquad j = 1, \ldots, n$$

(that is, τ is an *intertwining operator* for the tuples $A_1, \ldots, A_n$, $B_1, \ldots, B_n$). Then

$$\tau \circ f(\overset{1}{A_1}, \ldots, \overset{n}{A_n}) = f(\overset{1}{B_1}, \ldots, \overset{n}{B_n}) \circ \tau$$

for any function $f(x_1, \ldots, x_n)$.

Also, if

$$B_j = \gamma(A_j), \qquad j = 1, \ldots, n$$

for some morphism γ of operator algebras, then

$$\gamma(f(\overset{1}{A_1}, \ldots, \overset{n}{A_n})) = f(\overset{1}{B_1}, \ldots, \overset{n}{B_n}).$$

As R. Feynman mentioned, the presence of numbers over operators often allows to treat the latter as if they were commutative; the numbers govern the order of their action automatically; e.g., the following expansion is valid:

$$(13) \qquad e^A e^B = e^{\overset{2}{A} + \overset{1}{B}} = \sum_{n=0}^{\infty} \frac{(\overset{2}{A} + \overset{1}{B})^n}{n!}.$$

Of course, this does not mean that noncommutativity becomes insignificant, once the numbers over operators are introduced. In fact, a necessity arises to develop general noncommutative functional analysis, comprising a collection of rules and formulas, which supply researchers with necessary techniques to deal with functions of noncommuting operators.

Now let us give some examples.

Consider the notion of the differential. We define the differential of an operator function $f(A)$ in a usual way:

$$(14) \qquad < df(A), H > \overset{\text{def}}{=} \lim_{t \to 0} \frac{f(A + tH) - f(A)}{t}.$$

However, a simple example shows that the equality

$$(15) \qquad < df(A), H > = f'(A)H$$

is not true in general. In other words, one cannot merely compute the derivative $f'(x)$ and then substitute the operator A into the result. Indeed, let $f(x) = x^2$. Then

$$\frac{df}{dx}(A) = 2A;$$

on the other hand,

$$\begin{aligned}
(16) \qquad f(A+tH) - f(A) &= (A+tH)^2 - A^2 = (A+tH)(A+tH) - A^2 \\
&= A^2 + t(AH + HA) + t^2 H^2 - A^2 \\
&= t(AH + HA) + t^2 H^2 \\
&= t(AH + HA) + O(t^2),
\end{aligned}$$

so that if the commutator $[H, A] = HA - AH$ does not vanish, the relation (15) is not valid.

Thus we have the problem of computing $df(A)$ explicity.

Here is the solution of this problem. Let $\mathcal{A}$ be an algebra, and let $D : \mathcal{A} \to \mathcal{A}$ be a derivation of $\mathcal{A}$. Then

$$(17) \qquad Df(A) = \overset{2}{D}A\frac{\delta f}{\delta x}(\overset{1}{A}, \overset{3}{A}),$$

where

$$(18) \qquad \frac{\delta f}{\delta x}(x, y) \overset{\text{def}}{=} \frac{f(x) - f(y)}{x - y}$$

is the difference derivative of the function f.

Let us give a sketch of the proof. It suffices to consider the case $f(x) = e^{itx}$ (the general case follows then via the Fourier integral). We have

$$\frac{\delta e^{itx}}{\delta x}(x, y) = i \int_0^t e^{i((t-\tau)x + \tau y)} \, d\tau.$$

It is clear that

$$D(e^{iAt})\Big|_{t=0} = D(1) = 0.$$

Next,

$$\frac{d}{dt}(D(e^{iAt})) = D(\frac{d}{dt}e^{iAt}) = i(AD(e^{iAt}) + DA(e^{iAt})),$$

so that

$$D(e^{iAt}) = i \int_0^t e^{iA(t-\tau)}(DA)e^{iA\tau}\,d\tau = \overset{2}{D}A\frac{\delta e^{itx}}{\delta x}(\overset{1}{A}, \overset{3}{A}),$$

q.e.d.

Formula (17) is one of the fundamental formulas in noncommutative analysis. In particular, the theorem given below is a particular case of (17):

THEOREM. (Daletskii Yu.L. and Krein S.G. (1951)). *The equality*

$$\frac{d}{dt}f(A(t)) = A'(t)\frac{\overset{2}{\delta f}}{\delta x}(\overset{1}{A}(t), \overset{3}{A}(t))$$

holds.

Here the algebra $\mathcal{A}$ consists of operator families with the parameter t, $D = \dfrac{d}{dt}$.

We obtain an important corollary of (17), taking $D = \mathrm{ad}_B$, where ad_B is the operator of commutation with $B \in \mathcal{A}$, $\mathrm{ad}_B(X) = [B, X]$. Clearly, ad_B is a derivation, and we have

$$\mathrm{ad}_B(f(A)) = \overline{\overset{2}{\mathrm{ad}_B(A)}}\frac{\delta f}{\delta x}(\overset{1}{A}, \overset{3}{A});$$

in conventional notations the latter formula reads

$$[B, f(A)] = [\overset{2}{B}, A]\,\frac{\delta f}{\delta x}\,(\overset{1}{A}, \overset{3}{A}).$$

Another example arises, when we consider superposition of functions. Here even the problem of its proper definition and adequate notations is not trivial. In fact, the expression $f(g(\overset{1}{A}, \overset{2}{B}))$ is ambiguous: it may be considered either as the result of substituting the operators $\overset{1}{A}, \overset{2}{B}$ into the function $f(g(x,y))$ or as the result of substituting the operator $C = g(\overset{1}{A}, \overset{2}{B})$ into the function $f(x)$ ("true" superposition for functions of operators). If A and B do not commute and f is not a linear function, these two interpretations lead, as a rule, to different results. For example, let $f(x) = x^2$, $g(x,y) = x + y$. Then

$$h(x,y) = (x+y)^2, \qquad h(\overset{1}{A}, \overset{2}{B}) = (\overset{1}{A} + \overset{2}{B})^2 = A^2 + 2BA + B^2;$$

on the other hand,

$$C = \overset{1}{A} + \overset{2}{B} = A + B,$$

$$f(C) = (A+B)^2 = (A+B)(A+B) = A^2 + AB + BA + B^2 \neq h(\overset{1}{A}, \overset{2}{B}).$$

In order to avoid the ambiguity, we adopt the convention that the expression $f(g(\overset{1}{A}, \overset{2}{B}))$ always gets the former interpretation, while the "true" superposition will be denoted by $f(\ll g(\overset{1}{A}, \overset{2}{B}) \gg)$. That is,

$$(19) \qquad\qquad f(\ll g(\overset{1}{A}, \overset{2}{B}) \gg) \overset{\text{def}}{=} f(C), \qquad \text{where } C = g(\overset{1}{A}, \overset{2}{B}).$$

(Thus the "autonomous brackets" $\ll\gg$ (Maslov V.P. (1973)) define the order of computations in operator expressions — first the expression in these brackets is evaluated and then the resulting operator is used in the subsequent evaluations.)

It is natural to pose the problem of explicit evaluation for "true" superposition, namely, of its representation via functions of operators, which do not contain autonomous brackets.

2. Special Noncommutative Analysis

Now consider the application of general noncommutative analysis to the construction of functional calculus for a given fixed tuple of operators $\overset{1}{A_1}, \ldots, \overset{n}{A_n}$. Surely, there is no need to prove its importance, since particular examples of such calculus are the wellknown algebra of classical pseudo-differential operators (which are functions of the differentiation operators $-i\dfrac{\partial}{\partial x_j}$ and multiplication operators x_j, $j = 1, \ldots, n$) and also the calculus of Fourier-Maslov integral operators (see, e.g., Nazaikinskii V.E. et al. (1981)), which form the module over the algebra of pseudo-differential operators.

Special noncommutative analysis studies algebraic operations (in particular, multiplication) in the set

$$M = \{f(\overset{1}{A_1}, \ldots, \overset{n}{A_n})\}_{f \in \mathcal{F}},$$

whose elements are functions of the operators $\overset{1}{A_1}, \ldots, \overset{n}{A_n}$ with *symbols* $f(x_1, \ldots x_n)$ belonging to a given class $\mathcal{F}$.

The requirement that the set M be an algebra, i.e., that the product of any two its elements $f(\overset{1}{A_1}, \ldots, \overset{n}{A_n})$ and $g(\overset{1}{A_1}, \ldots, \overset{n}{A_n})$ be representable in the form of a function of $\overset{1}{A_1}, \ldots, \overset{n}{A_n}$,

$$(20) \qquad \ll f(\overset{1}{A_1}, \ldots, \overset{n}{A_n}) \gg \ll g(\overset{1}{A_1}, \ldots, \overset{n}{A_n}) \gg = h(\overset{1}{A_1}, \ldots, \overset{n}{A_n}),$$

imposes rigid restrictions on the operators $\overset{1}{A_1}, \ldots, \overset{n}{A_n}$.

In particular, it is clear that (20) implies that any commutator $[A_j, A_k]$ is a function of $\overset{1}{A_1}, \ldots, \overset{n}{A_n}$, i.e., the operators $\overset{1}{A_1}, \ldots, \overset{n}{A_n}$ form a *Poisson algebra* (Karasev M.V. and Maslov V.P. (1981)),

$$(21) \qquad [A_j, A_k] = \phi_{jk}(\overset{1}{A_1}, \ldots, \overset{n}{A_n}).$$

We assume that this Poisson algebra is in fact a nilpotent Lie algebra, i.e., the functions $\phi_{jk}(x)$ are linear,

$$(22) \qquad \phi_{jk}(x) = \sum_{l=1}^{n} c_{jkl} x_l,$$

and all the commutators of order N are equal to 0 for N large enough,

$$[A_{j_1}, [\ldots [A_{j_{N-1}}, A_{j_N}] \ldots]] = 0.$$

(This case was originally considered in Maslov V.P. (1973) and covers a number of applications. For general case see, e.g., Maslov V.P. and Nazaikinskii (1988), Karasev M.V, and Maslov V.P. (1981) and the papers cited therein.)

Thus let the operators $\overset{1}{A_1}, \ldots, \overset{n}{A_n}$ be fixed and suppose that they realize a representation of a nilpotent Lie algebra $\mathcal{G}$, i.e.,

$$[A_j, A_k] \equiv A_j A_k - A_k A_j = -i \sum_{l=1}^{n} C_{jk}^{l} A_l, \qquad j, k = 1, \ldots, n,$$

where C_{jk}^{l} are the structure constants of $\mathcal{G}$ in some basis $\{a_j\}$ (the factor $-i$ is introduced for the sake of convenience, so that the structure constants be real when all the A_j's are self-adjoint).

As we have already mentioned, we take an interest in the algebraic structure of the set $M = \{f(\overset{1}{A_1}, \ldots, \overset{n}{A_n})\}$, in particular, in the existence of products and inverse elements in M. First of all, let us specialize the definition of $f(\overset{1}{A_1}, \ldots, \overset{n}{A_n})$ in our particular case. We shall assume that the operators $\overset{1}{A_1}, \ldots, \overset{n}{A_n}$ are self-adjoint and consequently (under certain auxiliary assumptions) they realize the derived representation T_* of a unitary representation $T : G \to U(H)$ of the connected simply connected Lie group G, corresponding to the Lie algebra $\mathcal{G}$, in a Hilbert space H:

$$A_j = -i T_*(a_j).$$

The operators A_j are called the *generators* of the representation T.

Consider the coordinates of second genus on the group G. These are the coordinates $(x_1, \ldots, x_n)$, introduced via the mapping

$$\exp_2 : \mathcal{G} \longrightarrow G$$

$$x = \sum_{i=1}^{n} x_i a_i \longmapsto \exp_2(x) \overset{\text{def}}{=} \exp(x_n a_n) \ldots \exp(x_1 a_1),$$

where $\exp : \mathcal{G} \to G$ is the exponential mapping.

The mapping $\exp_2$ is a diffeomorphism of a neighborhood of 0 in the Lie algebra onto a neighborhood of unity in the Lie group; in our case (nilpotent Lie algebra) it is a global diffeomorphism of $\mathcal{G}$ onto G (for a special choice of the basis $(a_1, \ldots, a_n)$).

Using the introduced coordinates, we may write

$$f(\overset{1}{A_1}, \ldots, \overset{n}{A_n}) = \left(\frac{1}{\sqrt{2\pi}}\right)^n \int \tilde{f}(x_1, \ldots, x_n) e^{ix_n A_n} \ldots e^{ix_1 A_1} \, dx_1 \ldots dx_n$$

$$= \left(\frac{1}{\sqrt{2\pi}}\right)^n \int \tilde{f}(x_1, \ldots, x_n) T(\exp(x_n a_n)) \ldots T(\exp(x_1 a_1)) \, dx_1 \ldots dx_n$$

$$= \left(\frac{1}{\sqrt{2\pi}}\right)^n \int \tilde{f}(x_1, \ldots, x_n) T(\exp_2(x_1 a_1 + \cdots + x_n a_n)) \, dx_1 \ldots dx_n$$

$$= \left(\frac{1}{\sqrt{2\pi}}\right)^n \int_G \check{f}(g) T(g) \, d\mu(g).$$

Here $d\mu(g)$ is the Haar measure on G, the Jacobian $\dfrac{d\mu}{dx}$ is equal to 1 because the group is nilpotent, $\check{f}(g)$ is the "group" Fourier transform of f, defined by the equality $\check{f}(\exp_2(x)) = \tilde{f}(x)$.

In what follows, we write $f(A)$ instead of $f(\overset{1}{A_1}, \ldots, \overset{n}{A_n})$ for brevity.

Now we are able to compute the product of two elements of M, say, $f_1(A)$ and $f_2(A)$, via integrals over the group:

$$f_1(A)f_2(A) = \left(\frac{1}{2\pi}\right)^n \int\limits_{G\times G} \check{f}_1(g)\check{f}_2(h)T(g)T(h)\,d\mu(g)\,d\mu(h)$$

$$= \left(\frac{1}{2\pi}\right)^n \int\limits_{G\times G} \check{f}_1(g)\check{f}_2(h)T(gh)\,d\mu(g)\,d\mu(h)$$

$$\text{(making change of variables}\quad gh = k,\quad \mu(g^{-1}k) = d\mu(k))$$

$$= \left(\frac{1}{2\pi}\right)^n \int\limits_{G\times G} \check{f}_1(g)\check{f}_2(g^{-1}k)T(k)\,d\mu(g)\,d\mu(k)$$

$$= \left(\frac{1}{\sqrt{2\pi}}\right)^n \int\limits_{G} \check{f}(k)T(k)\,d\mu(k) = f(A),$$

where

$$\check{f}(k) = (\check{f}_1 * \check{f}_2)(k) = \left(\frac{1}{\sqrt{2\pi}}\right)^n \int\limits_{G} \check{f}_1(g)\check{f}_2(g^{-1}k)\,d\mu(g)$$

is the convolution of $\check{f}_1$ and $\check{f}_2$ with respect to the Haar measure.

Let $\mathcal{L}$ be the left regular representation of the group g, i.e.,

$$\mathcal{L}(g)h(k) = h(g^{-1}k)$$

for any function h on the group G. Using the representation $\mathcal{L}$, the function $\check{f}$ may be put into the form

$$\check{f} = \left[\left(\frac{1}{\sqrt{2\pi}}\right)^n \int\limits_{G} \check{f}_1(g)\mathcal{L}(g)\,d\mu(g)\right]\check{f}_2 = f_1(L)(\check{f}_2),$$

where $L = (L_1, \ldots, L_n)$ are the generators of the representation $\mathcal{L}$ (L_j possesses a simple explicit description: it is a right-invariant vector field on G satisfying the condition $\left.L_j\right|_e = -i\frac{\partial}{\partial x_j}$).

Consequently,

$$f = [\mathcal{F}^{-1} \circ f_1(L) \circ \mathcal{F}](f_2) = f_1(l)(f_2),$$

where $\mathcal{F}$ is the group Fourier transformation,

$$l_j = \mathcal{F}^{-1} \circ L_j \circ \mathcal{F}$$

are pseudo-differential operators (in our case, these operators will be differential ones provided that a special basis in the Lie algebra is chosen).

Thus we have shown that the product $f_1(A)f_2(A)$ belongs to the set M and its symbol (a function f such that $f_1(A)f_2(A) = f(A)$) is given by

$$f = f_1(l)(f_2),$$

where $l = (l_1, \ldots, l_n)$ are the operators of the left regular representation, acting in the space of symbols.

More accurate considerations show that M is in fact only a *module* over the algebra $M_0 \subset M$ of the operators $f(\overset{1}{A_1}, \ldots, \overset{n}{A_n})$ with classical symbols. This module is an analog of the module of Fourier-Maslov integral operators over the algebra of pseudo-differential operators mentioned above.

How to calculate the inverse element? Suppose that we try to invert the operator $f(A)$. It is clear from the above considerations that it suffices to solve the equation

$$f(l)g = 1,$$

where g is the (unknown) symbol of the inverse operator. This equation is equivalent to a pseudo-differential equation

$$(23) \qquad\qquad f(L)\check{g} = \delta_e$$

on the group G (here δ_l is the Dirac δ-function at the point $e \in G$).

Thus the inversion problem in M reduces to a certain pseudo-differential equation on the group G.

3. Some applications to algebra and differential equations

3.1. Campbell-Hausdorff-Dynkin formula. There is a following important problem in the theory of Lie groups and Lie algebras:

PROBLEM. Given two elements A and B of a noncommutative algebra, find an element C such that the product of exponentials of A and B be expressed as the exponential of C,

$$e^A e^B = e^C.$$

This problem was partially solved yet in 19th century (see Campbell J.E. (1898)). Namely, it was discovered that the element $C = \ln(e^B e^A)$ may be expressed as

$$(24) \qquad C = B + A + \frac{1}{2}[B, A] + \ldots,$$

where dots stand for sum of higher-order commutators of elements A and B.

However, the explicit expression for the terms of series (24) was found more than fifty years later, when Dynkin E.B. (1949) proved that

$$(25) \qquad C = \sum_{m=1}^{\infty} \frac{(-1)^{m-1}}{m} \sum_{\substack{k_i + l_i \geq 1 \\ k_i, l_i \geq 0}} \frac{[B^{k_1} A^{l_1} \ldots B^{k_m} A^{l_m}]}{k_1! l_1! \ldots k_m! l_m!},$$

where

$$[S_1 S_2 \ldots S_N] = \frac{1}{N}[S_1, [S_2, \ldots [S_{N-1}, S_N] \ldots].$$

Noncommutative analysis makes it possible to obtain a new formula for $\ln(e^B e^A)$, namely

$$(26) \qquad \ln(e^B e^A) = \int_0^1 \varphi(e^{-t\, a\, d_A} e^{-t\, a\, d_B}) e^{-t\, a\, d_A}(A + B)\, dt,$$

where $\varphi(t) = \ln(t)/(t-1)$ (Mosolova M.V. (1978)). Neither this formula, nor its proof, very short and elegant, use expansions into cumbersome series; it follows directly from (26) that $\ln(e^B e^A)$ belongs to the Lie algebra, generated by A and B, and the series (25) may be obtained from (26) as its Taylor expansion in the powers of t.

3.2. Jacobi condition and Poincare-Birkhoff-Witt theorem.

The famous Poincare-Birkhoff-Witt theorem states that the ordered monomials

$$a^\nu \stackrel{\text{def}}{=} a_n^{\nu_n} \dots a_1^{\nu_1}, \qquad |\nu| = 0, 1, 2, \dots,$$

where $a_1, \dots, a_n$ is a linear base of a Lie algebra L, form a linear base of the enveloping algebra U of the algebra L (i.e., these monomials are linearly independent and span the whole U).

This turns out to be a consequence of the following general statement:

THEOREM. *Let a system of relations Σ be given (i.e., a subset of the free associative algebra $\mathcal{A}$ with 1, generated by noncommuting letters $a_1, \dots, a_n$). Suppose that there exists a tuple $l_1, \dots, l_n$ of operators acting on functions such that*

$$\ll f(\overset{1}{A_1}, \dots, \overset{n}{A_n}) \gg \ll g(\overset{1}{A_1}, \dots, \overset{n}{A_n}) \gg = [f(\overset{1}{l}_1, \dots, \overset{n}{l}_n)g](\overset{1}{A_1}, \dots, \overset{n}{A_n})$$

once the operator tuple $A_1, \dots, A_n$ satisfies Σ, and

$$l_j f(x) = x_j f(x)$$

once f doesn't depend on $x_{j+1}, \dots, x_n$ (($l_1, \dots, l_n$) are called left regular representation operators for the system Σ).

Then the mapping $f \longmapsto f(\overset{1}{A_1}, \dots, \overset{n}{A_n})$ is injective for some tuple of operators $A_1, \dots, A_n$ satisfying Σ if and only if the operators of left regular representation themselves satisfy Σ.

The latter condition is called *the generalized Jacobi condition*, since in the case of Lie algebras its validity follows from the Jacobi identity for the structure constants.

3.3. Unification of asymptotic expansions

To study a solution via certain asymptotic expansions is a common idea in the theory
of differential equations. These expansions appear to be quite different — asymptotics
with respect to large (small) parameter, smoothness asymptotics, mixed asymptotics
(see 3.4 below) etc. However, special noncommutative analysis often allows to reduce
various asymptotic problems to a "standard" one, namely to the problem of asymptotics with respect to smoothness for a pseudodifferential equation on a Lie group G.

Here is the sketch of such a reduction. Consider a differential equation

$$(27) \qquad P(\overset{2}{x}, \overset{1}{D})u = f.$$

What does it mean to construct its asymptotic solution? This means that some
filtration
$$\cdots \subset H^k \subset H^{k-1} \subset \cdots \subset H^{k-l} \subset \cdots$$
of functional spaces is given, and we wish to find a function u_l for any l such that

$$P(\overset{2}{x}, \overset{1}{D})u_l - f \in H^l$$

(Most known asymptotics fall under this scheme; in particular, H^l is a usual Sobolev
space for asymptotics with respect to smoothness).

Suppose that we managed to find a tuple of operators $A_1, \ldots, A_n$ such that

i) For some symbol f

$$P(\overset{2}{x}, \overset{1}{D}) = f(\overset{1}{A_1}, \ldots, \overset{n}{A_n});$$

ii) The filtration $\{H^l\}$ coincides for some fixed k with the scale generated by the
operators $A_1, \ldots, A_k$ (in the same way as the operators $\dfrac{\partial}{\partial x_1}, \ldots, \dfrac{\partial}{\partial x_n}$ generate the
Sobolev scale);

iii) The operators $A_1, \ldots, A_n$ form a representation of a nilpotent Lie algebra.

Equation (27) reduces in this situation to the inversion problem for the operator $f(A)$ which in turn reduces to the equation (23). As a rule, one cannot obtain the precise solution of this equation. Fortunately, it is sufficient to obtain *asymptotic solutions*. In our case (asymptotics with respect to powers of the operators $A_1, \ldots, A_k$) the problem of construction of the asymptotically inverse operator for $f(A)$ leads to the problem of smooth asymptotics for solutions of (23). (More precisely, we should speak of partial smoothness in directions defined by vector fields $L_1, \ldots, L_k$).

Note that the phase space of our problem turns out to be the cotangent space T^*G, while the Hamilton function is given by the left shifts of the principal part of the symbol f.

Thus we may conclude that the notions of the phase space and the Hamiltonian depend not only on the equation itself but also on the type of the required asymptotics.

3.4. Example: Electromagnetic Waves in Plasma

Consider the following Cauchy problem:

$$(28) \qquad \begin{cases} -\dfrac{\partial^2 u}{\partial t^2} + c^2 \dfrac{\partial^2 u}{\partial x^2} - \lambda^2 b^2(x) u = \lambda \delta(x) r(t) e^{-i\lambda q(t)} \overset{\text{def}}{=} F(x, \lambda, t), \\[2mm] u\big|_{t=0} = u_t\big|_{t=0} = 0. \end{cases}$$

Equation (28) describes propagation of electromagnetic waves in plasma. Here λ is the average plasma frequency, which may be considered as a large parameter, $b(x) > 0$ everywhere.

The right-hand side of the equation (28) describes an oscillating point source with the amplitude $r(t)$ and the instantaneous frequency $\lambda q'(t)$. $F(x, \lambda, t)$ is a singular function (distribution) which oscillates rapidly as $\lambda \to \infty$, so it is natural to pose a problem:

PROBLEM. *Construct the simultaneous asymptotics of the Cauchy problem (28) with respect to smoothness and the large parameter λ.*

One of the possible ways to solve this asymptotic problem is to represent the operator in the left-hand side of (28) as a function of the operators

$$\overset{1}{A_1} = -i\frac{\partial}{\partial x}, \qquad A_2 = \lambda, \qquad \overset{2}{B} = x,$$

which gives an opportunity to find asymptotic solution in the scale generated by the norms

$$\|u\|_s = \|(1 + A_1^2 + A_2^2)^{s/2}u\|_{L^2}.$$

References

Maslov V.P. (1973). Operational Methods. Nauka, Moscow.

Maslov V.P. (1987). Asymptotic Methods of Solution of Pseudo-differential Equations. Nauka, Moscow.

Maslov V.P. and Nazaikinskii V.E. (1988). Asymptotics of Operator and Pseudo-differential Equations. Consultants Bureau. New York.

Karasev M.V. and Maslov V.P. (1981). Global asymptotic operators of regular representation. Dokl. Acad. Nauk SSSR, vol. 257, N 1, 33-38.

Feynman R.P. (1951). An operator calculus having applications in quantum electrodynamics. Phys. Rev., vol. 84, N 2, 108-128.

Nazaikinskii V.E. et al. (1981). Fourier integral operators and canonical operator. Uspekhi Mat. Nauk, vol. 36 N 2, 81-140.

Daletskii Yu.D. and Krein S.G. (1951). A formula for differentiating with respect to parameter of functions of Hermitian operators. Dokl. Acad. Nauk SSSR, vol. 76, N 1, 13-66.

Campbell J.E. (1898). Introductory Treatise on Lie Theory of Finite Continuous Transformation Groups. London.

Dynkin E.B. (1949). On a representation of the series $\log(e^x e^y)$ in non-commuting x and y via commutators. Mat. USSR Sbornik, vol. 25(67), N 1, 155-162.

Mosolova M.V. (1978). A new formula for $\log(e^B e^A)$ via commutators of the elements A and B. Mat. Zametki, vol. 23, N 6, 817-823.

Nazaikinskii V.E. et al. (1990a). Introduction to Maslov's operational method (Non-commutative analysis and differential equation). In: Algebraic Questions of Analysis and Topology. Voroneš, 52 – 60.

Nazaikinskii V.E. et al. (1990b). Applications of noncommutative operators method to diffractional problems. In: Wave and Diffraction, Moscow, vol. 2, 33 – 40.

Nazaikinskii V.E. et al. (1990c). Maslov's operational method and diffraction problems, Proc. XXIII General Assembly of the URSI, Prague, 369.

Nazaikinskii V.E. et al. (1991). On the application of Maslov's operational method to a diffraction problem. Docl. Acad. Nauk, vol. 317, N 4, 832– 834.

Author's address

Dr. Nazaikinskii Vladimir, prof. Sternin Boris and prof. Shatalov Victor

Dept. of Computer Math.

and Cybernetics

Moscow State University

Lenin Hills, Moscow 119899

R U S S I A

e.mail: nonlin @ cs.msu.su

T e l e x: 411483 MGU SU, VMK

F a x: (7) (095) 227 28 07

Operator Theory:
Advances and Applications, Vol. 57
© 1992 Birkhäuser Verlag Basel

Relative Time Delay and Trace Formula for Long Range perturbations of Laplace Operator

Didier Robert

Département de Mathématiques, Université de Nantes,
2, rue de la Houssinière, 44072-Nantes-Cédex 03, France

1 Introduction

Let be L_1, L_2, two long range perturbations of $L_0 = -h^2\Delta$ where Δ is the usual Laplace operator on $\mathbb{R}^n$, h is a small positive number proportional to the Planck constant.

We assume that L_1, L_2 are obtained from L_0 by perturbation of the gravitational field and the electromagnetic field. Let us consider an electric potential V, a magnetic potential $A = (A_1, \ldots, A_n)$ and a Riemaniann metric $g = \{g_{j,k}\}$. We use the usual notations: $G = det(g)$, $\{g^{j,k}\} = \{g_{j,k}\}^{-1}$. The natural quantum Hamiltonian to compare with L_0 in $L^2(\mathbb{R}^n)$ (with Lebesgue measure) is:

$$L(g, A, V) = -G^{-1/4} \cdot \sum_{\substack{1 \leq j \leq n \\ 1 \leq k \leq n}} (h\partial_j + iA_j)G^{1/2}g^{j,k}(h\partial_k + iA_k)G^{-1/4} + V$$

We assume that the data are smooth and satisfy the following decreasing assumption: $\exists \delta > 0$ such that :

$$\forall \alpha, \text{multiindex}, \exists C_\alpha \quad \text{such that:} \quad \forall x \in \mathbb{R}^n,$$

$$|\partial^\alpha(g(x) - (\delta_{j,k}))| + |\partial^\alpha A(x)| + |\partial^\alpha V(x)| \leq C_\alpha(1 + |x|)^{-\delta-|\alpha|} \quad ((DC)_\delta)$$

It is well known that L is essentially self-adjoint on $L^2(\mathbb{R}^n)$ with domain the usual Sobolev space $H^2(\mathbb{R}^n)$. Let us denote by $U(t)$ the propagator: $U(t) = exp(-\frac{it}{h}L)$. Using the commutator technics (see [14]) it is not difficult to prove the following:

Theorem 1 *For $L = L(g, A, V)$ satisfying $(DC)_\delta, \delta > 0$ and $I = [a, b] \in$ $]0, +\infty[$ we have:*

i) The spectrum of L in I is absolutely continuous with at most a finite number of eigenvalues.

ii) If L has no eigenvalues in I, then for every $s > k + \frac{1}{2}(k \in \mathbb{N})$ $\lambda \mapsto (1 + |x|)^{-s}(L - \lambda \mp i0)^{-1}(1 + |x|)^{-s}$ exists and is C^k in I and its derivatives are given by the expected formula.

iii) Let I be as in ii). Then for every $0 < \tau < s$ and every $\varphi \in C_0^\infty(]a, b[)$ we have:

$$\|(1 + |x|)^{-s}\varphi(L)U(t)(1 + |x|)^{-s}\| \le C(1 + |t|)^{-\tau} \quad \forall t \in \mathbb{R}.$$

The estimates given in Theorem 1 are not controled in h for $h \searrow 0$ nor for $\lambda \nearrow +\infty$. To get uniform estimates we have to consider the classical dynamical system in the phase space $\mathbb{R}_x^n \times \mathbb{R}_\xi^n$ generated by the Hamiltonian:

$$\ell(x, \xi) = g(x)(\xi - A(x)) \cdot (\xi - A(x)) + V(x)$$

Let us denote by Φ_ℓ^t the corresponding Hamiltonian flow.

Definition 1 We say that $J \subset \ ]0, +\infty[$ is non trapping for ℓ if for every $\lambda \in J$ and every $(x, \xi) \in \mathbb{R}_x^n \times \mathbb{R}_\xi^n$ such that $\ell(x, \xi) = \lambda$ then: $\lim_{t \to \pm\infty} |\Phi_\ell^t(x, \xi)| = +\infty$.

We have the following improvement of theorem 1 ([13])

Theorem 2 *Let $J \subset \ ]0, +\infty[$ be a non trapping compact interval for ℓ such that L has no eigenvalue in J. Then we have:*

i) $\forall s > k - \frac{1}{2}, \quad k \in \mathbb{N}, \exists c > 0$ such that:

$$\|(1 + |x|)^{-s}(L - \lambda \mp i0)^{-k}(1 + |x|)^{-s}\| \le Ch^{-k} \quad \forall \lambda \in J \quad \text{and} \quad \forall h \in 0, 1].$$

ii) $\forall s > \tau > 0, \forall \varphi \in C_0^\infty(J) \ \exists C > 0$ such that:

$$\|(1 + |x|)^{-s}\varphi(L)U(t)(1 + |x|)^{-s}\| \le C(1 + |t|)^{-\tau} \quad \forall t \in \mathbb{R} \quad \text{and} \quad \forall h \in]0, 1]$$

iii) Conversely if i) holds for $k = 0$ or if ii) holds for some $s > \tau > 0$ then J is a non trapping interval for ℓ.

Remark. We have an analogue statement for the high energy behaviour (ie for $\lambda \nearrow +\infty$) assuming the non trapping condition for the principal symbol $\ell_0(x, \xi) = g(x)\xi \cdot \xi$. By a scaling this case can be deduce from Theorem 2 admitting lower order perturbations in $h := \lambda^{\frac{-1}{2}}$ see ([13]).

Let us introduce now the relative time-delay for the pair (L_1, L_2). Let us consider $L_\kappa = L(g^\kappa, A^\kappa, V^\kappa)$ for $\kappa = 1 \, or \, 2$. We assume that L_2 is a short range perturbation of L_1:

$$\forall \alpha, \text{multiindex}, \exists C_\alpha \quad \text{such that} \quad \forall x \in \mathbb{R}^n,$$

$$|\partial^\alpha(g^2(x) - g^1(x))| + |\partial^\alpha(A^2(x) - A^1(x))| + |\partial^\alpha(V^2(x) - V^1(x))|$$
$$\leq C_\alpha(1 + |x|)^{-\rho - |\alpha|} \qquad (SR)_\rho$$

Let us denote by $U_\kappa(t) = e^{(-i\frac{t}{\hbar}L_\kappa)}$ the propagator generated by L_κ.
From Theorem 1 we get easily in standard way existence and completeness for the wave operators: $\Omega_\pm = \lim_{t \to \pm\infty}(U_2(-t) \cdot U_1(t) \cdot \Pi_{ac}(L_1)))$ where $\Pi_{ac}(L)$ denotes the spectral projection on the absolutely continuous subspace of L.
Following Jauch-Sinha-Martin ([10]) define now the local time delay. For K compact set in $\mathbb{R}^n$, $\mathbb{1}_K$ denotes the caracteristic function of K. The time-delay in K for the pair (L_1, L_2) is defined as the quadratic form:

$$\langle T_K^D \psi, \psi \rangle = \int_{-\infty}^{+\infty} \left(\|\mathbb{1}_K U_2 \Omega_- \psi\|^2 - \|\mathbb{1}_K U_1 \psi\|^2 \right) dt \qquad (1)$$

for $\psi \in L^2(\mathbb{R}^n)$ such that $\varphi(L_1)\psi = \psi$ for some $\varphi \in C_0^\infty(J)$ with $J \subset \,]0, +\infty[$ doe'snt contain eigenvalues of L_1. Each term in the r.h.s of (1) is t-integrable by smoothness, in the Kato sense, of $\mathbb{1}_K$ (see Theorem 1). The global time-delay is defined by:

$$\lim_{K \nearrow \mathbb{R}^n} \langle T_K^D \psi, \psi \rangle = \langle T^D \psi, \psi \rangle \qquad (2)$$

when this limit exists!

The existence of T^D as a self-adjoint, dense defined operator in $L^2(\mathbb{R}^n)$ is a non trivial question. This was solved for $L_1 = L_0$, independently, by Nakamura ([9]) and Wang ([17]). For the general case considered here the question is open. Nevertheless it is possible to define the average global time-delay on the energy shell of L_1, following Rauch-Sinha-Misra ([10]). We can write:

$$T_K^D = \int_{-\infty}^{+\infty} U_1(-t)\left(\Omega_-^* . \mathbb{1}_K . \Omega_- - \mathbb{1}_K\right) U_1(t) dt \qquad (3)$$

In particular T_K^D commute with $U_1(t)$ and T_K^D can be decomposed in the spectral representation of L_1. Let φ be as above (energy cutoff) such that $\varphi = 1$ on a compact interval I. We have:

$$T_K^D = \int_{-\infty}^{+\infty} U_1(-t).\Gamma_K.U_1(t) dt$$

$$\text{where } \Gamma_K = (\Omega_-^* . \mathbb{1}_K . \Omega_- - \mathbb{1}_K).$$

By Fourier transform we get:

$$(\varphi(L_1))^2 . T_K^D = 2\pi . \int \frac{\partial E_1(\lambda)}{\partial \lambda} . \Gamma_K . \frac{\partial E_2(\lambda)}{\partial \lambda} d\lambda \qquad (4) \bullet$$

Here E_1 denotes the spectral resolution of L_1. It is well known that we have:

$$\frac{\partial E_1(\lambda)}{\partial \lambda} = \mathcal{F}_1^*(\lambda) . \mathcal{F}_1(\lambda)$$

where the unitary operator $\mathcal{F}_1$ gives the spectral representation of L_1 ,

$$\mathcal{F}_1 : R(E_1(I)) \mapsto \int_I^\oplus \mathcal{H}(\lambda) d\lambda$$

the integral being an Hilbertian integral (see [11]) and $\mathcal{F}_1(\lambda)$ is the trace of $\mathcal{F}_1$ on $\mathcal{H}(\lambda)$. It results that on the energy shell space $\mathcal{H}(\lambda)$, T_K^D is expressed by: $T_K^D(\lambda) = 2\pi \mathcal{F}_1(\lambda) . \Gamma_K . \mathcal{F}_1(\lambda)^*$, $\quad \forall \lambda \in J$. Furthermore we can prove that $T_K^D(\lambda)$ is a trace-class operator in $\mathcal{H}(\lambda)$. Let us denote: $\tau_K(\lambda) = tr(T_K^D(\lambda)$. $\tau_K(\lambda)$ can be called the local average time delay for (L_1, L_2). By an easy computation using cyclicity of traces and intertwining property of wave operators, we get the following identity:

$$\forall \varphi \in C_0^\infty(J) \quad \text{we have: } \int \varphi(\lambda) . \tau_K(\lambda) d\lambda = 2\pi tr \Big(\mathbb{1}_K . \big(\varphi(L_2) - \varphi(L_1)\big) \Big) \quad (5)$$

So we have:

$$\tau_K(\lambda) = 2\pi . tr \big(\mathbb{1}_K (\frac{\partial E_2(\lambda)}{\partial \lambda} - \frac{\partial E_1(\lambda)}{\partial \lambda}) \mathbb{1}_K \big) \qquad (6)$$

Remark that the l.h.s of (6) is well defined and smooth for $\lambda \in I$ by Theorem 1. To be able to define the global average time-delay **we assume** $\rho > \mathbf{n}$ (see condition $(SR)_\rho$). Then we see from (6) that $\lim_{K \nearrow \mathbb{R}^n} tau_K = \tau$ exists in the distribution sense and then we have:

$$\tau(\lambda) = 2\pi tr \Big(\frac{\partial E_2(\lambda)}{\partial \lambda} - \frac{\partial E_1(\lambda)}{\partial \lambda} \Big) \qquad (7)$$

The main results presented here concern the asymptotic behavior of $\tau(\lambda)$ as $\lambda \nearrow +\infty$ or as $h \searrow 0$.

2 A Trace Formula

Before giving the main result of this part we need some notations and definitions.

Definition 1 A h-dependent symbol $q(h, x, \xi)$ will be called an h-admissible pseudodifferential operator of weight $< x >^\mu < \xi >^{-\infty}$ if there exists an asymptotic expansion: $q(h) \asymp \sum_{j \geq 0} h^j q_j$ such that for every N, M we have:

$$|\partial_x^\alpha \partial_\xi^\beta q_j| \leq C < x >^\mu < \xi >^{-M}$$

$$\left| \partial_x^\alpha \partial_\xi^\beta \left(q(h) - \sum_{j=0}^{j=N-1} q_j \right) \right| \leq C h^N < x >^{\mu - N} < \xi >^{-M}$$

We recall the following:

Definition 2 For an h-dependent symbol $q(h)$ we denote by $q^w(h)$ the Weyl h-quantization of $q(h)$ whose definition for $u \in \mathcal{S}(\mathbb{R}^n)$ is:

$$q^w(h)u(x) = (2\pi h)^{-n} \iint_{\mathbb{R}^{2n}} e^{\frac{i}{h} < x - y, \xi >} q\left(h, \frac{(x+y)}{2}, \xi\right) u(y) dx d\xi$$

We denote also: $R_j(z) = (L_j - z)^{-1}$

Now we can state our trace formula which is very useful for our purpose because it essentially reduces the study of gobal average time-delay to the local one.

Theorem 3 *There exists $R_0 > 0$ large enough such that for every $\theta \in C_0^\infty(\mathbb{R}^n)$ satisfying: $\theta(x) = 1$ for $|x| \leq 1$ we can find h-admissible symbols $k_j(h)$ for $j = 1$ or 2 such that:*

$$\tau(\lambda) = 2\pi tr\left(\theta \frac{\partial E_2(\lambda)}{\partial \lambda} - \frac{\partial E_1(\lambda)}{\partial \lambda} \theta \right) + tr\left(k_1^w(h) . \frac{\partial E_0}{\partial \lambda} \right) + tr\left(k_2^w(h) \mathcal{R}e(R_0(\lambda \pm i0)) \right)$$

$$+ tr\left(X_1^\pm R_1(\lambda \pm i0) . Y_1^\pm . R_0(\lambda \pm i0) . Z_1^\pm \right)$$

$$+ tr\left(X_2^\pm R_2(\lambda \pm i0) . Y_2^\pm . R_0(\lambda \pm i0) . Z_2^\pm \right) \tag{8}$$

in the last line we mean that we have a (+) and a (-) term, and $X_j^\pm, Y_j^\pm, Z_j^\pm$, operators are negligeable operators in the following sense: $\forall M, \quad \forall N,$ we have:

$$\| < x >^M Y_j^\pm R_0(\lambda \pm i0) Z_j^\pm < x >^M \|_{tr} = O(h^N)$$

$$\| < x >^M X_j^\pm < x >^M \| = O(h^N) \tag{9}$$

$O(h^N)$ *being uniform in the energy parameter* λ. *Furthermore formula (8) can be differentiated in* λ *at any order and we have also estimates like (9)*

sketch of proof: For $L = L_j$ we construct a long time parametrix for the associated propagator: $U(t) = exp(-\frac{it}{h}L)$. This construction was first introduced by Isozaki-Kitada (ref. in[13]) for pure electric potential perturbations and extended to more general perturbations by the author ([13]). Here we recall briefly the main steps of this construction. For that let us introduce outgoing and incoming areas in the classical phase space. For $R > 0$, $\sigma \in]-1, 1[$, J a compact interval, $J \subset]0, +\infty[$, we define:

$$\Gamma^{\pm}(R, \sigma, J) = \left\{ (x, \xi), \quad |x| > R, \quad \pm < x, \xi > \geq \pm\sigma|x|.|\xi|, \quad |\xi|^2 \in J \right\}$$

An important step in the construction is a solution of a stationary Hamilton-Jacobi equation:

Proposition 1 *For every* $\mu, \sigma \in]-1, 1[, \mu < \sigma$, *there exists* $R > 0$ *large enough and* $\Phi \in C^{\infty}(\mathbb{R}_x^n \times \mathbb{R}_\xi^n)$ *such that:*
    i) $\ell(x, \partial_x \Phi(x, \xi)) = |\xi|^2$ *in* $\Gamma^+(R, \sigma, J) \bigcup \Gamma^-(R, \mu, J)$
    ii) For each multiidex α, β *we have:*

$$|\partial_x^\alpha \partial_\xi^\beta(\Phi(x, \xi) - < x, \xi >)| \leq C < x >^{1-|\alpha|-\delta} \forall (x, \xi) \in \mathbb{R}_x^n \times \mathbb{R}_\xi^n$$

Although not unic there exists a canonical procedure to find the phase function Φ of proposition 1.(the details are given in [15]) So for L_1 and L_2 we get canonical associated phases Φ_1 and Φ_2. Using assumption $(SR)_\rho$ we can prove:

Proposition 2 *For each multiidex* α *and* β *there exists* $C > 0$ *such that:*

$$|\partial_x^\alpha \partial_\xi^\beta(\Phi_2(x, \xi) - \Phi_1(x.\xi))| \leq C < x >^{1-|\alpha|-\rho} \qquad \forall (x, \xi) \in \mathbb{R}_x^n \times \mathbb{R}_\xi^n).$$

To the phase Φ such as in proposition 1 and to some amplitude a we associate the Fourier integral operator:

$$\mathcal{J}(\Phi, a)u(x) = (2\pi h)^{-n}. \iint_{\mathbb{R}^{2n}} e^{\frac{i}{h}.(\Phi(x,\xi) - <y,\xi>)}.a(x, \xi)u(y)dyd\xi$$

Let us consider a cutoff χ supported in a outgoing area Γ^+. Then we have the following approximation for the propagator $U(t)$ for $t \geq 0$.Of course the same statement is true for $t \leq 0$ if χ is supported in Γ^-.

Theorem 4 *There exists h-admissible symbols a(h) and b(h) of weight 1 such that for every N we have:*

$$U(t).\chi^w(h) = \mathcal{J}(\varPhi, a^{(N)}).U_0(t).\mathcal{J}(\varPhi, b^{(N)})^* + h^N.U(t).\chi_N^w(h) + h^{N+1}.R_N(t)$$

where

$$R_N(t) = \int_0^t U(t-s)\mathcal{J}(\varPhi, r_N).U_0(s).\mathcal{J}(\varPhi, b^{(N)})^* ds$$

and χ_N , r_N are uniformly bounded symbols, with respect to $h \in]0,1]$, in $S(-N-1, -\infty)$.

Now to prove Theorem 3 we remark that $\frac{\partial E_j}{\partial \lambda}$ is the Fourier transform of $U_j(t)$. So using proposition 2 and theorem 4 we can obtain Theorem 3 (see [15] for the details)

3 Semi-Classical Asymptotics for Average Time-Delay

At first we give a result for the local average time-delay . We assume $(DC)_\delta$ with $\delta \geq 0$ and $(SR)_\rho$ with $\rho > 1$ for the pair (L_1, L_2). Then we have:

Theorem 5 *Let J be a compact interval, $J \subset]0, +\infty[$ and K a compact set in $\mathbb{R}^n$ such that $K \supset \{|x| \leq R_0\}$ for some R_0 large enough. Assume that J is non trapping for ℓ_1 and ℓ_2. Then we have:*

$$\tau_K(\lambda) \asymp h^{-n}. \sum_{j \geq 0} \gamma_{K,j}(\lambda) \qquad as \quad h \searrow 0 \quad uniformly \ for \quad \lambda \in J$$

Remark. This theorem is proved by studying asymptotics for:

$$tr\left(\mathbb{1}_K.\frac{\partial E_j}{\partial \lambda}\mathbb{1}_K\right)$$

which is much easier than to study a difference, like it is necessary to do for $\tau(\lambda)$.

Now we come to the global average time-delay.

Theorem 6 *i) τ is a C^∞ function in $]0, +\infty[$.*
ii) If J is a compact interval, $J \subset]0, +\infty[$ non trapping for ℓ_j, $\quad j = 1, j = 2$ then we have:

$$\tau(\lambda) \asymp h^{-n}. \sum_{j \geq 0} c_j(\lambda)h^j \qquad as \quad h \searrow 0 \quad uniformly \quad for \quad \lambda \in J \qquad (10)$$

Furthermore (10) can be differentiated in λ at any order.
iii) For $h > 0$ fixed, assuming that the metrics g^1 and g^2 have no trapped geodesics, then we have the high energy asymptotics:

$$\tau(\lambda) \asymp \lambda^{-\frac{n}{2}-1}. \sum_{j \geq 0} \alpha_j.\lambda^{-j} \qquad as \quad \lambda \nearrow +\infty \qquad (11)$$

Furthermore (11) can be differentiated in λ at any order

In the above results the non trapping condition play a very important rôle. We want to give now results without this condition. We state results only for the $h \searrow 0$ case. For $\gamma > 0$ let us introduce the Riesz means of order γ for the function $\tau(\lambda)$ which is defined by:

$$\tau_\gamma(\lambda) = \int_{-\infty}^{\lambda} (\lambda - \mu)^\gamma \tau(\mu) d\mu$$

In what follows we use the notation:
$[\gamma]_+$ is the smallest integer $\geq \gamma$.

Theorem 7 *Let us assume that J is a non critical compact interval for ℓ_j, $j=1, 2$.*

i) For every $\gamma > 0$ we have the finite asymptotic expansion:

$$\tau_\gamma(\lambda) = h^{-n} . \sum_{j=0}^{j=[\gamma]_+} c_{j,\gamma}(\lambda) h^j + O(h^{-n+\gamma+1}) \tag{12}$$

as $h \searrow 0$ uniformly for λ

Furthermore if on the energy surfaces: $\{\ell_j = \lambda\}$ for $j=1,2$, the set of closed trajectories is of measure 0 for the Liouville measure, then the remainder term in (12) can be improved in $o(h^{-n+\gamma+1})$ and we get a term more in the expansion if $\gamma \in \mathbb{N}$.

ii) For $\gamma = 0$ we need the following technical assumption: there exists positive numbers s_0, S, k, C, such that:

$$\| < x >^{-s_0} R_j(\lambda + i\tau) < x >^{-s_0} \| \leq C . e^{\left(\frac{S}{h^k}\right)} \tag{13}$$

for $0 < |\tau| \leq 1$, $\lambda \in J$, $h \in]0,1]$.
If (13) holds then we have for τ_0:

$$\tau_0(\lambda) = C_{0,0}(\lambda) h^{-n} + O(h^{1-n}) \tag{14}$$

Remark. 1. For $L_1 = L_0 = -h^2 \triangle$ we have proved in [15] that (14) holds without assumption (13). Anyway (13) seems to be a very weak condition and I do'nt know if it exists Hamiltonians like L which do'nt satisfies it.In particular it is satisfied for a potential wells in an island.(see remark 2).

Remark. 2. Let us consider the particular case $L_1 = -h^2 \triangle$, $L_2 = -h^2 \triangle + V$ and assume that we are in the situation of the "well in an island" in a neighborhood of the energy level λ.   In a joint paper with C.Gérard and A. Martinez [4], we proved that $\tau(\lambda)$ can be approximated by the Breit-Wigner formula:

$$\tau(\lambda) = 2\frac{\mathcal{R}e(r(h))}{\lambda}\mathcal{I}m\left(\frac{1}{r(h)-\lambda}\right) + O_\varepsilon\left(\frac{e^{-\frac{(2S_0-\varepsilon)}{h}}}{|r(h)-\lambda|}\right) + O(h^{-n}) \qquad (15)$$

Here $r(h)$ denotes a resonance for L_2 such that $\lim_{h\searrow 0} r(h) = \lambda$ and $S_0 > 0$ is the Agmon distance from the wells to the sea (we refer to the paper [4] for the details).In particular for the bottom of the wells we have: $r(h) \approx e^{\frac{-2S_0}{h}}$ ([6]). So we see from (15) that $\tau(\lambda) \approx e^{\frac{2S_0}{h}}$ when λ is close to $\mathcal{R}e(\lambda)$.To prove (15) , besides analyticity assumption on V, we assumed that condition $(DC)_\rho$ holds with $\rho > n + 1$. Using for this case our trace formula (8) and the principal result of Gérard-Martinez ([4]) we can prove easily that (15) holds under the natural assumption $\rho > n$.

References

1   W.O. Amrein, M.B. Cibils: Helv. Phys. Acta, **60** 481 (1987)

2   M.S. Birman, M.G. Krein: Dokl Akad Nauk SSSR5.**5** 475 (1962)

3   V.S. Buslaev: Soviet Math Dokl **12** 591 (1971)

4   C. Gérard, A. Martinez; D. Robert: Comm. Math. Phys **121** 323 (1989)

5   A. Jensen: Comm. Math. Phys. **82** 435 (1981)

6   B.Helffer, J. Sjstrand: Bull. S.M.F. Mémoire n24-25, tome **114** (1986)

7   A. Majda, J. Ralston: Duke Math. J. ; **45** 183 (1978); **45** 513 (1978); **46** 725 (1979)

8   Ph. Martin: Acta Phys. Austriarca, Supp. **23** 157 (1981)

9   S. Nakamura: Comm. Math. Phys. **109** 397 (1987)

10   J. M. Rauch, K. B. Sinha, B. N. Misra: Helv. Phys. Acta **45** 398 (1972)

11   M. Reed, B. Simon: Scattering Theory, Academic Press (1979)

12   D. Robert: Autour de l'Approximation Semi-Classique **PM 68** Bikhäuser (1987)

13   D. Robert: Asympt. Anal. **3** 301 (1991); Ann. ENS. Paris to appear; Conf. in Honour of S. Agmon, to appear in J.d'Anal. Math.

14   D. Robert, H. Tamura: J. of Funct. Anal. **80:1** 124 1988)

15   D. Robert: preprint.University of Nantes (1992)

16   R. Schrader: Z. Phys.C. Part. and Fields **4** 27 (1980)

17   X. P. Wang: Helv. Phys. Acta **60** 501 (1987)

18   E. P. Wigner: Phys. Rev. **98:1** 145 (1955)

Operator Theory:
Advances and Applications, Vol. 57
© 1992 Birkhäuser Verlag Basel

Functional Calculus and Fredholm Criteria
for Boundary Value Problems on Noncompact Manifolds

Elmar Schrohe

Abstract. A Boutet de Monvel type calculus is developed for boundary value problems on (possibly) noncompact manifolds. It is based on a class of weighted symbols and Sobolev spaces. If the underlying manifold is compact, one recovers the standard calculus. The following is proven:

(1)   The algebra $\mathcal{G}$ of Green operators of order and type zero is a spectrally invariant Frechet subalgebra of $L(H)$, H a suitable Hilbert space, i.e.

$$\mathcal{G} \cap L(H)^{-1} = \mathcal{G}^{-1}.$$

(2)   There is a necessary and sufficient criterion for the Fredholm property of boundary value problems, based on the invertibility of a symbol, and

(3)   There is a holomorphic functional calculus for the elements in $\mathcal{G}$ in several complex variables.

Introduction. Boutet de Monvel's calculus, established in 1971, showed a new way of treating boundary value problems by pseudodifferential methods. In particular, it gave necessary and sufficient conditions for the Fredholm property of boundary value problems on smooth compact manifolds. For earlier work in this area, cf. Višik & Eskin (1967). Functional calculus for boundary problems on compact manifolds was treated by G. Grubb in 1986.

The present paper deals with both questions in the context of manifolds that may be noncompact, also with noncompact boundaries - in a situation where the standard methods fail. It offers a new approach and a solution based on a combination of pseudodifferential and operator theoretical methods.

Instead of a direct analysis of the boundary problems, I am developing an extension of Boutet de Monvel's calculus adapted to the noncompact situation. I am focusing on the algebra $\mathcal{G}$ of elements of order and type zero. It turns out to be a Fréchet-*-subalgebra of $L(H)$, where H is a Hilbert space these operators are naturally acting on. Moreover, $\mathcal{G}$ is *spectrally invariant*, i.e. $\mathcal{G} \cap L(H)^{-1} = \mathcal{G}^{-1}$.

The importance of spectral invariance in Fréchet algebras was first observed by Gramsch

(1984) who introduced the notion of Ψ^*-*algebras*: A Ψ^*-subalgebra of L(H), H a Hilbert space, is a spectrally invariant, symmetric, continuously embedded Fréchet subalgebra with the same unit.

Establishing the Ψ^*-property is the crucial step towards a variety of interesting results. Here, I first show a Fredholm criterion for boundary problems on noncompact manifolds, which is new even for differential problems on the half-space $\mathbb{R}^n_+$. Together with the results in Erkip & Schrohe (1990), one also obtains a uniform analog of the classical Lopatinski-Shapiro conditions. Based on the spectral invariance and general Fredholm theory, the proof is much simpler than earlier concepts that have been used e.g. in the case of classical pseudodifferential operators, relying on variants of Gohberg's lemma.

In connection with results of Waelbroeck, the spectral invariance also yields a holomorphic functional calculus for the elements in $\mathcal{G}$ in several complex variables.

If the underlying manifold is compact, the algebra $\mathcal{G}$ will coincide with the standard Boutet de Monvel algebra based on symbols in the Hörmander class S^0_{10}.

Spectral invariance and the Ψ^*-property were open questions also in this case. The present result on functional calculus not only extends Grubb's 1986 results to several complex variables; it gives a stronger version even for one variable only. Moreover, this follows here directly - without the need to first establish a parameter-dependent calculus.

Specializing further to the algebra of classical (pseudohomogeneous) elements, one recovers Grubb's theorem for one variable and gets an extension to several. For this situation, Schulze (1989) has proven spectral invariance.

In addition, the Ψ^*-property gives access to remarkable results in perturbation theory, on non-abelian cohomology and Oka principle, and on analytic Fréchet submanifolds in the algebra (Gramsch 1984, 1990, Lorentz 1990) or for the division problem for operator-valued distributions (Gramsch & Kaballo 1989). Connes and Bost have shown that the K-theory of a Ψ^*-algebra coincides with that of its C*-closure.

Spectral invariance for pseudodifferential operators was first proven by Beals (1977), cf. Ueberberg (1988). Since then it has been shown to hold in many interesting cases (Cordes 1985, Schrohe 1988, 1990, 1991a, Leopold & Schrohe 1991), although it fails in slightly different situations (Widom 1960, Davies et al. 1988).

In many algebras of pseudodifferential operators there is a close connection between the facts that the algebra is spectrally invariant and that the Fredholm property can be characterized by ellipticity in a suitable sense. This has been observed and exploited in Schrohe (1991a and 1991b). Already in 1989, Schulze has shown how to deduce spectral invariance in an abstract setting, provided that only elliptic operators are Fredholm.

Compared to the Banach algebra techniques established by Cordes and his associates (cf. Cordes 1979, 1987, Cordes & Erkip 1980), the present approach has the advantage that it yields existence and regularity results at the same time: An elliptic operator is a Fredholm operator, and whenever an operator in the calculus is Fredholm, then there is a Fredholm inverse which is a parametrix in the calculus. Since all these operators respect the whole scale of Sobolev spaces, this allows conclusions on the regularity of solutions in the spirit of Weyl's lemma: As soon as the Fredholm property is established between one fixed pair of Sobolev spaces, the application of the parametrix will give a hold on the regularity of the solution given the regularity of the right hand side.

1. The SG-Calculus for Pseudodifferential Operators. SG-Manifolds.

In order to overcome the basic difficulties stemming from the non-compactness of the underlying manifold, we are going to use symbol classes and Sobolev spaces with a very controlled behavior near infinity. On $\mathbb{R}^n$, this concept is due to Shubin (1971), Parenti (1972a), and Cordes (1976).

1.1 Definition. For $m = (m_1, m_2) \in \mathbb{R} \times \mathbb{R}$, $SG^m(\mathbb{R}^n)$ is the space of all smooth functions p on $\mathbb{R}^n \times \mathbb{R}^n$ such that $D_\xi^\alpha D_x^\beta p(x,\xi) = O(<\xi>^{m_1-|\alpha|} <x>^{m_2-|\beta|})$; $<x> = (1+|x|^2)^{\frac{1}{2}}$.

We will call m the *(double) order* of the symbol p. The intersection $\cap\, SG^m$, is the space of *regularizing symbols*.

The pseudodifferential operator $Op\, p$ or $p(x,D)$ associated with p is defined by

$$[Op\, p]\, f(x) = (2\pi)^{-n/2} \int e^{ix\xi}\, p(x,\xi)\, \hat{f}(\xi)\, d\xi.$$

Here, f is a rapidly decreasing function. The hat $\hat{\cdot}$ denotes the Fourier transform, also written $\mathscr{F}$, and p is called the *symbol* of $Op\, p$.

1.2 Theorem. (Shubin, Parenti, Cordes) The SG classes are closed under compositions and adjoints: If p, q are SG-symbols of orders m and m', resp., then $Op\, p \circ Op\, q = Op\, r$ for a symbol r of order $m+m'$, and $(Op\, p)^* = Op\, s$, where s also has order m. The pseudodifferential operators *(pdo)* with regularizing symbols *(regularizing pdo)* are precisely the integral operators with rapidly decreasing kernels.

The SG pseudodifferential operators naturally act on weighted Sobolev spaces.

1.3 Definition. For $s = (s_1, s_2) \in \mathbb{R} \times \mathbb{R}$, let $H^s = H^s(\mathbb{R}^n) = \{u \in \mathscr{S}': <x>^{s_2}(1-\Delta)^{s_1}u \in L^2\}$.

A symbol p of order m yields a bounded linear operator $Op\, p: H^s \longrightarrow H^{s-m}$ for all s.

1.4 Definition. A symbol $p \in SG^m$ is called elliptic, if $p(x,\xi)$ is invertible for large $|x|+|\xi|$ and $p(x,\xi)^{-1} = O(<\xi>^{-m_1}<x>^{-m_2})$.

Ellipticity allows the construction of a parametrix modulo regularizing operators: Given an elliptic $p \in SG^m$, there is a $q \in SG^{-m}$ such that both $Op\ p \circ Op\ q - I$ and $Op\ q \circ Op\ p - I$ are regularizing.

It is obvious that we will only be able to transfer these symbol classes to manifolds if the manifold has a special structure near infinity.

1.5 Definition (Schrohe 1987, Erkip & Schrohe 1989) Let Ω be an n-dimensional manifold without boundary. Call Ω *SG-compatible*, if conditions (SG1) - (SG3) hold.

(SG1)   There are finitely many coordinate neighborhoods that cover Ω, say $\Omega = \bigcup\limits_{j=1}^{J} \Omega_j$.

(SG2)   This cover has a good shrinking.

(SG3)   All the changes of coordinates χ satisfy $\partial^\alpha \chi(x) = O(<x>^{1-|\alpha|})$.

Let X be an n-dimensional submanifold of Ω with boundary $\partial X = Y$, where Y is a smooth (n-1)-dimensional submanifold without boundary. Assume additionally that

(SG4)   the coordinate charts $\kappa_j : \Omega_j \longrightarrow \mathbb{R}^n$ map $X \cap \Omega_j$ to $\mathbb{R}^n_+$, $Y \cap \Omega_j$ to $\mathbb{R}^{n-1} \times \{0\}$, and $\Omega_j \backslash \overline{X}$ to $\mathbb{R}^n_-$, and

(SG5)   there is a fixed Riemannian metric g on Ω whose metric tensor (g_{ij}) satisfies (in local coordinates) $\partial^\alpha g_{ij}(x) = O(<x>^{-|\alpha|})$, $g^{-1}(x) = O(1)$.

We then call the tuple (Ω, X, Y, g) an *SG-manifold with boundary*.

1.6 Remarks. (i) The existence of a good shrinking means that Ω is the union of open sets $\Omega_j' \subseteq \Omega_j$, and there is an $\varepsilon > 0$ such that $B(x, \varepsilon <x>) \subseteq \kappa_j(\Omega_j)$ for all $x \in \kappa_j(\Omega_j')$.
This is a typical condition for SG-manifolds. More generally, for open subsets U, U' of $\mathbb{R}^n$, we will say that U is a *conic neighborhood* of U' if there is an $\varepsilon > 0$ such that $B(x, \varepsilon <x>) \subseteq U$ for every $x \in U'$. This notion also makes sense on SG-manifolds.
(ii) A metric with (SG5) always exists. I just want to fix one in order to fix normal coordinates. Any choice of such a metric, however, makes the manifold Ω asymptotically flat: the Christoffel symbols satisfy $\partial^\alpha \Gamma(x) = O(<x>^{-1-|\alpha|})$.

1.7 Theorem. (Schrohe 1987) (a) On an SG compatible manifold Ω, there is a partition of unity together with cut-off functions subordinate to the cover $\{\Omega_j\}$ in SG^0.
(b) The symbol classes SG^m are invariant under changes of coordinates with (SG2) and (SG3). Hence SG-pdo may be defined on Ω, using a partition of unity $\{\varphi_j\}$ and cut-off functions $\{\psi_j\}$ subordinate to the cover $\{\Omega_j : j = 1, ..., J\}$, and asking that the non-local terms be integral operators with rapidly decreasing kernels and that the local terms be defined by an SG symbol. More precisely: For $A: \mathscr{S}(\Omega) \longrightarrow \mathscr{S}(\Omega)$ write

$$A = \sum_{j=1}^{J} \varphi_j A \psi_j + \sum_{j=1}^{J} \varphi_j A (1 - \psi_j).$$

The operators $\varphi_j A \psi_j$ induce operators A_j on $\mathbb{R}^n$ by $A_j f(x) = [\varphi_j A \psi_j (f \circ \kappa_j)](\kappa_j^{-1}(x))$. We shall say that A belongs to $SG^m(\Omega)$, if each A_j has a symbol in $SG^m(\mathbb{R}^n)$, and each of the operators $\varphi_j A(1-\psi_j)$ is - in local coordinates - an integral operator with a rapidly decreasing kernel.

1.8 Theorem. (Erkip & Schrohe 1989, 1990). If (Ω,X,Y,g) is SG-compatible, then one can switch to normal coordinates near the boundary *within the calculus*: One can introduce additional charts in a conic neighborhood of Y such that locally, the manifold there looks like $\mathbb{R}^{n-1} \times \mathbb{R}_+$, and the changes of coordinates are of the form $(y,t) \mapsto (\overline{\chi}(y),t)$ with a function $\overline{\chi}$ satisfying (SG3).

Moreover, the normal derivative (defined in a neighborhood of the boundary) is an operator with a symbol in $SG^{(1,0)}$.

2. The Algebra of Green Operators on SG-Manifolds

2.1 Standard notation. Let (Ω,X,Y,g) be SG-compatible with boundary.

(a) r^+ denotes restriction of functions or distributions to X, e^+ denotes extension (by zero) from X to Ω.

(b) Given a pseudodifferential operator P on Ω, define P_X by $P_X = r^+Pe^+$.

(c) The weighted Sobolev spaces on X are defined by restriction: $H^s(X) = r^+H^s(\Omega)$.

(d) $H = H^+ \oplus H_0^- \oplus H'$, where
$$H^+ = \{(e^+f)^\wedge : f \in \mathscr{S}(\mathbb{R}_+)\}, \quad H_0^- = \{(e^-f)^\wedge : f \in \mathscr{S}(\mathbb{R}_-)\}$$
(e^- is extension by zero from $\mathbb{R}_-$ to $\mathbb{R}$). H' is the space of all polynomials. Let $H^- = H_0^- \cup H'$, $H_d = \{ f \in H^- : f(\xi) = O(<\xi>^{d-1})\}$, $d \in \mathbb{N}_0$, $H_d^- = H_d \cap H^-$.

(e) For a function on $\mathbb{R}$ let $r'f = \lim_{t \to 0+} f(t)$. Define $\Pi' : H \to \mathbb{C}$ by $\Pi' = (2\pi)^{-\frac{1}{2}} r' \mathscr{F}^{-1}$.

For the definition of the symbol classes let us first assume that (Ω,X,Y,g) is $(\mathbb{R}^n, \mathbb{R}^n_+, \mathbb{R}^{n-1}$, Euclidean metric). Write $\mathbb{R}^n_+$ as $\{(x',x_n) \in \mathbb{R}^n : x_n > 0\}$.

2.2 Definition. A symbol $p \in SG^m$ has the *transmission property*, if for every $k \in \mathbb{N}_0$,

$$(1) \qquad p_{[k]}(x',\xi,v) = D_{x_n}^k p(x',x_n,\xi,<\xi>v)\big|_{x_n=0} \in SG_{x',\xi}^{m-(0,k)} \otimes H_{d,v}$$

Here, $d = [m_1]+1$, and $\otimes$ is an abbreviation for the completed tensor product $\hat{\otimes}_\pi$.

Write $p \in \mathscr{A}^m$. The indices x', ξ, v refer to the arguments of the functions. Together with the Fréchet topology on SG^m, (1) yields a Fréchet topology for $\mathscr{A}^m$.

2.3 Definition. (a) A function $g \in C^\infty(\mathbb{R}^{n-1} \times \mathbb{R}^{n-1} \times \mathbb{R} \times \mathbb{R})$ is a *singular Green symbol of order* m *and type* d, written $g \in \mathscr{B}^{m,d}$, provided

(1) $$g(x',\xi',<\xi'>v,<\xi'>\eta) \in SG^m_{x',\xi'} \otimes H^+_v \otimes H^-_{d,\eta} \; ;$$

(b) $t \in C^\infty(\mathbb{R}^{n-1} \times \mathbb{R}^{n-1} \times \mathbb{R})$ is a *trace symbol of order* m *and type* d, written $t \in \mathcal{T}^{m,d}$, if

(2) $$t(x',\xi',<\xi'>v) \in SG^m_{x',\xi'} \otimes H^-_{d,v} \; , \text{ and}$$

(c) $k \in C^\infty(\mathbb{R}^{n-1} \times \mathbb{R}^{n-1} \times \mathbb{R})$ is a *potential symbol of order* m, written $k \in \mathcal{K}^m$, if

(3) $$k(x',\xi',<\xi'>v) \in SG^m_{x',\xi'} \otimes H^+_v.$$

(d) Relations (1), (2), and (3) define Fréchet topologies on $\mathcal{B}^{m,d}$, $\mathcal{T}^{m,d}$, and $\mathcal{K}^m$.

2.4 Definition and Theorem. (a) The various symbols induce boundary symbol operators (acting in the normal direction only) in the standard way:

Let $f \in \mathcal{S}(\mathbb{R}_+)$, $c \in \mathbb{C}$. For fixed x', ξ' define

$$[P_{\mathbb{R}_+}(x,\xi',D_n)f](x_n) = r^+p(x',x_n,\xi',D_n)e^+f \; ;$$

$$[g(x',\xi',D_n)f](x_n) = (2\pi)^{-\frac{1}{2}} \int e^{ix_n\xi_n} \Pi'_{\eta_n} g(x',\xi',\xi_n,\eta_n)(e^+f)^\wedge(\eta_n) \, d\xi_n$$

$$t(x',\xi',D_n)f = \Pi'_{\xi_n} \{t(x',\xi)(e^+f)^\wedge(\xi_n)\};$$

$$[k(x',\xi',D_n)c](x_n) = (2\pi)^{-\frac{1}{2}} \int e^{ix_n\xi_n} k(x',\xi) \, d\xi_n \cdot c$$

(b) The following mappings are bounded

(i) $p_{\mathbb{R}_+}(x,\xi,D_n), \; g(x',\xi',D_n)$: $\mathcal{S}(\mathbb{R}_+) \longrightarrow \mathcal{S}(\mathbb{R}_+)$,

(ii) $t(x',\xi',D_n)$: $\mathcal{S}(\mathbb{R}_+) \longrightarrow \mathbb{C}$, and

(iii) $k(x',\xi',D_n)$: $\mathbb{C} \longrightarrow \mathcal{S}(\mathbb{R}_+)$.

(c) The full operators are defined from the boundary symbol operators by pseudo-differential action in the (x',ξ')-variables, denoted here by Op' :

pseudodifferential operators $\qquad Op_X p = (Op\ p)_X = r^+Op\ p\ e^+ = Op'\ p_{\mathbb{R}_+}(x,\xi,D_n)$

singular Green operators $\qquad\qquad Op_G g = Op'\ g(x',\xi',D_n)$

trace operators $\qquad\qquad\qquad\quad Op_T t = Op'\ t(x',\xi',D_n)$

potential operators $\qquad\qquad\qquad Op_K k = Op'\ k(x',\xi',D_n)$

(d) The following mappings are bounded for every choice of order and type

(i) $Op_X p, \; Op_G g$: $\mathcal{S}(\mathbb{R}^n_+) \longrightarrow \mathcal{S}(\mathbb{R}^n_+)$

(ii) $Op_T t$: $\mathcal{S}(\mathbb{R}^n_+) \longrightarrow \mathcal{S}(\mathbb{R}^{n-1})$

(iii) $Op_K k$: $\mathcal{S}(\mathbb{R}^{n-1}) \longrightarrow \mathcal{S}(\mathbb{R}^n_+)$

2.5 Definition. A *Green operator of order* m *and type* d is a matrix A of operators

$$(1) \qquad A = \begin{bmatrix} Op_X p + Op_G g & Op_K k \\ Op_T t & Op\ s \end{bmatrix} : \begin{matrix} \mathscr{S}(\mathbb{R}_+^n) \\ \oplus \\ \mathscr{S}(\mathbb{R}^{n-1}) \end{matrix} \longrightarrow \begin{matrix} \mathscr{S}(\mathbb{R}_+^n) \\ \oplus \\ \mathscr{S}(\mathbb{R}^{n-1}) \end{matrix}$$

where $p \in \mathscr{A}^m$, $g \in \mathscr{B}^{m-(1,0),d}$, $t \in \mathscr{T}^{m,d}$, $k \in \mathscr{K}^{m-(1,0)}$, $s \in SG^m(\mathbb{R}^{n-1})$.
Write $A \in \mathscr{G}^{m,d}$. The *boundary symbol operator* associated with A is the operator

$$a(x',\xi',D_n) = \begin{bmatrix} p_{\mathbb{R}_+}(x,\xi',D_n) + g(x',\xi',D_n) & k(x',\xi',D_n) \\ t(x',\xi',D_n) & s(x',\xi') \end{bmatrix} : \begin{matrix} \mathscr{S}(\mathbb{R}_+) \\ \oplus \\ \mathbb{C} \end{matrix} \longrightarrow \begin{matrix} \mathscr{S}(\mathbb{R}_+) \\ \oplus \\ \mathbb{C} \end{matrix} \ .$$

All the entries may be matrix-valued, fitting together appropriately, i.e. p, g are $n_1 \times n_2$
matrices, k is $n_1 \times n_3$, t is $n_4 \times n_2$, s is $n_4 \times n_3$. Call this an $(n_1,n_4) \times (n_2,n_3)$-matrix.

2.6 Theorem. Let $A \in \mathscr{G}^{m,d}$, $A' \in \mathscr{G}^{m',d'}$ with matrix sizes so that the composition
AA' makes sense. Then $AA' \in \mathscr{G}^{m'',d''}$, where $m'' = m+m'$, $d'' = \max\{m'+d,d'\}$. If $d = 0$, then the formal adjoint A^* also belongs to $\mathscr{G}^{m,0}$.
For the various compositions of operators, the classical asymptotic expansion formulas
(cf. sections 2.6, 2.7 in Grubb 1986) hold with respect to the SG-calculus.
In particular, $\mathscr{G}^{0,0}$ is a *-algebra, if the matrix size is $(n_1,n_2) \times (n_1,n_2)$.

2.7 Theorem. Let $A \in \mathscr{G}^{m,d}$ be an $(n_1,n_4) \times (n_2,n_3)$ matrix. Then

$$A : \begin{matrix} H^s(X)^{n_2} \\ \oplus \\ H^{s-(\frac{1}{2},0)}(Y)^{n_3} \end{matrix} \longrightarrow \begin{matrix} H^{s-m}(X)^{n_1} \\ \oplus \\ H^{s-m-(\frac{1}{2},0)}(Y)^{n_4} \end{matrix}$$

is bounded, provided $s_1 > d - \frac{1}{2}$. For lower values of s_1 one has to use H_0^s - spaces.
The Fréchet topologies on the symbol spaces are stronger than the corresponding topologies of bounded operators.

2.8 Definition and Theorem. (a) The regularizing Green operators of type zero (those in $\cap_m \mathscr{G}^{m,0}$) are precisely the integral operators with rapidly decreasing kernels over the respective spaces, i.e. in $\mathscr{S}(X \times X)$ for $Op_G g$, in $\mathscr{S}(Y \times X)$ for $Op_T t$, etc.

(b) The regularizing Green operators of type d differ from those in (a) only in that a
regularizing singular Green operator G of type d has the form

$$G = \sum_{j=0}^{d-1} K_j \, D_n^j \big|_Y + G_0.$$

where K_j are regularizing potential operators; G_0 is a regularizing singular Green operator of type zero. Similarly, a regularizing trace operator T of type d can be written

$$T = \sum_{j=0}^{d-1} S_j \, D_n^j \big|_Y + T_0,$$

where S_j are regularizing pdos on Y, and T_0 is a regularizing trace operator of type 0.
(c) Let $\varphi \in C_0^\infty(\mathbb{R})$, $\varphi \equiv 1$ near zero, and let G, K, T be singular Green, potential and

trace operators of order m and type d. By M denote for the moment the operator of multiplication with the function $1-\varphi(x_n/<x'>)$. Then MK is a regularizing potential operator, GM and TM are regularizing singular Green and trace operators, resp., of type zero, and MG is a regularizing singular Green operator of type d.

We can now start to define Green operators on an *arbitrary SG-manifold* (Ω,X,Y,g).

2.9 Definition. (a) Call a vector bundle over Ω an *SG-(vector) bundle*, if it is trivial in all local coordinates and the transition matrices (a_{ij}) satisfy $\partial^\alpha a_{ij}(x) = O(<x>^{-|\alpha|})$, $(a_{ij}(x))^{-1} = O(1)$. SG-bundles over X are simply restrictions of SG-bundles over Ω.
(b) It is some what lengthy but straightforward to introduce rapidly decreasing sections into a bundle E and weighted Sobolev spaces of distribution sections. The notation is $\mathscr{S}(\Omega,E)$, $\mathscr{S}(X,E)$, and $H^s(\Omega,E)$, $H^s(X,E)$ respectively. Identify $H^0(\cdot)$ and $L^2(\cdot)$.
(c) If φ is a C^∞-function on Ω, and A is a matrix of operators acting on sections over X and Y, respectively, then the multiplication φA of A with φ is to be understood as multiplication of the matrix A with the matrix $\begin{bmatrix} \varphi & 0 \\ 0 & \varphi|_Y \end{bmatrix}$.

2.10 Remark. By 1.8 there is a conic neighborhood of Y (cf. 1.6) with normal coordinates. We can and will assume now that, given any coordinate neighborhood, it will either not intersect a conic neighborhood of Y or else be one of those with normal coordinates. Then choose functions $\varphi_1,...,\varphi_k \in SG^0(\Omega)$, supported in the neighborhoods intersecting Y, such that $\sum_j \varphi_j = 1$ in a conic neighborhood of Y. Choose cut-off functions $\psi_j \in SG^0(\Omega)$, supported in the same coordinate neighborhoods with $\varphi_j\psi_j = \varphi_j$.

2.11 Definition. Let E, F be SG-bundles over X and Y, resp.. A *Green operator of order* m *and type* d is a matrix $\begin{bmatrix} P_x+G & K \\ T & S \end{bmatrix}$ acting on $\mathscr{S}(X,E) \oplus \mathscr{S}(Y,F)$, where

(i)   P is a pdo of order m on Ω, $P_x = r^+Pe^+$,

(ii)   S is a pdo of order m on Y, and where

(iii)   G, T, and K are given locally near the boundary by singular Green operators of order m-(1,0) and type zero, trace operators of order m and type d, and potential operators of order m-(1,0), cf. 2.5, while the nonlocal terms behave accoording to 2.9.

Let me state this explicitly only for G: Write $G = \sum \varphi_j G\psi_j + \sum \varphi_j G(1-\psi_j) + (1-\sum \varphi_j) G = \sum G_j + \sum R_j + R'$, and ask that each G_j is - in local coordinates - a singular Green operator of order m-(1,0) and type d, each R_j is an integral operator with a rapidly decreasing kernel on X×X, and R' can be written $R' = \sum_{j=0}^{d-1} K_j D_n^j|_Y + R''$, where K_j and

R" are integral operators with rapidly decreasing kernels on X×Y and X×X resp..

2.12 Theorem. The $\mathbb{R}^n_+$ results of 2.6/2.7 extend to manifolds and bundles with one restriction: The boundary symbol operators are only defined in the neighborhood of Y with normal coordinates. However, if we are dealing with φA where φ is an SG^0-function supported in this neighborhood, then there is no problem.

2.13 Definition. Let $A \in \mathscr{G}^{m,d}$, $d \leq \max\{m_1,0\}$.

(a) A *parametrix* to A is an operator $B \in \mathscr{G}^{-m,d'}$, $d' \leq \max\{-m_1,0\}$, such that $AB - I$ and $BA - I$ are both regularizing .

(b) Call A *elliptic of order* m, if the following holds:

(i)   the pseudodifferential operator P is elliptic on X, i.e. $p(x,\xi)^{-1} = O(<\xi>^{-m_1}<x>^{-m_2})$ for all large $|x|+|\xi|$, $x \in X$.

(ii)   Near the boundary, the boundary symbol of A is locally invertible by boundary symbols of order $-m$ and type d', i.e. given the functions φ_j, ψ_j as in 2.10 and denoting the boundary symbols of $\varphi_j A \psi_j$ by a_j, there are b_j such that

$$a_j b_j - \varphi_j = g_{1j} \quad \text{and} \quad b_j a_j - \varphi_j = g_{2j}$$

are regularizing boundary symbols.

2.14 Remark. The ellipticity condition in 2.13(ii) is slightly stronger than that given in the present version of Schrohe (1991b), Def. II.6.19. It follows from the theorem, below, and Theorem III.6.1 in Schrohe (1991b) that both formulations are equivalent.

2.15 Theorem. Let $A \in G^{m,d}$, $d \leq \max\{m_1, 0\}$.

(a) There is a parametrix B to A if and only if A is elliptic.

(b) For $s_1 > d-\frac{1}{2}$, ellipticity of A implies the Fredholm property of

$$A : H^s(X,E) \oplus H^{s-(\frac{1}{2},0)}(Y,F) \longrightarrow H^{s-m}(X,E) \oplus H^{s-m-(\frac{1}{2},0)}(Y,F).$$

2.16 Remark. In the half-space case, the boundary symbol operators are globally defined. It turns out that then the existence of a parametrix B to A is equivalent to the existence of a boundary symbol operator b that inverts the boundary symbol operator a of A modulo regularizing singular Green boundary symbols, g_1, g_2: The operators

$$a(x',\xi',D_n) \circ b(x',\xi',D_n) - 1 = g_1(x',\xi',D_n)$$

and

$$b(x',\xi',D_n) \circ a(x',\xi',D_n) - 1 = g_2(x',\xi',D_n)$$

are regularizing boundary symbol operators.

3. Spectral Invariance

Consider first the case of the half-space $X = \mathbb{R}^n_+$ with boundary $\partial X = Y = \mathbb{R}^{n-1}$. Denote by $\mathscr{G} = \mathscr{G}^{0,0}$ the algebra of all Green operators of order and type zero, of matrix size $(n_1,n_2) \times (n_1,n_2)$. These operators define bounded maps on the Hilbert space

$$H = L^2(X)^{n_1} \oplus H^{-(\frac{1}{2},0)}(Y)^{n_2}.$$

In fact, $\mathcal{G}$ is a $*$-subalgebra of $L(H)$, the bounded operators on H, cf. 2.6.

Each element $A \in \mathcal{G}$ can be written as a matrix of the form 2.5(1). This representation is not unique, but starting from the Fréchet topology on the symbol spaces and going over to suitable quotient topologies, one obtains

3.1 Theorem. $\mathcal{G}$ is a Fréchet-$*$-subalgebra of $L(H)$ with a stronger topology.

The principal theorem of this section is the following.

3.2 Theorem. $\mathcal{G}$ is also *spectrally invariant* in $L(H)$: $\mathcal{G} \cap L(H)^{-1} = \mathcal{G}^{-1}$.

Hence $\mathcal{G}$ is a Ψ^*-subalgebra of $L(H)$, cf. Gramsch (1984), Definition 5.1.

Details of the proof are given in Schrohe (1991b). Here is a rough sketch, broken up into several theorems and lemmas. First note the corresponding result for SG-manifolds *without* boundary.

3.3 Theorem. (Schrohe 1988) The algebra of pdo with symbols in $SG^0(Y)$ is a Ψ^*-subalgebra of $L(H^{(-\frac{1}{2},0)}(Y))$.

Clearly, the choice of n_1 and n_2 does not matter, so assume $n_1 = n_2 = 1$. Moreover, the proof can be reduced to considering a neighborhood of the identity:

3.4 Lemma. Let $(\mathcal{C}, \|\cdot\|)$ be a unital C^*-algebra, and let $\mathcal{A}$ be a symmetric Fréchet subalgebra with the same unit e. If there is an $\varepsilon > 0$ such that

(1) $\qquad\qquad\qquad (e-x)^{-1} \in \mathcal{A}$ for all $x \in \mathcal{A}$ with $\|x\| < \varepsilon$,

then $\mathcal{A}$ is spectrally invariant in $\mathcal{C}$.

Proof. $\mathcal{A}$ is dense in its C^* closure $\mathcal{B} = C^*(\mathcal{A})$. Suppose $a \in \mathcal{A}$ is invertible in $\mathcal{C}$ with inverse b. Then $b \in \mathcal{B}$. Choose a sequence $\{b_j\}$ in $\mathcal{A}$ with $b_j \to b$. Then $c_j = e - ab_j \in \mathcal{A}$ tends to zero in $\mathcal{B}$. Thus $(e+c_j)^{-1} \in \mathcal{A}$ for large j by (1), and $a^{-1} \in \mathcal{A}$.

3.5 Lemma. It is sufficient to show that there is an $\varepsilon > 0$ such that $I-(P_x+G)$ is invertible within the calculus for all P_x+G in a neighborhood of zero in $L(L^2(X))$.

Proof. Write a matrix $A \in \mathcal{G}$ as $A = \begin{bmatrix} g_{11} & g_{12} \\ g_{21} & g_{22} \end{bmatrix}$. Then invertibility of g_{11} lets us write

$$A = \begin{bmatrix} I & 0 \\ g_{21}g_{11}^{-1} & I \end{bmatrix} \begin{bmatrix} g_{11} & 0 \\ 0 & g_{22} - g_{21}g_{11}^{-1}g_{12} \end{bmatrix} \begin{bmatrix} I & g_{11}^{-1}g_{12} \\ 0 & I \end{bmatrix}.$$

For g_{11} and g_{22} close to I and g_{12}, g_{21} close to zero in the corresponding operator norms, 3.3 shows that A is invertible within the calculus whenever this is true for g_{11}.

We have therefore reduced the problem to showing the spectral invariance of the algebra

$$\mathscr{C} = \{Op_X p + Op_G g \colon p \in \mathscr{A}^0, g \in \mathscr{B}^{(-1,0),0}\}$$

in $L(L^2(X))$. The calculus says that $\{Op_G g \colon g \in \mathscr{B}^{(-1,0),0}\}$ is a two-sided ideal in $\mathscr{C}$.

3.6 Theorem. $\mathscr{B} = \{\lambda I + Op_G g \colon \lambda \in \mathbb{C}, g \in \mathscr{B}^{(-1,0),0}\}$ is a Ψ^*-algebra in $L(L^2(X))$.

Sketch of the proof. By 3.4, we may assume $\lambda = 1$, G small. Now $G = Op' g(x',\xi',D_n)$. Proving the operator-valued version of a result by Hörmander (1967, Thm. 3.3) one concludes that for large $|x'| + |\xi'|$, $\|g(x',\xi',D_n)\|$ is small.

On the other hand, one can show the following: If $g \in \mathscr{B}^{(-1,0),0}$ satisfies $\|g(x',\xi',D_n)\|$ < 0.1 for all (x',ξ') (all norms in $L(L^2(\mathbb{R}_+))$), then the boundary symbol operator $1 - g(x',\xi',D_n)$ is invertible in $L(L^2(\mathbb{R}_+))$, and the inverse is of the form $1 - k(x',\xi',D_n)$ with a $k \in \mathscr{B}^{(-1,0),0}$.

This allows us to find a $g' \in \mathscr{B}^{(-1,0),0}$ such that $(I-Op_G g')(I-G) = I-Op_G h$, $h \in \mathscr{B}^{(-\infty,-\infty),0}$. For $H = Op_G h$ the mapping properties in 2.8 show that $(I-H)^{-1} = I+H+ H(I-H)^{-1}H = I-Op_G h'$ for some $h' \in \mathscr{B}^{(-\infty,-\infty),0}$, provided the inverse exists.

Now an operator theoretic construction by Gramsch&Kaballo (1989) shows the theorem.

The proof of the important final proposition, below, is rather technical.

3.7 Proposition. There is a $C > 0$ such that for all $P_x + G \in \mathscr{C}$ with $\|P_x + G\|_{L(L^2(X))} < 1$ we can find operators $\overline{P} \in Op \mathscr{A}^0$ (with symbol defined on $\mathbb{R}^n$) and $\overline{G} \in Op_G \mathscr{B}^{(-1,0),0}$ with $\overline{P}_x + \overline{G} = P_x + G$, and $\|\overline{P}\|_{L(L^2(\mathbb{R}^n))} < C$, $\|\overline{G}\|_{L(L^2(X))} < C$.

3.8 Conclusion. Suppose $\|P_x + G\|$ in $L(L^2(X))$ is small. Find small representatives $\overline{P}$, $\overline{G}$, according to 3.7. Then $I-\overline{P}$ is invertible in $L(L^2(\mathbb{R}^n))$. By 3.3 and the calculus, its inverse, say $I-Q$, has a symbol in $\mathscr{A}^0$. Moreover, $(I-Q_x)(I-\overline{P}_x) = I-L$, with a small singular Green operator L. I-L is invertible within the calculus by 3.6, thus also $I-\overline{P}_x$.

We finally obtain the spectral invariance of $\mathscr{C}$ by writing

$$I-(P_x+G) = I-(\overline{P}_x+\overline{G}) = (I-\overline{P}_x)\,(I-(I-\overline{P}_x)^{-1}G')$$

and applying again 3.6, since the second factor on the right belongs to $\mathscr{B}$.

Let us now have a look at the manifold case. The spectral invariance result holds again:

3.9 Theorem. Let (Ω,X,Y,g) be SG-compatible, E, F fixed SG-bundles over Ω, Y, resp.. Then the algebra $\mathscr{G}$ of Green operators of order and type zero is a Ψ^*-subalgebra of L(H), where $H = H^0(X,E) \oplus H^{(-\frac{1}{2},0)}(Y,F)$.

Proof. Again the crucial point is spectral invariance. By Lemma 3.4 we can confine ourselves to small perturbations of the identity. We start with two observations.

3.10 Lemma. A is a regularizing Green operator in $\mathcal{G}$ if and only if A has a continuous extension A: $\mathcal{S}'(X,E) \oplus \mathcal{S}'(Y,F) \longrightarrow \mathcal{S}(X,E) \oplus \mathcal{S}(Y,F)$. If, in addition, I-A is invertible in L(H), then $(I-A)^{-1} = I - A - A(I-A)^{-1}A = I-A'$ for a regularizing $A' \in \mathcal{G}$.

3.11 Lemma. Let $A \in \mathcal{G}$, and let φ, ψ be SG^0-functions supported in the same coordinate neighborhood. Suppose that for the induced operator in local coordinates, we have $\|(\varphi A\psi)_*\| < 1$. Then $I - \varphi A\psi$ is invertible, and $(I-\varphi A\psi)^{-1} = I-\varphi B\psi$ for some $B \in \mathcal{G}$

The proof of 3.11 uses the spectral invariance theorem for Euclidean space.

3.12 Conclusion. Choose an SG-partition of unity $\{\varphi_j\}$ on X and cut-off functions $\{\psi_j\}$ subordinate to the coordinate charts, $\{\Omega_j: j = 1,...,J\}$. Then write for small A:

$$I-A = I - \varphi_1 A - \varphi_2 A - ... -\varphi_J A = (I - \varphi_2 A(I-\varphi_1 A)^{-1}- ...)(I-\varphi_1 A),$$

and note that $I-\varphi_1 A = (I-\varphi_1 A\psi_1)(I-(I-\varphi_1 A\psi_1)^{-1}\varphi_1 A(1-\psi_1))$, where both factors are invertible within the calculus by 3.10 and 3.11. Iteration yields the assertion.

4. Applications: Fredholm Criteria, Functional Calculus, Compact Manifolds

Use the notation of section 3. The theorem, below, is essential for the Fredholm criteria:

4.1 Theorem. Let $A \in \mathcal{G}$ Then A: $H \to H$ is a Fredholm operator iff A is elliptic.

Proof. Ellipticity implies the Fredholm property by 2.15. So assume that A is Fredholm. The Ψ^*-property yields that there is a $B \in \mathcal{G}$ such that $F_1 = BA-I$ and $F_2 = AB-I$ are finite rank operators (Gramsch 1984, Bem. 5.7). In particular, F_1 and F_2 are seen to be integral operators with rapidly decreasing kernels, thus regularizing Green operators. Hence B is a parametrix, and A is elliptic.

4.2 Theorem. (Reduction of the Order) Let E be an SG-bundle over X. For every $m \in \mathbb{Z}\times\mathbb{R}$ there is a family $\{\Lambda_-^m(\mu): \mu \geq 1\}$ of pdo such that for all $s \in \mathbb{R}\times\mathbb{R}$, $s_1 > d-\frac{1}{2}$,

$$\Lambda_-^m(\mu)_x: H^s(X,E) \longrightarrow H^{s-m}(X,E)$$

is invertible for large μ.

The proof uses ideas of Rempel & Schulze (1982, 1983/84) and Grubb (1990).

4.3 Corollary. Together with the fact that (cf. Erkip 1987, Erkip & Schrohe 1989, 1990)

$$A = ((I-\Delta)^m_x, \gamma_0,...,\gamma_{m-1}) : H^{(2m,0)}(X) \longrightarrow L^2(X) \oplus \prod_{j=0}^{m-1} H^{(2m-j-\frac{1}{2},0)}(Y)$$

is a Fredholm operator (Δ denotes the Laplace-Beltrami operator, $\gamma_j(f) = D_n^j f|Y$), 4.1 and 4.2 furnish Fredholm criteria for boundary value problems of arbitrary order and size.

4.4 Theorem. There is a holomorphic functional calculus for the elements in $\mathcal{G}$ in several complex variables.

Proof. Given k commuting elements $A_1,...,A_k \in \mathcal{G}$, choose $\mathcal{A}$ as a maximal commutative subalgebra of $\mathcal{G}$ containing $I, A_1,...,A_k$. $\mathcal{G}$ is a Ψ^*-subalgebra of $L(H)$, thus has an open group of invertible elements. As a closed subalgebra of $\mathcal{G}$, $\mathcal{A}$ is Fréchet with an open group of invertible elements, hence inversion is continuous, cf. Waelbroeck (1971). Again by Waelbroeck (VI, prop. 4), there is a functional calculus for the holomorphic functions on the joint spectrum of $A_1,...,A_k$ in $\mathcal{A}$ with values in $\mathcal{A} \subseteq \mathcal{G}$.

4.5 Corollary. (a) On a compact manifold with boundary, the algebra $\mathcal{G}$ coincides with the algebra of elements of order and type zero in Boutet de Monvel's algebra for (non-classical) symbols based on the Hörmander class $S^0_{1,0}$. The notion of SG-bundles reduces to usual vector bundles, $H^s(X)$, $H^s(Y)$, $s \in \mathbb{R}^2$, coincide with $H^{s_1}(X)$, $H^{s_1}(Y)$.

In this situation, Theorem 4.4 not only gives an extension of Grubb's theorem on functional calculus (Grubb 1986, Thm. 3.4.4) to several variables. Even in the case of one variable only, one obtains a stronger result, since in Theorem 4.4 we have no additional restrictions on the choice of the paths, cf. Grubb (1986), p. 356f.

(b) The *classical* elements in Boutet de Monvel's algebra on a compact manifold are those where all symbols have asymptotic expansions into homogeneous terms in the respective classes ('polyhomogeneous'). Denote the corresponding algebra of classical elements of order and type zero by $\mathcal{G}^{cl}$. Following an idea of Guillemin, it is possible to endow $\mathcal{G}^{cl}$ with a Fréchet topology. It is then a consequence of 3.9 and the calculus that $\mathcal{G}^{cl}$ is a Ψ^*-sublagebra of $L(H)$, too.

This also gives a functional calculus for the elements of $\mathcal{G}^{cl}$. In the case of one variable, one recovers Grubb's result (1986, Thm. 3.4.4). For $\mathcal{G}^{cl}$, spectral invariance was shown earlier by Schulze (1989).

Acknowledgements. The present paper contains a concise version of the results in Schrohe (1991b). I would like to thank B. Gramsch for many valuable and productive discussions. I am grateful to B.-W. Schulze, H.O. Cordes, and G. Grubb for their support.

References

Beals, R. (1977/79) Characterization of Pseudodifferential Operators and Applications,

Duke Math.J.44, 45 - 57, ibid. 46, p.215

Boutet de Monvel, L. (1971) Boundary Problems for Pseudo-Differential Operators, Acta Math. 126, 11 - 51

Cordes, H.O. (1976) A Global Parametrix for Pseudo-Differential Operators over $\mathbb{R}^n$, with Applications, preprint no. 90, SFB 72, Bonn

Cordes, H.O. (1979) Elliptic Pseudo-Differential Operators - An Abstract Theory, Springer LNM 756, Berlin, Heidelberg, New York: Springer 1979

Cordes, H.O. (1985) On Some C*-Algebras and Fréchet-*-Algebras of Pseudodifferential Operators, Proc. Symp. Pure Math. 43, 79 - 104

Cordes, H.O. (1987) Spectral Theory of Linear Differential Operators and Comparison Algebras, London Math. Soc. Lecture Notes 76, Cambridge, London, New York 1987

Cordes, H.O., and Erkip, A.K. (1980) The N-th Order Eliptic Boundary Problem for Noncompact Boundaries, Rocky Mountain J. Math. 10, 7 - 24

Davies, E. et al. (1988) L^p Spectral Theory of Kleinian Groups, J.Funct.Anal.78, 116-136

Erkip, A.K. (1987) Normal Solvability of Boundary Value Problems in Half Space, in: Pseudo-Differential Operators, Proceedings Oberwolfach 1986, Springer LNM 1256, 123 - 134, Berlin, Heidelberg, New York, Tokyo 1987

Erkip, A.K., and Schrohe, E. (1989) Pseudodifferential Operators on Weighted Sobolev Spaces for Manifolds with Noncompact Boundary, Symp. "Partial Differential Equations" Holzhau 1988. Teubner-Texte zur Mathematik 112, Leipzig 1989

Erkip, A.K., and Schrohe, E. (1990/preprint) Normal Solvability of Boundary Value Problems on Asymptotically Flat Manifolds, to appear in Journal Funct. Anal.

Gramsch, B. (1984) Relative Inversion in der Störungstheorie von Operatoren und Ψ-Algebren, Math. Ann. 269, 27 - 71

Gramsch, B. (1990) Analytische Bündel mit Fréchet-Faser in der Störungstheorie von Fredholmfunktionen zur Anwendung des Oka-Prinzips in F-Algebren von Pseudo-Differentialoperatoren, 120p., Arbeitsgruppe Funktionalanalysis, Univ. Mainz 1990

Gramsch, B., and Kaballo, W. (1989) Decompositions of Meromorphic Fredholm Resolvents and Ψ*-Algebras, Int.Eq.Op.Th. 12, 23 - 41

Gramsch, B. et al. (1991/preprint) Spectral Invariance and Submultiplicativity for Fréchet Algebras with Applications to Rings of Pseudo-Differential Operators

Grubb, G. (1986) Functional Calculus of Pseudo-Differential Boundary Problems, Progress in Math., vol.65, Boston, Basel: Birkhäuser 1986

Grubb, G. (1990) Pseudo-Differential Boundary Problems in L_p Spaces, Comm. PDE 15, 289 - 340

Hörmander, L. (1967) Pseudodifferential Operators and Hypoelliptic Equations, AMS Proceedings Symp. Pure Math. 10 (Singular Integrals) 138 - 183

Leopold,H.-G., and Schrohe,E.(1991) Spectral Invariance for Algebras of Pseudodifferential Operators on Besov Spaces of Variable Order of Differentiation, to app. Math. Nachr.

Leopold, H.-G., and Schrohe, E. (1992/preprint) Spectral Invariance for Algebras of Pseudodifferential Operators on Besov-Triebel-Lizorkin Spaces

Lorentz, K. (1991) On the Rational Homogeneous Manifold Structure of the Similarity Orbit of Jordan Elements in Operator Algebras, Operator Theory, Advances and Applications, vol. 50, 293 - 306

Melrose, R. (1981) Transformation of Boundary Problems, Acta Math. 147, 149 - 236

Parenti, C. (1972a) Operatori pseudodifferenziali in $\mathbb{R}^n$ e applicazioni, Annali Mat.Pura ed Appl. 93, 359 - 389

Parenti, C. (1972b) Un problema ai limiti ellittico in un dominio non limitato, Annali Mat. Pura ed Appl. 93, 391 - 406

Rempel, S., and Schulze, B.-W. (1982) Index Theory of Elliptic Boundary Problems, Berlin: Akademie-Verlag 1982

Rempel, S., and Schulze, B.W. (1983/84) Complex Powers for Pseudo-Differential Boundary Problems, part I, Math. Nachr. 111, 41-109, part II ibid. 116, 269-314

Rempel, S., and Schulze, B.-W. (1989) Asymptotics for Elliptic Mixed Boundary Problems, Berlin: Akademie-Verlag 1989

Schrohe, E. (1987) Spaces of Weighted Symbols and Weighted Sobolev Spaces on Manifolds, in: Pseudo-Differential Operators, Ed. Cordes, Gramsch, Widom, Springer LNM 1256, 360-377, Berlin, New York, Tokyo 1987

Schrohe, E. (1988) A Ψ^*-Algebra of Pseudodifferential Operators on Noncompact Manifolds, Arch. Math. 51, 81 - 86

Schrohe, E. (1990) Boundedness and Spectral Invariance for Standard Pseudodifferential Operators on Anisotropically Weighted L_p-Sobolev Spaces, Int. Eq. Op. Th. 13, 271-284

Schrohe, E. (1991a/prepr.) Spectral Invariance, Ellipticity, and the Fredholm Property for Pseudodifferential Operators on Weighted Sobolev Spaces, to app. Ann.Glob.Anal.Geom.

Schrohe, E. (1991b/preprint) A Pseudodifferential Calculus for Weighted Symbols and a Fredholm Criterion for Boundary Value Problems on Noncompact Manifolds

Schulze, B.-W. (1989) Topologies and Invertibility in Operator Spaces with Symbolic Structure, Proc. 9. TMP, Karl-Marx-Stadt, Teubner-Texte zur Mathematik, 111, 257 - 270

Schulze, B.-W. (1991) Pseudodifferential Operators on Manifolds with Singularities, Amsterdam: North-Holland 1991

Schulze, B.-W. (1989-91) Pseudodifferential Operators and Asymptotics on Manifolds with Corners, vols. I - IX, Reports of the Karl-Weierstraß-Institut, Berlin

Shubin, M.A. (1971) Pseudodifferential Operators in $\mathbb{R}^n$, Dokl.Akad.Nauk SSSR 196, N2, 316-319 = Sov.Math.Dokl. 12, N1, 147-151

Ueberberg, J. (1988) Zur Spektralinvarianz von Algebren von Pseudodifferentialoperatoren, manuscripta math. 61, 459 -475

Višik, M.I., and Eskin, G.I. (1967) Normally Solvable Problems for Elliptic Systems of Equations in Convolutions, Math. USSR Sbornik 14 (116), 326-356

Waelbroeck, L. (1971) Topological Vector Spaces and Algebras, Springer Lecture Notes Math. 230, Berlin, Heidelberg, New York 1971

Widom, H. (1960) Singular Integral Equations in L^p, Trans. AMS 97, 131 - 160

Author's address:

    Elmar Schrohe
    FB Mathematik
    Johannes Gutenberg-Universität
    Saarstraße 21
  D-W-6500 Mainz / Germany

Operator Theory:
Advances and Applications, Vol. 57
© 1992 Birkhäuser Verlag Basel

The variable discrete asymptotics of solutions of singular boundary value problems

By B.-W. Schulze

Max–Planck–Arbeitsgruppe

"Partielle Differentialgleichungen und Komplexe Analysis"

Universität Potsdam

December 14, 1990

Contents

1   Preliminary Remarks

The theory of pseudo–differential operators ($\psi DO's$) on manifolds with higher-dimensional singularities (e.g. edges) was developed in [S3], [S6] in such a way that the standard elliptic (pseudo–differential) boundary value problems can be regarded as "edge problems". In this approach the boundary is interpreted as the edge and the inner normal $I\!R_+$ as the model cone of the wedge. In this case the "wedge" is of the form $I\!R_+ \times \Omega$ with some open set $\Omega \subseteq \{$ boundary $\}$. The idea is to perform a pseudo–differential calculus along Ω with operator-valued amplitude functions (symbols) acting in distribution spaces along $I\!R_+$. Elements of the technique may also be found in [R2], [S3]. The point of view of boundary value problems has been elaborated in detail also in [S4, Chapter 2].

The present exposition is part of a continuation of [S4]. The typical aspects of general edge problems will already appear here, but the final result (to be published elsewhere) is in fact an extension of the algebra of edge problems for general variable discrete asymptotics (cf. the notations and definitions below). This requires from the very beginning arbitrary interior symbols, i.e. symbols without any sort of transmission property. The case with the transmission property was studied by many authors, cf. Boutet de Monvel [B1], Eskin [E1], Rempel, Schulze [R1]. The asymptotics in that situation are the constant ones (namely Taylor expansions close to the boundary). For non-trivial asymptotics (general interior symbols) there can be red off practically nothing from the latter case.

The violated transmission property is by no means an artificial assumption, also if we have not in mind the edge problems. Every classical symbol $a(\eta)$ of order η with $|\eta|$ as homogeneous principal part ($\eta \in I\!R^q$ being the covariable to $y \in I\!R^q$) is of this kind. Elliptic mixed problems, i.e. boundary value problems for elliptic operators with jumping elliptic conditions, lead to such symbols after a reduction to the boundary. In case of the Zaremba problem for the Laplacian in a $(q+1)$-dimensional domain (where Dirichlet and Neumann conditions are posed on different parts of the boundary with the jump along a submanifold of the boundary Y of codimension 1), we just obtain such an $a(\eta)$ (up to a constant factor), η being the covariable along Y. The solution of the Zaremba problem then needs the solution of a boundary value problem on the "Neumann side" of Y for the operator with symbol $a(\eta)$. More subtle examples follow by reducing problems with jumping oblique derivatives to Y (cf. [S5], [S6]). In other words another motivation for our investigations are mixed and transmission problems. The technique also plays a role in crack theory (cf. Morozov [M1], Schulze [S7]). Furthermore the work of Višik, Eskin [V1], [E1] belongs to the background of our calculus.

The essential new phenomena, compared with the transmission property, concern the nature of the elliptic regularity of solutions $u(t,y)$, $(t,y) \in I\!R_+ \times \Omega$, close to $t = 0$.

After the experience with conical singularities (cf., in particular, Kondrat'ev's technique [K1]) one may expect asymptotics of solutions $u(t,y) \in C^\infty(I\!R_+ \times \Omega)$ of the form

$$u(t,y) \sim \sum_{j=0}^{\infty} \sum_{k=0}^{m_j} c_{jk}(y)\, t^{-p_j}\, log^k t \tag{1.1}$$

as $t \to 0$, with exponents $p_j \in \mathbb{C}$, $j \in I\!N$, $Re\ p_j < \frac{1}{2} - \gamma$ for some weight $\gamma \in I\!R$, $Re\ p_j \to -\infty$ as $j \to \infty$. This is actually true when the p_j are independent of y. Note that the Taylor expansion of a solution $u(t,y) \in C^\infty(\bar{I\!R}_+ \times \Omega)$ is also like (1.1) with $p_j = -j$, $m_j = 0$ for all j. It is also natural to look at finite expansions in the sense that j runs through $\{0,1,...,N\}$ for some N. For instance, $u(t,y) \in H^s(I\!R_+ \times I\!R^q)$ for

$s > -\frac{1}{2}$, $s \neq \frac{1}{2}$ mod $\mathbb{Z}$, $N = [s + \frac{1}{2}] - 1$, implies

$$u(t,y) = \sum_{j=0}^{N} \mathcal{F}_{\eta \to y}^{-1}\{\hat{c}_j(\eta)[\eta]^{\frac{1}{2}}(t[\eta])^j \omega(t[\eta])\} \, mod H_0^s(\bar{\mathbb{R}}_+ \times \mathbb{R}^q) \qquad (1.2)$$

with $c_j(y) \in H^s(\mathbb{R}^q)$. Here $\omega(t)$ is a cut–off function (i.e. $\omega \in C_0^\infty(\bar{\mathbb{R}}_+)$, $\omega(t) = 1$ close to $t = 0$) and $\eta \to [\eta]$ means a strictly positive C^∞ function in $\eta \in \mathbb{R}^q$ with $[\eta] = | \eta |$ for $| \eta | > const > 0$. Further $\mathcal{F}_{y \to \eta}$ is the Fourier transform, and we also set $\hat{c}(\eta) = (\mathcal{F}_{y \to \eta} c)(\eta)$. $H^s(\mathbb{R}^n)$ is the standard Sobolev space on $\mathbb{R}^n$ of smoothness s, and $H^s(\Omega) = H^s(\mathbb{R}^n)|_\Omega$ for every open sufficiently regular $\Omega \subseteq \mathbb{R}^n$; further $H_0^s(\bar{\Omega})$ is the closure of $C_0^\infty(\Omega)$ in $H^s(\mathbb{R}^n)$ .The asymptotic terms of (1.2) suggest a Sobolev space generalization of (1.1). Under the condition of y–independent exponents p_j we get

$$u(t,y) \sim \sum_{j=0}^{\infty} \sum_{k=0}^{m_j} \mathcal{F}_{\eta \to y}^{-1}\{\hat{c}_{jk}(\eta)[\eta]^{\frac{1}{2}}(t[\eta])^{-p_j} log^k(t[\eta]) \omega(t[\eta])\}, \qquad (1.3)$$

$c_{jk}(y) \in H^s(\mathbb{R}^q)$. In [R2] it was proved for the first time that the elliptic regularity with asymptotics of pseudo–differential boundary value problems for violated transmission condition is actually of the sort (1.3), provided the p_j are y–independent. This was considerably generalized in [S4, part IX] to more singular boundary value problems as well as to the continuous asymptotics. The latter ones were explained in detail in [S6] (cf. also [S2], [S4, IV], [S3]). Among the singular differential operators on $\mathbb{R}_+ \times \Omega$ for open $\Omega \subseteq \mathbb{R}^q$ are those of the from

$$A(t,y,t\frac{\partial}{\partial t},tD_y)\, u(t,y) = t^{-\mu} \sum_{k+|\alpha| \leq \mu} \{\prod_{j=0}^{k}(d_{j\alpha}(t,y) + t\frac{\partial}{\partial t})\}(tD_y)^\alpha u(t,y) \qquad (1.4)$$

$\mu \in \mathbb{N}$, with coefficients $d_{j\alpha}(t,y) \in C^\infty(\bar{\mathbb{R}}_+ \times \Omega)$. The exponents p_j of the asymptotics are the solutions $z \in \mathbb{C}$ of

$$\prod_{r=0}^{\mu}(d_{r0}(0,y) - z) = 0. \qquad (1.5)$$

Then the $p_j(y)$ will depend in general on y as well as $m_j(y)$, where $m_j(y) + 1$ is the multiplicity of the zero $p_j(y)$.

It is not obvious at first glance how to formulate the y–dependent asymptotics in a general functional analytic framework. But the idea is to employ y–dependent families of analytic functionals in the complex Mellin plane which correspond to a finite linear combination of derivatives of the Dirac measure, concentrated at p_j. In Section 2 of the present paper we shall formulate elements of the structure of these y–dependent asymptotics. This extends and improves corresponding results from [S1]. Section 3 will present the nature of smoothing operators with y–dependent asymptotics. They

occur as remainders in operations between pseudo–differential operators. A precise
description of the smoothing operators is also needed for obtaining the elliptic regularity
with asymptotics when we apply a left parametrix to a given equation in order to get
the identity modulo a smoothing object of that sort. The theory of [S4], Section
2.4., will also show that the smoothing operators belong to an operator algebra of
pseudo–differential boundary value problems, "extending" Boutet de Monvel's algebra.
Analogously we shall also talk about Green operators instead of smoothing ones, where
here the general asymptotics are the substitute of the Taylor asymptotics from Boutet
de Monvel's theory.

2   The variable discrete asymptotics

Let

$$(Mu)(z) = \int_0^\infty t^{z-1} u(t) dt$$

be the Mellin transform, first for $u(t) \in C_0^\infty(I\!\!R_+), z \in \mathbb{C}$, and then extended to an
isomorphism

$$M : L^2(I\!\!R_+) \to L^2(\Gamma_{\frac{1}{2}}). \tag{2.1}$$

Here $\Gamma_\rho = \{z \in \mathbb{C} : Re\ z = \rho\}$ for any real ρ. Spaces on Γ_ρ are understood as
corresponding ones on $I\!\!R$ under the identification $\Gamma_\rho \ni z \to Im\ z$. In particular, it
makes sense to talk about $\hat{H}^s(\Gamma_{\frac{1}{2}})$, where hat indicates the Fourier image of the given
space on $\Gamma_{\frac{1}{2}} \cong I\!\!R$. The Mellin transform can be applied for introducing "Mellin"
–Sobolev spaces $\mathcal{H}^s(I\!\!R_+), s \in I\!\!R$. For every real s there is a well–defined subspace
$\mathcal{H}^s(I\!\!R_+) \subset \mathcal{D}'(I\!\!R_+)$ such that (2.1) restricts (or extends) to an isomorphism

$$M : \mathcal{H}^s(I\!\!R_+) \to \hat{H}^s(\Gamma_{\frac{1}{2}})$$

(cf. [S6]). Let us set $\mathcal{H}^{s,\gamma}(I\!\!R_+) = t^\gamma \mathcal{H}^s(I\!\!R_+)$ for $s, \gamma \in I\!\!R$, and

$$\mathcal{K}^{s,\gamma}(I\!\!R_+) = \{\omega u + (1 - \omega)v : u \in \mathcal{H}^{s,\gamma}(I\!\!R_+), v \in H^s(I\!\!R_+)\}, \tag{2.2}$$

with some fixed cut–off function $\omega(t)$. In view of $\mathcal{H}^{s,\gamma}(I\!\!R_+) \subset H_{loc}^s(I\!\!R_+)$ the latter
space is independent of the concrete choice of ω. The space (2.2) can be endowed with
an adequate Banach structure (even Hilbert). Then $\mathcal{K}^{\infty,\gamma}(I\!\!R_+) = \bigcap_{s\in I\!\!R} \mathcal{K}^{s,\gamma}(I\!\!R_+)$ is a
Fréchet space. γ is also called a weight. If we set

$$M_\gamma u(z) = M(t^{-\gamma}u)(z + \gamma), \tag{2.3}$$

first for $u(t) \in C_0^\infty(I\!\!R_+)$, $z \in \Gamma_{\frac{1}{2}-\gamma}$, then M_γ induces isomorphisms
$M_\gamma : \mathcal{H}^{s,\gamma}(I\!\!R_+) \to \hat{H}^s(\Gamma_{\frac{1}{2}-\gamma})$. Since the weight γ will usually be clear, we shall
often write for abbreviation $M = M_\gamma$, again.

Next we want to describe subspaces of $\mathcal{K}^{s,\gamma}(\mathbb{R}_+)$ or more generally of $C^\infty(\Omega, \mathcal{K}^{s,\gamma}(\mathbb{R}_+))$ of elements with asymptotics in $t \to 0$. To this end we first fix a compact set $K \subset \mathbb{C}$ and denote by $\mathcal{A}'(K)$ the space of all analytic functionals carried by K. Let $G \subseteq \mathbb{C}$ be an open set and $\mathcal{A}(G)$ be the space of all holomorphic functions in G, equipped with the (nuclear Fréchet) topology of uniform convergence on all compact subsets of G. Then, there is a canonical isomorphism

$$\mathcal{A}'(K) \cong \mathcal{A}(\mathbb{C} \setminus K)/\mathcal{A}(\mathbb{C}).$$

Every $\zeta \in \mathcal{A}'(K)$ can be represented in the form

$$\langle \zeta, h \rangle = \frac{1}{2\pi i} \int_C f(z)h(z)dz,$$

$h \in \mathcal{A}(\mathbb{C})$, for some $f(z) \in \mathcal{A}(\mathbb{C} \setminus K)$, C being a sufficiently regular curve surrounding K. The space $\mathcal{A}'(K)$ is nuclear Fréchet in the corresponding quotient topology.

For a given weight $\gamma \in \mathbb{R}$ we will always assume

$$K \subset \{z \in \mathbb{C} : Re\ z < \frac{1}{2} - \gamma\}.$$

Then $\zeta \in \mathcal{A}'(K)$ implies $\langle \zeta_w, t^{-w} \rangle \omega(t) \in \mathcal{K}^{\infty,\gamma}(\mathbb{R}_+)$ for every cut–off function $\omega(t)$ (sub w at ζ means that the pairing refers to $w \in \mathbb{C}$). A simple calculation shows that for every sequence of pairs

$$P = \{(p_j, m_j)\}_{j=0,\dots,N} \subset \mathbb{C} \times \mathbb{N} \tag{2.4}$$

($\mathbb{N} = \{0,1,2,3,\dots\}$), with $\pi_{\mathbb{C}}P := \bigcup_{j=0}^{N}\{p_j\} \subset \{\frac{1}{2} - \gamma + \delta < Re\ z < \frac{1}{2} - \gamma\}$ for some $\delta < 0$ and for every choice of constants c_{jk}, $0 \le k \le m_j$, $j = 0,..,N$, there is a $\zeta \in \mathcal{A}'(\pi_{\mathbb{C}}P)$ such that

$$\langle \zeta, t^{-w} \rangle = \sum_{j=0}^{N} \sum_{k=0}^{m_j} c_{jk} t^{-p_j} log^k t. \tag{2.5}$$

The associated $f(z) \in \mathcal{A}(\mathbb{C} \setminus \pi_{\mathbb{C}}P)$ is just meromorphic with poles at p_j of multiplicities $m_j + 1$, $j = 0,\dots,N$. Let us set

$$m(\zeta) = \sum_{j=0}^{N}(1 + m_j), \quad sg^\bullet(\zeta) = P. \tag{2.6}$$

Fix a compact set $K \subset \{z \in \mathbb{C} : \frac{1}{2} - \gamma + \delta < Re\ z < \frac{1}{2} - \gamma\}$, $\delta < 0$, and denote by

$$C^\infty(\Omega, \mathcal{A}'(K))^\bullet \tag{2.7}$$

the subspace of all $\zeta(y) \in C^\infty(\Omega, \mathcal{A}'(K))$ which are of the form (2.5) for every fixed $y \in \Omega$, where the data c_{jk}, p_j, m_j, N are allowed to vary along $y \in \Omega$, but satisfy the condition

$$\sup_{y \in A} m(\zeta(y)) < \infty \text{ for every compact subset } A \subset \Omega.$$

Note that the associated families of meromorphic functions may have poles of variable multiplicities. The following result was proved in [S6].

PROPOSITION: **2.1** $\zeta(y) \in C^\infty(\Omega, \mathcal{A}'(K))^\bullet$ *implies* $D_y^\alpha \zeta(y) \in C^\infty(\Omega, \mathcal{A}'(K))^\bullet$ *for every* $\alpha \in I\!\!N^q$, *and we have*

$$\pi_{\mathbb{C}} \, sg^\bullet(D_y^\alpha \zeta)(y) \subseteq \pi_{\mathbb{C}} \, sg^\bullet(\zeta)(y)$$

for all $y \in \Omega$.

By construction every $\zeta(y) \in C^\infty(\Omega, \mathcal{A}'(K))^\bullet$ is associated with a system of sequences $P^\alpha(y)$ of y–dependent pairs like (2.4), $\alpha \in I\!\!N^q$, such that

$$sg^\bullet(D^\alpha \zeta)(y) \subseteq P^\alpha(y) \ for \ all \ y \in \Omega, \ \alpha \in I\!\!N^q. \tag{2.8}$$

Set $P = \{P^\alpha(y)\}_{y \in \Omega, \alpha \in I\!\!N^q}$ and denote by $C^\infty(\Omega, \mathcal{A}'_P(K))$ the subspace of all $\zeta(y) \in (2.7)$ with (2.8).

PROPOSITION: **2.2** *The space* $C^\infty(\Omega, \mathcal{A}'_P(K))$ *is Fréchet in the topology induced by* $C^\infty(\Omega, \mathcal{A}'(K))$.

PROOF: As noted above the space $\mathcal{A}'(K)$ is Fréchet. Then $C^\infty(\Omega, \mathcal{A}'(K))$ is also a Fréchet space in the topology of uniform convergence of all y–derivatives over compact subsets of Ω. The convergence of a sequence $\{\zeta_k(y)\}_{k \in I\!\!N}$ of elements in $C^\infty(\Omega, \mathcal{A}'_P(K))$ with respect to the topology of $C^\infty(\Omega, \mathcal{A}'(K))$ implies the convergence of $\{(D_y^\alpha \zeta_k)(y_0)\}_{k \in I\!\!N}$ in $\mathcal{A}'(K)$ for every fixed $y_0 \in \Omega$. But it is known from the space $\mathcal{A}'(K)$ that the limit $\zeta(y)$ has the property $sg^\bullet(D_y^\alpha \zeta)(y_0) \subseteq P^\alpha(y_0)$ for all $\alpha \in I\!\!N^q$, $y_0 \in \Omega$. This was just the point to be verified. $\square$

REMARK: **2.3** We have for $u(t,y) = \langle \zeta(y)_w, t^{-w} \rangle \omega(t)$ with $\zeta(y) \in C^\infty(\Omega, \mathcal{A}'_P(K))$

$$(Mu)(z,y) \in \mathcal{A}(\mathbb{C} \setminus \pi_{\mathbb{C}} P^0(y))$$

for ever $y \in \Omega$.

The set $\pi_{\mathbb{C}} P^0(y)$ will also be called the carrier of asymptotics of $u(t,y)$ at y. More generally it may happen for a $u(t,y) \in C^\infty(\Omega, \mathcal{K}^{\infty,\gamma}(I\!\!R_+))$ that $(M\,\omega u)(z,y)$ is holomorphic outside a larger subset of $\mathbb{C}$ than a compact discrete one. In any such case we shall talk about the carrier of asymptotics at a given y.

The occurring sets V of carriers can be specified by the following properties. $V \subset \mathbb{C}$ is closed,

$$V \cap \{z : c \leq Re\ z \leq c'\}$$

is compact for all $c, c' \in \mathbb{R}$, and $V = V^c$, where upper c indicates the complement of the union of all unbounded connected components of $\mathbb{C} \setminus V$. Denote by $\mathcal{V}$ the system of those V. If V belongs to a weight $\gamma \in \mathbb{R}$ we will assume

$$V \subset \{z : Re\ z < \frac{1}{2} - \gamma\}. \tag{2.9}$$

Let us fix a (half-open) weight interval $\Delta = (\delta, 0]$, $\delta < 0$, and set

$$K = V \cap \{z : \frac{1}{2} - \gamma + \delta \leq Re\ z\} \tag{2.10}$$

for a $V \in \mathcal{V}$ with (2.9). Furthermore we fix a covering $\mathcal{U}(\Omega)$ of Ω by open sets U with compact $\bar{U} \subset \Omega$, such that every compact subset $A \subset \Omega$ does intersect only finitely many elements of $\mathcal{U}(\Omega)$.

Now the raw material for all (usually more sophisticated) notions of asymptotics which are variable along $\Omega \ni y$ is the following

DEFINITION: **2.4** *Let* $\gamma \in \mathbb{R}$ *and* $\Delta = (\delta, 0]$ *be finite. Then* $u(t, y) \in C^{\infty}(\Omega, \mathcal{K}^{s,\gamma}(\mathbb{R}_+))$ *is said to have variable discrete asymptotics, associated with the weight data* (γ, Δ), *if there is a* $V \in \mathcal{V}$ *with (2.9) and a map*

$$\zeta : (\delta, 0) \times \mathcal{U}(\Omega) \to C^{\infty}(\Omega, \mathcal{A}'(K))^{\bullet}$$

with (2.10) such that for any cut-off function $\omega(t)$

$$u(t, y) - \langle \zeta_b(y)_w,\ t^{-w} \rangle \omega(t) \in C^{\infty}(U, \mathcal{K}^{s,\gamma-\beta}(\mathbb{R}_+)) \tag{2.11}$$

for arbitrary $b = (\beta, U) \in (\delta, 0) \times \mathcal{U}(\Omega)$, *and* ζ_b *being the value of* ζ *at* b. *The pairing* $\langle \cdot, \cdot \rangle$ *refers to the complex variable* w.

For $\Delta = (-\infty, 0]$ we impose analogous relations, now for $K = V \cap \{\frac{1}{2} - \gamma - (m+1) \leq Re\ z\}$ for all $m \in \mathbb{N}$.

Set

$$\mathbb{H}_{\rho} := \{z \in \mathbb{C} : Re\ z > \frac{1}{2} - \rho\}.$$

If $u(t, y)$ has variable discrete asymptotics in the sense of Definition 2.4 then $(M_{t \to z} \omega u)(z, y)$ for any cut-off function ω is a y-dependent family of meromorphic functions in $\mathbb{H}_{\gamma-\delta}$. The essential properties of those families with respect to the y-dependent pattern of poles can also be described as follows. First, if $K \subset \mathbb{C}$

is a set, a K–excision function $\chi(z)$ is a C^∞ function in $\mathbb{C}$ with $\chi(z) = 0$ for $\mathrm{dist}(z, K) < c_0$, $\chi(z) = 1$ for $\mathrm{dist}(z, K) > c_1$ for certain $0 < c_0 < c_1 < \infty$.

Let $\Omega \subseteq \mathbb{R}^q$ be open and fix the weight data (γ, Δ), $\Delta = (\delta, 0]$. Then $\mathcal{F}(\Omega, (\gamma, \Delta))$ is the space of all functions $f(y, z)$ with the following properties. There exists a $V \in \mathcal{V}$ with (2.9), dependent on f, such that

$$f(y, z) \in C^\infty(\Omega, \mathcal{A}(\mathbb{H}_{\gamma-\delta} \setminus K))$$

for the compact set (2.10), and further

$$\chi(z) f(y, z)|_{z=\beta+i\tau} \in C^\infty(\Omega, \mathcal{S}(\mathbb{R}_\tau))$$

for every K–excision function $\chi(z)$ and all $\beta > \frac{1}{2} - \gamma + \delta$, uniformly in β for $c \leq \beta \leq c'$ for all $\frac{1}{2} - \gamma + \delta < c < c' < \infty$. Denote by $\mathcal{F}_V(\Omega, (\gamma, \Delta))$ the subspace of those elements of $\mathcal{F}(\Omega, (\gamma, \Delta))$ which belong to a given fixed V.

Moreover $\mathcal{F}(\Omega, (\gamma, \Delta))^\bullet$ will denote the subspace of all $f(y, z) \in \mathcal{F}(\Omega, (\gamma, \Delta))$ which extend to a meromorphic function in $\mathbb{H}_{\gamma-\delta}$ for every fixed $y \in \Omega$, with a finite number of poles $p_0(y), ..., p_N(y) \in K \setminus \Gamma_{\frac{1}{2}-\gamma+\delta}$, with (2.10)$= K(f)$, and multiplicities $m_j(y) + 1$, $j = 0, ..., N = N(y, f)$, such that

$$m(f)(y) := \sum_{j=0}^{N} (1 + m_j(y)) \tag{2.12}$$

satisfies

$$\sup_{y \in A} m(f)(y) < \infty \text{ for every compact subset } A \subset \Omega.$$

THEOREM: 2.5 *For every $f(y, z) \in F(\Omega, (\gamma, \Delta))^\bullet$, $\Delta = (\delta, 0]$, $-\infty < \delta < 0$, there is a map*

$$\zeta : (\delta, 0) \times \mathcal{U}(\Omega) \to C^\infty(\Omega, \mathcal{A}'(K))^\bullet,$$

where (2.10) belongs to f in the sense of the above notation, such that

$$\langle (\zeta_b(y) - \zeta_{b'}(y))_w, \ t^{-w} \rangle \omega(t) \in C^\infty(U \cap U', \mathcal{K}^{\infty, \gamma - \max(\beta, \beta')}(\mathbb{R}_+))$$

for arbitrary $b = (\beta, U)$, $b' = (\beta', U') \in (\delta, 0) \times \mathcal{U}(\Omega)$. Moreover

$$f(y, z) - M_{t \to z}\langle \zeta_b(y)_w, \ t^{-w} \rangle \omega(t) \in F_O(U, (\gamma, B))$$

for $B = (\beta, 0]$, where O indicates the space $F_V(...)$ for $V = \varnothing$.

PROOF: By assumption $f(y, z)$ has finitely many poles $p_0(y), ..., p_N(y)$ in $\mathbb{H}_{\gamma-\delta}$ of multiplicities $m_j(y) + 1$, $j = 0, ..., N = N(y)$, and we have (2.12)$<$ *const* uniformly in $y \in \bar{U}$ for every $U \in \mathcal{U}(\Omega)$. Let us fix $y_0 \in \bar{U}$ and denote the poles of real part $< \frac{1}{2} - \gamma + \beta$ by $q_0(y_0), ..., q_M(y_0)$, $M = M(y_0)$, $\delta < \beta < 0$. Then there exists an $\varepsilon = \varepsilon(y_0) > 0$ such that

$$dist\{((\{p_0(y_0), ..., p_N(y_0)\} \setminus \{q_0(y_0), ..., q_M(y_0)\}), \Gamma_{\frac{1}{2} - \gamma + \beta}\} > \varepsilon.$$

There is an open neighbourhood $U_0(y_0)$ of y_0 in Ω, U_0 also dependent on ε and β, such that $f(y, z)$ is holomorphic in $\frac{1}{2} - \gamma + \beta - \frac{\varepsilon}{2} < Re\ z < \frac{1}{2} - \gamma + \beta + \frac{\varepsilon}{2}$ for all $y \in U_0$. Set

$$R_0 = \{z \in \mathbb{C} : |\ Im\ z\ | \leq c,\ \frac{1}{2} - \gamma + \beta - \frac{\varepsilon}{4} \leq Re\ z \leq \frac{1}{2} - \gamma\},$$

where $c > 0$ is so large that $K \subset \{z \in \mathbb{C} : |\ Im\ z\ | \leq c\}$. Then $C_0 = \partial R_0$ does not intersect the poles of $f(y, z)$ for all $y \in U_0$. Let us form the family of analytic functionals

$$\mathcal{A}(\mathbb{C}) \ni h(z) \to \frac{1}{2\pi i} \int_{C_0} f(y, z) h(z) dz =: \langle \zeta_0(y), h \rangle.$$

Then $\zeta_0(y) \in C^\infty(U_0, \mathcal{A}'(R_0))^\bullet$, and $\langle \zeta_0(y), t^{-w} \rangle \omega(t) \in C^\infty(U_0, \mathcal{K}^{\infty, \gamma}(\mathbb{R}_+))$ has variable discrete asymptotics. If y_0 runs over $\bar{U}$ the associated sets U_0 form an open covering of $\bar{U}$. Since $\bar{U}$ is compact, we can choose a finite subcovering $U_0, ..., U_L$, belonging to certain $y_0, ..., y_L \in \bar{U}$. Choose $\varphi(y) \in C_0^\infty(U_j)$ with $\sum_{j=0}^{L} \varphi_j \equiv 1$ over $\bar{U}$. Denote by $\zeta_j(y)$ the family of analytic functionals in $C^\infty(U_j, \mathcal{A}'(R_j))^\bullet$, constructed analogously to the above $\zeta_0(y)$, $j = 0, ..., L$. Then it follows a family

$$\zeta_b(y) := \sum_{j=0}^{L} \varphi_j(y) \zeta_j(y) \in C^\infty(\Omega, \mathcal{A}'(K))^\bullet,$$

$b = (\beta, U)$ which has obviously the asserted properties. The last statement of Theorem 2.5 is a consequence of the remarks in the beginning. $\square$

COROLLARY: 2.6 *Let* $u(t, y) \in C^\infty(\Omega, \mathcal{K}^{\bullet, \gamma}(\mathbb{R}_+))$ *have variable discrete asymptotics in the sense of Definition 2.4 for the weight data* (γ, Δ). *Then, there is an* $f(y, z) \in F(\Omega, (\gamma, \Delta))^\bullet$ *such that*

$$(M_{t \to z} \omega u)(z, y) - f(y, z) \in C^\infty(\Omega, \mathcal{A}(\mathbb{H}_{\gamma - \delta}))$$

for any cut-off function $\omega(t)$. *Conversely every* $f(y, z) \in \mathcal{F}(\Omega, (\gamma, \Delta))^\bullet$ *is the Mellin image of* ωu *with some* $u(t, y) \in C^\infty(\Omega, \mathcal{K}^{\infty, \gamma}(\mathbb{R}_+))$ *of variable discrete asymptotics, modulo* $C^\infty(\Omega, \mathcal{A}(\mathbb{H}_{\gamma - \delta}))$.

REMARK: 2.7 The generalization of (1.3) to the variable discrete case is not obvious, since the smoothness of items on the right of (1.3) in y does depend on $Re\, p_j$. Therefore the jumping exponents p_j lead to a rather complex behaviour of smoothness of those asymptotic terms in y. We shall neglect the details here. They will be published in [S4, Chapter 2.4.] (cf. also [S5]).

It is possible to introduce variable discrete asymptotic types P for classifying corresponding distribution spaces, similarly to the constant discrete case (cf. [S4, 1.1.2.Definition 1]). It suffices to talk about an arbitrary fixed system

$$P = \{P^\alpha(y)\}_{y\in\Omega,\alpha\in I\!N^q} \tag{2.13}$$

with $P^\alpha(y) = \{(p_j^\alpha(y), m_j^\alpha(y))\}_{j=1,\ldots,N(y)}$, and to say that an $f(y,z) \in \mathcal{F}(\Omega,(\gamma,\Delta))^\bullet$ belongs to the space $\mathcal{F}_P(\Omega,(\gamma,\Delta))$ if $(D^\alpha f)(y,z)$ has poles at $p_j^\alpha(y)$ of multiplicities $m_j^\alpha(y) + 1$, for all $y \in \Omega$, $\alpha \in I\!N^q$. Analogously to Proposition 2.2 it follows that $\mathcal{F}_P(\Omega,(\gamma,\Delta))$ is a Fréchet space.

3 The nature of smoothing operators

This section will study pseudo–differential operators with operator–valued symbols which are "Green" of finite order and smoothing, respectively, in the sense of the variable discrete asymptotics. We shall mainly look at the operator–valued symbols. The operators themselves follow by applying the standard operator convention, based on the Fourier transform $\mathcal{F} = \mathcal{F}_{y\to\eta}$ in y, i.e. $Op(\cdot) = \mathcal{F}^{-1}(\cdot)\mathcal{F}$. Remember that it was already useful in Boutet de Monvel's algebra (of pseudo–differential boundary value problems with the transmisson property) to invent and to describe the role of ("singular") Green operators there.

Let us briefly recall the Green boundary symbols of class zero and order zero, here in the formulation of the theory of [S4, VII].

First, let E be a Banach space, and $\{\kappa_\lambda\}_{\lambda\in I\!R_+}$ be a group of isomorphisms in E, with $\kappa_\lambda\kappa_\mu = \kappa_{\lambda\mu}$, $\kappa_{\lambda^{-1}} = (\kappa_\lambda)^{-1}$ for all $\lambda,\mu \in I\!R_+$, and $\{\kappa_\lambda\}_{\lambda\in I\!R_+} \in C(I\!R_+,\mathcal{L}_\sigma(E))$ (σ indicates the strong operator topology in $\mathcal{L}(E)$). Let further $\tilde{E}$, $\{\tilde{\kappa}_\lambda\}_{\lambda\in I\!R_+}$ be another pair of data of that sort. Then we can talk about the space

$$S^\nu(U \times I\!R^q; E, \tilde{E}) \tag{3.1}$$

of operator–valued amplitude functions ("symbols") with respect to an open $U \subseteq I\!R^p$. The space (3.1) is defined as the set of all $a(y,\eta) \in C^\infty(U \times I\!R^q, \mathcal{L}(E; \tilde{E}))$ satisfying

$$\| \tilde{\kappa}_{[\eta]}^{-1}\{D_y^\alpha D_\eta^\beta a(y,\eta)\}\kappa_{[\eta]} \|_{\mathcal{L}(E,\tilde{E})} \leq c[\eta]^{\nu-|\beta|} \tag{3.2}$$

for all $\alpha \in \mathbb{N}^p$, $\beta \in \mathbb{N}^q$, $y \in K$ for every compact $K \subset\subset U$, $\eta \in \mathbb{R}^q$, with constants $c = c(\alpha, \beta, K) > 0$.

An example is $E = \tilde{E} = L^2(\mathbb{R}_+)$, $(\kappa_\lambda u)(t) = \lambda^{\frac{1}{2}} u(\lambda t)$, $\lambda > 0$.

If $a_{(\nu)}(y, \eta) \in C^\infty(U \times (\mathbb{R}^q \setminus \{0\}), \mathcal{L}(E, \tilde{E}))$ satisfies

$$a_{(\nu)}(y, \lambda\eta) = \lambda^\nu \tilde{\kappa}_\lambda a_{(\nu)}(y, \eta) \kappa_\lambda^{-1} \tag{3.3}$$

for all $\lambda > 0$, $(y, \eta) \in U \times (\mathbb{R}^q \setminus \{0\})$, and if $\chi(\eta)$ is any η–excision function, then $\chi(\eta) a_{(\nu)}(y, \eta) \in (3.1)$. Define the subspace

$$S_{cl}^\nu(U \times \mathbb{R}^q; E, \tilde{E}) \tag{3.4}$$

of classical symbols $a(y, \eta)$ by the condition $a(y, \eta) \sim \sum_{j=0}^\infty \chi(\eta) a_{(\nu-j)}(y, \eta)$, where $a_{(\nu-j)}$ is homogeneous in η of order $\nu - j$ in the sense of a relation (3.3), with $\eta - j$ instead of ν.

These notions allow a straightforward extension to Fréchet spaces $\tilde{E}$, written as projective limits $\tilde{E} = \lim_{\longleftarrow} \tilde{E}^{(j)}$ of Banach spaces $\tilde{E}^{(j)}$, with $\{\tilde{\kappa}_\lambda\}_{\lambda \in \mathbb{R}_+} \in C(\mathbb{R}_+, \mathcal{L}_\sigma(\tilde{E}^{(j)}))$ for all $j \in \mathbb{N}$. We then define

$$S^\nu(U \times \mathbb{R}^q; E, \tilde{E}) = \lim_{\longleftarrow} S^\nu(U \times \mathbb{R}^q; E, \tilde{E}^{(j)}) \tag{3.5}$$

and similary (3.4).

Analogously to the standard scalar symbol spaces (i.e. when $E = E = \mathbb{C}$, $\kappa_\lambda = \tilde{\kappa}_\lambda = identity$ for all λ) the spaces (3.1), (3.4) are Fréchet, with systems of seminorms following immediately from the definitions, both for Banach and Fréchet spaces $\tilde{E}$. The groups $\{\kappa_\lambda\}$, $\{\tilde{\kappa}_\lambda\}$ are fixed in every concrete case. Clearly (3.1), (3.4) depend on $\{\kappa_\lambda\}$, $\{\tilde{\kappa}_\lambda\}$, but for abbreviation that was not indicated in the notations.

Let $\mathcal{S}(\mathbb{R})$ be the space of Schwartz functions on $\mathbb{R}$ and set $\mathcal{S}(\bar{\mathbb{R}}_+) = \mathcal{S}(\mathbb{R}) \mid_{\mathbb{R}_+}$, equipped with the standard (nuclear) Fréchet topology. Then $\mathcal{S}(\bar{\mathbb{R}}_+)$ may be written as projective limit of (Hilbert–) spaces $E^{(j)} \subset L^2(\mathbb{R}_+)$ with the induced actions of $\{\kappa_\lambda\}_{\lambda \in \mathbb{R}_+}$ on $E^{(j)}$ from $L^2(\mathbb{R}_+)$. Then we obtain the spaces

$$S_{cl}^\nu(\Omega^2 \times \mathbb{R}^q; L^2(\mathbb{R}_+), \mathcal{S}(\bar{\mathbb{R}}_+)) \tag{3.6}$$

for open $\Omega \subseteq \mathbb{R}^q$, $\Omega^2 := \Omega \times \Omega$, $\nu \in \mathbb{R}$. The space $L^2(\mathbb{R}_+)$ is endowed with the scalar product

$$(u, v) = \int_0^\infty u(t) \overline{v(t)} dt \tag{3.7}$$

Points in Ω^2 will also be denoted by (y, y'). If we define $a^*(y, y', \eta)$ for an $a(y, y', \eta) \in (3.6)$ by the point–wise adjoint with respect to (3.7), we can talk about the conditions

$$a(y, y', \eta), a^*(y, y', \eta) \in S_{cl}^\nu(\Omega^2 \times \mathbb{R}^q; L^2(\mathbb{R}_+), \mathcal{S}(\bar{\mathbb{R}}_+)). \tag{3.8}$$

A consequence of results of one of the forthcoming parts of [S4, Section 2.2.7.] will be the following

THEOREM: **3.1** *An operator G is a ("singular") Green operator of class and order zero in Boutet de Monvel's algebra over $\mathbb{R}_+ \times \Omega$, $\Omega \subseteq \mathbb{R}^q$ open, if and only if $G = Op(a)$ for some operator-valued amplitude function $a(y, y', \eta)$ satisfying (3.8) for $\nu = 0$. Here $Op(a)u(y) = \mathcal{F}_{\eta \to y}^{-1} a(y, y', \eta) \mathcal{F}_{y' \to \eta} u(y')$, with $u(y)$ being regarded as function on Ω (of compact support) with values in the appropriate functions over $\mathbb{R}_+ \ni t$ ($Op(a)$ is uniquely determined by the actions over $C_0^\infty(\Omega, L^2(\mathbb{R}_+)))$.*

REMARK: **3.2** The Green operators in Boutet de Monvel's algebra of arbitrary class d and order ν can similarly be characterized, namely as finite linear combinations

$$G = \sum_{j=1}^{N} Op(a_j) D_j$$

where D_j is a differential operator (in t or likewise in (t, y) of order $d_j \leq d$ and $a_j(y, y', \eta)$, $a_j^*(y, y', \eta) \in S^{\nu - d_j}(\Omega^2 \times \mathbb{R}^q; L^2(\mathbb{R}_+), \mathcal{S}(\bar{\mathbb{R}}_+))$ for all j. Analogous characterizations hold for the trace and potential operators in Boutet de Monvel's algebra (cf. also [S4, VIII]). In the latter case $\tilde{E}$ (or E) equals some $\mathbb{C}^N$ and $\tilde{\kappa}_\lambda$ (or κ_λ) acts as the identity on $\mathbb{C}^N$ for all $\lambda > 0$.

Let us discuss for convenience for a moment the case of order zero, again. Let $p(t, y, \tau, \eta)$ be in the usual scalar Hörmander symbol class $S_{cl}^0(\bar{\mathbb{R}}_+ \times \Omega \times \mathbb{R}_{\tau, \eta}^{q+1})$, here over $\bar{\mathbb{R}}_+ \times \Omega$, defined as the restriction to $\bar{\mathbb{R}}_+ \times \Omega$ of the corresponding space over $\mathbb{R} \times \Omega \ni (t, y)$. Analogously we can also look at $p(t, y, y', \tau, \eta) \in S_{cl}^0(\bar{\mathbb{R}}_+ \times \Omega^2 \times \mathbb{R}^{q+1})$. Assume for simplicity that p is independent of t for $t \geq 1$. Consider the operators

$$op(p)(y, \eta) \quad : \quad L^2(\mathbb{R}) \to L^2(\mathbb{R}),$$
$$op(p)(y, \eta)u(t) \quad = \quad (2\pi)^{-1} \int \int e^{i(t-t')\tau} p(t, y, \tau, \eta) u(t') dt' d\tau,$$

which depend on (y, η) as parameters, and

$$e^+ : L^2(\mathbb{R}_+) \to L^2(\mathbb{R}), \quad r^+ : L^2(\mathbb{R}) \to L^2(\mathbb{R}_+),$$

where e^+ extends functions by zero to the negative half–axis, r^+ restricts to $\mathbb{R}_+$. Then, we get the operator family

$$op_\psi(p)(y, \eta) := r^+ op(p)(y, \eta) e^+ \quad : \quad L^2(\mathbb{R}_+) \to L^2(\mathbb{R}_+). \tag{3.9}$$

In an analogous manner we can form $op_\psi(p)(y, y', \eta)$ when p depends on y, y'. It is not hard to show that

$$op_\psi(p)(y, \eta) \in S^0(\Omega \times \mathbb{R}^q; L^2(\mathbb{R}_+), L^2(\mathbb{R}_+)) \tag{3.10}$$

where $\{\kappa_\lambda\}_{\lambda \in \mathbb{R}_+}$ acts on $L^2(\mathbb{R}_+)$ as mentioned. In addition, if p is independent of t, then $op_\psi(p)(y, \eta)$ belongs to the corresponding classical symbol subspace (cf. [S6, Chapter 3]). Analogous results hold in the (y, y')–dependent case.

Now a question of general importance in the theory of pseudo–differential boundary value problems is to look at the subalgebra of $S^0(\Omega \times \mathbb{R}^q; L^2(\mathbb{R}_+), L^2(\mathbb{R}_+))$, generated by the symbols of the form (3.10), when p runs over $S^0_{cl}(\bar{\mathbb{R}}_+ \times \Omega \times \mathbb{R}^{q+1})$ (independent of t for $t \geq 1$) or over a subclass of such p. This problem was solved in full generality in [S4], cf. in particular parts $VII, VIII$. A crucial point was to characterize the remainders under compositions, i.e. the operator families

$$op_\psi(p_1)(y, \eta) op_\psi(p_2)(y, \eta) - op_\psi(p_1 \#_t p_2)(y, \eta), \tag{3.11}$$

for fixed y, η, where $\#_t$ denotes the Leibniz product with respect to the t-variable. Partial results have been obtained by Eskin in [E1] (namely a description in terms of Mellin operators of highest conormal order, modulo a Hilbert Schmidt operator). If p_1, p_2 have the transmission property (3.11) is just a Green symbol of order and class zero, cf. the above Theorem 3.1 .

If we want to find a "minimal" algebra containing also inverses, multiplied by an excision function $\chi(\eta)$, under the condition that, say, $op_\psi(p)(y, \eta)$ is invertible for $| \, \eta \, | > const$, then we need (apart from a certain kind of smoothing Mellin operators which we do not discuss here) analogues of the mentioned Green symbols of Theorem 3.1, where roughly speaking the space $\mathcal{S}(\bar{\mathbb{R}}_+)$, in (3.6) is to be replaced by $\mathcal{S}_P(\mathbb{R}_+)$. Here $P = P(y)$ indicates a y–dependent discrete asymptotic type in the sense that $\omega(t) \mathcal{S}_{P(y)}(\mathbb{R}_+) \subset C^\infty(\Omega, \mathcal{K}^{\infty,0}(\mathbb{R}_+))$ is a subspace of variable discrete asymptotics (cf. Definition 2.4), $\omega(t)$ being an arbitrary cut–off function, further $(1 - \omega(t)) \mathcal{S}_{P(y)}(\mathbb{R}_+) \subset C^\infty(\Omega, \mathcal{S}(\bar{\mathbb{R}}_+))$. For y-independent asymptotic types (2.13) the corresponding operator-valued symbols have been described in [S4, VII]. The associated operators are also responsible for the regularity of solutions of singular boundary value problems with interior symbols from (1.4) when the solutions of (1.5) are independent of y (i.e. when the d_{j0} are constants). The precise formulation for varying discrete asymptotic types is more difficult. We shall employ a generalization of ideas from Section 2, but avoid here a direct use of spaces like $\mathcal{S}_{P(y)}(\mathbb{R}_+)$.

Let us first introduce the space

$$\underline{\mathcal{S}}^\rho(\mathbb{R}_+) = \{t^\rho \omega(t) u(t) + (1 - \omega(t)) v(t) : u(t) \in M^{-1}_{z \to t} \mathcal{S}(\Gamma_{\frac{1}{2}}), \ v(t) \in \mathcal{S}(\bar{\mathbb{R}}_+)\} \tag{3.12}$$

for fixed $\rho \in I\!\!R$ and any cut–off function ω. This space is independent of the concrete choice of ω and nuclear, Fréchet. It can be written as a projective limit of Hilbert spaces.

For arbitrary $\gamma, \mu, \nu \in I\!\!R$ we can form the space of all

$$g(y, y', \eta) \in \bigcap_{s \in I\!\!R} S_{cl}^{\nu}(\Omega^2 \times I\!\!R^q; \, \mathcal{K}^{s,\gamma}, \underline{\mathcal{S}}^{\gamma-\mu}), \tag{3.13}$$

with

$$g^*(y, y', \eta) \in \bigcap_{s \in I\!\!R} S_{cl}^{\nu}(\Omega^2 \times I\!\!R^q; \mathcal{K}^{s,-\gamma+\mu}, \underline{\mathcal{S}}^{-\gamma}). \tag{3.14}$$

Here and in the sequel for abbreviation we shall often write $\mathcal{K}^{s,\gamma}$ instead of $\mathcal{K}^{s,\gamma}(I\!\!R_+)$, and so on. * means in (3.14) the point–wise formal adjoint with respect to an extension of $(\cdot, \cdot)_{L^2(I\!\!R_+)} : C_0^{\infty}(I\!\!R_+) \times C_0^{\infty}(I\!\!R_+) \to \mathbb{C}$ to a non–degenerate sesqui–linear pairing

$$(\cdot, \cdot) : \mathcal{K}^{s,\gamma}(I\!\!R_+) \times \mathcal{K}^{-s,-\gamma}(I\!\!R_+) \to \mathbb{C}$$

for all $s, \gamma \in I\!\!R$. Next we shall generalize the notation (2.7) to the vector–valued case. In other words

$$C^{\infty}(\Omega, \mathcal{A}'(K, E))^{\bullet}$$

for some Fréchet space E will be defined in an analogous manner, where the only difference is that analytic functionals are replaced by E–valued ones,
$\mathcal{A}'(K, E) \cong \mathcal{A}'(K) \bigotimes_{\pi} E$ ($\bigotimes_{\pi}$ is the (completed) projective tensor product). We shall also introduce the space $\mathcal{F}(\Omega, (\gamma, \Delta); E)^{\bullet}$ of families $f(y, z)$ of E–valued meromorphic functions by a straightforward generalization of the conditions for $\mathcal{F}(\Omega, (\gamma, \Delta))^{\bullet}$ from the previous section. Then we get an immediate analogue of Theorem 2.5 .

DEFINITION: **3.3** *Denote by*

$$R_G^{\nu}(\Omega^2 \times I\!\!R^q, (\gamma, \gamma - \mu, \Delta))^{\bullet} \tag{3.15}$$

for $\nu, \mu, \gamma \in I\!\!R$, $\Delta = (\delta, 0]$, $-\infty < \delta < 0$, the subspace of all $g(y, y', \eta) \in (3.13)$ with (3.14) such that for some $f(y, z) \in \mathcal{F}(\Omega, (\gamma - \mu, \Delta); E)^{\bullet}$ for the Fréchet space

$$E = \{e(y, y', \eta) \in \bigcap_{s \in I\!\!R} S_{cl}^{\nu+\frac{1}{2}}(\Omega^2 \times I\!\!R_{\eta}^q; \mathcal{K}^{s,\gamma}, \mathbb{C}) : e^*(y, y', \eta) \in S_{cl}^{\nu+\frac{1}{2}}(\Omega^2 \times I\!\!R_{\eta}^q; \mathbb{C}, \underline{\mathcal{S}}^{-\gamma})\}$$

$$\tag{3.16}$$

the associated map

$$\zeta : (\delta, 0) \times \mathcal{U}(\Omega) \to C^{\infty}(\Omega, \mathcal{A}'(K, E))^{\bullet} \tag{3.17}$$

(obtained by the E–valued analogue of Theorem 2.5) satisfies the following relations. If we set

$$h(\zeta_b)(y, y', \eta)u(t) = \omega(t[\eta])\langle \zeta_b(y; y, y', \eta)u, (t[\eta])^{-w}\rangle \tag{3.18}$$

(where the first y-variable in ζ_b comes from $\Omega \ni y$ in (3.17), and $;y, y', \eta$ from (3.16)) *then we get for all $b = (\beta, U) \in (\delta, 0) \times \mathcal{U}(\Omega)$*

$$(g - h(\zeta_b))(y, y', \eta) \in \bigcap_{s \in \mathbb{R}} S^{\nu}_{cl}(U^2 \times \mathbb{R}^q; \mathcal{K}^{s, \gamma}, S^{\gamma - \mu}_{(\beta, 0]}). \qquad (3.19)$$

Here, for any $\rho \in \mathbb{R}$,

$$S^{\rho}_{(\beta, 0]}(\mathbb{R}_+) = \{\omega u + (1 - \omega)v : \ u \in \bigcap_{\varepsilon > 0} \mathcal{K}^{\infty, \rho - \beta - \varepsilon}(\mathbb{R}_+), \ v \in S(\bar{\mathbb{R}}_+)\}$$

for a cut-off function $\omega(t)$ (the space is independent of the concrete ω).

$\Delta = (\delta, 0]$ plays the role of a weight interval, γ of a weight, and μ is a weight shift relative to γ. In the abstract definition we could write at once $\gamma - \mu = \rho$ with some real ρ. The elements of Definition 3.3 are motivated by a machinery that will be published in [S4, Section 2.4.]. By posing $y' = y$ we get a definition of the analogue of (3.15) for Ω instead of Ω^2. In this case the definition is actually a result which gives a description of elements of $S^0(\Omega \times \mathbb{R}^q; L^2(\mathbb{R}_+), L^2(\mathbb{R}_+))$ belonging to the algebra generated by $op_\psi(p)(y, \eta)$ ($p(t, y, \tau, \eta)$ running over a sufficiently rich space of interior symbols), together with inverses. An analogous result holds for arbitrary orders. Here, in the zero order case, we have, of course, $\nu = 0$, $\gamma = \mu = 0$. In particular, it follows

REMARK: 3.4 The Green boundary symbols $a(y, y', \eta)$ of Theorem 3.1 belong to $R^0_G(\Omega^2 \times \mathbb{R}^q, (0, 0, \Delta))^\bullet$ for arbitrary $\Delta = (\delta, 0]$, $\delta < 0$.

For illustrating the role of Green boundary symbols as an ideal in the algebra containing the operator families $op_\psi(p)(y, y', \eta)$ we want to announce from the corresponding chapter of [S4] the following

THEOREM: 3.5 *Let $p(t, y, y', \tau, \eta) \in S^0_{cl}(\bar{\mathbb{R}}_+ \times \Omega^2 \times \mathbb{R}^{q+1})$ be independent of t for $t \geq$ const. Then $g(y, y', \eta) \in R^0_G(\Omega^2 \times \mathbb{R}^q, (0, 0, \Delta))^\bullet$ implies*

$$op_\psi(p)(y, y', \eta)g(y, y', \eta), g(y, y', \eta)op_\psi(p)(y, y', \eta) \in R^0_G(\Omega^2 \times \mathbb{R}^q, (0, 0, \Delta))^\bullet$$

for arbitrary $\Delta = (\delta, 0]$, $-\infty < \delta < 0$. Analogous relations hold for Ω instead of Ω^2.

From [S6, Chapter 3] it is known that the $\psi DO's$ with operator-valued symbols of the class (3.1) for $U = \Omega^2$ fit to the "abstract" wedge Sobolev spaces

$W^s_{comp}(\Omega, E)$ and $W^s_{loc}(\Omega, E)$, respectively. Remember that for $\Omega = I\!R^q$ we have the scale $\{W^s(I\!R^q, E)\}_{s\in I\!R}$, where $W^s(I\!R^q, E)$ is the completion of $S(I\!R^q, E)$ with respect to the norm

$$\| u \|_{W^s(I\!R^q, E)} = \{ \int [\eta]^{2s} \| \kappa^{-1}_{[\eta]}(F_{y\to\eta}u)(\eta) \|^2_E \, d\eta \}^{\frac{1}{2}}. \tag{3.20}$$

The concrete choice of $\{\kappa_\lambda\}_{\lambda\in I\!R_+}$ was not explicitly indicated in the notation, since it is fixed once and for all. An exception is the case $\kappa_\lambda = identity$ for all $\lambda > 0$. We then write $H^s(I\!R^q, E)$.

An obvious modification of definitions from the scalar case gives us the comp, loc-versions of the spaces $W^s(...)$ over $\Omega \subseteq I\!R^q$ (cf. also [S4, 2.2.1. Definition 27]). Note that

$$W^\infty_{loc}(\Omega, E) = H^\infty_{loc}(\Omega, E) = C^\infty(\Omega, E). \tag{3.21}$$

Now

$$Op(a) : W^s_{comp}(\Omega, E) \to W^{s-\nu}_{loc}(\Omega, \tilde{E}) \tag{3.22}$$

is continuous for all $s \in I\!R$, for every $a(y, y', \eta) \in (3.1)$, $U = \Omega^2$ (cf.[S6], [H1]).

As noted above, for establishing the regularity of a solution u of the equation $Au = f$ with f being of described regularity and u of finite order (in the distributional sense) we compose from the left by a parametrix P. It follows $PAu = (1+G)u = Pf$, where G is smoothing within the operator algebra. That scheme in the wedge pseudo–differential algebra, in particular, in singular boundary value problems, gives rise to Green operators G of order $-\infty$, i.e. $\psi DO's$ along Ω with symbols of the class (3.15) for $\nu = -\infty$, $\mu = 0$, with appropriate γ. The conclusion on the regularity of u, here with asymptotics, does employ the following

THEOREM: **3.6** *Let $G = Op(g)$ for $g(y, y', \eta) \in R^\infty_G(\Omega^2 \times I\!R^q, (\gamma, \gamma, \Delta))^\bullet$, $\Delta = (\delta, 0]$. Then G induces continuous operators*

$$G : W^s_{comp}(\Omega, \mathcal{K}^{s,\gamma}(I\!R_+)) \to C^\infty(\Omega, \mathcal{K}^{\infty,\gamma}(I\!R_+)) \tag{3.23}$$

for all $s \in I\!R$, where the image consists of elements with variable discrete asymptotics in the sense of Definition 2.4 .

PROOF: The space (3.15) for $\nu = -\infty$, $\mu = 0$ is contained in $S^{-\infty}(\Omega^2 \times I\!R^q; \mathcal{K}^{s,\gamma}, \underline{S}^\gamma)$ for all $s \in I\!R$. In virtue of (3.22) it follows in the present case

$$Op(g) : W^s_{comp}(\Omega, \mathcal{K}^{s,\gamma}(I\!R_+)) \to W^\infty_{loc}(\Omega, \underline{S}^\gamma(I\!R_+)) \tag{3.24}$$

for all $s \in I\!R$. From (3.21) we obtain the continuity (3.23). Let $U \in \mathcal{U}(\Omega)$ be fixed and $(y, y') \in U^2$. Then, by definition, $g(y, y', \eta)$ can be written as a sum

$$g(y, y', \eta) = g_0(y, y', \eta) + g_1(y, y', \eta)$$

for every given $\beta \in (\delta, 0)$, $b = (\beta, U)$, namely

$$g_0(y, y', \eta) \in S^{-\infty}(U^2 \times \mathbb{R}^q; \mathcal{K}^{s,\gamma}, S^\gamma_{(\beta,0]}),$$
$$g_1(y, y', \eta) = h(\zeta_b)(y, y', \eta)$$

with $\zeta_b(y; y, y', \eta) \in C^\infty(\Omega, \mathcal{A}'(K, E^{-\infty}))^\bullet$, $E^{-\infty}$ being given by (3.16), here for $\nu = -\infty$. The mapping properties of $Op(g_0)$ follow similarly to (3.24) with the only difference that $\underline{S}^\gamma(\mathbb{R}_+)$ is to be replaced by $S^\gamma_{(\beta,0]}(\mathbb{R}_+)$. But that means

$$Op(g_0) : W^s_{comp}(U, \mathcal{K}^{s,\gamma}(\mathbb{R}_+)) \to C^\infty(U, \mathcal{K}^{\infty,\gamma-\beta-\varepsilon}(\mathbb{R}_+))$$

for arbitrary $\varepsilon > 0$. Since we could start as well with $\tilde{\beta} = \beta - \varepsilon$, $\varepsilon > 0$ being so small that $\beta - \varepsilon > \delta$, we obtain after modifying g_0 to a corresponding $\tilde{g}_0$ a map to $C^\infty(U, \mathcal{K}^{\infty,\gamma-\beta}(\mathbb{R}_+))$. Then we just obtain the flat remainder (of flatness $-\beta$ relative to the weight γ) in the sense of (2.11), Definition 2.4 . The final step of the proof is to recognize that for every $b = (\beta, U) \in (\delta, 0) \times \mathcal{U}(\Omega)$ there is a map

$$e_b : C^\infty(\Omega, \mathcal{A}'(K, E^{-\infty}))^\bullet \to C^\infty(\Omega, \mathcal{A}'(K))^\bullet \tag{3.25}$$

(where K is the compact set from Definition 3.3) such that for $e_b(\zeta_b)(y) =: \alpha_b(y)$

$$(Op(g_1)u)(t, y) - \langle \alpha_b(y), t^{-w} \rangle \omega(t) \in C^\infty(U, \mathcal{K}^{\infty,\gamma-\beta}(\mathbb{R}_+)).$$

The latter consideration is in a sense straightforward, though lenghty. Therefore some arguments will be sketched here. First the y'–dependence of $g_1(y, y', \eta)$ can be removed by applying a representation of C^∞ functions of y' in terms of elements of a projective tensor product as convergent sums of products. The convergence refers to the corresponding analogues of Fréchet topologies from Proposition 2.2 . This reduces the discussion to factors consisting of C^∞ functions in y' and y'–independent symbols. The elements in $W^s_{comp}(\Omega_{y'}, E)$ allow multiplications by $\psi(y') \in C^\infty(\Omega)$ within that distribution class. In other words the main point is the discussion of y'–independent symbols.

A further observation is that for orders $\nu = -\infty$ it is allowed to replace $\omega(t[\eta])$ in (3.18) by $\omega(t)$, modulo a flat remainder in the sense of the relations (3.19). This reduces (3.18) to

$$\omega(t)\langle \zeta_b(y; y, \eta)u, (t[\eta])^{-w} \rangle. \tag{3.26}$$

The factor $[\eta]^{-w}$ may be drawn to the analytic functional, since it is allowed to multiply families of analytic functionals by holomorphic functions without destroying the character of the variable discrete pattern of carriers. Because of $\nu = -\infty$ it follows

$$[\eta]^{-w}\zeta_b(y; y, \eta) \in C^\infty(\Omega, \mathcal{A}'(K, E_0))^\bullet$$

with $E_0 = C^\infty(\Omega, \mathcal{S}(\mathbb{R}^q_\eta, \mathcal{L}(\mathcal{K}^{s,\gamma}(\mathbb{R}_+), \mathbb{C})))$. Therefore it is clear that $u \in \mathcal{W}^s_{comp}(\Omega, \mathcal{K}^{s,\gamma}(\mathbb{R}_+))$ (where comp only refers to y–variables) implies

$$\mathcal{F}^{-1}_{\eta \to y}\{[\eta]^{-w}\zeta_b(y; y, \eta)u\} =: \alpha_b(u) \in C^\infty(\Omega, \mathcal{A}'(K))^\bullet.$$

In this way we have constructed (3.25). The above convergence argument in connection with projective tensor products does require in the y'–independent case to verify that (3.25) is continuous. But this is obvious by the construction. $\square$

References

[B1] L.Boutet de Monvel
Boundary problems for pseudo–differential operators.
Acta Math. $\underline{126}$ (1971), 11–51

[E1] G.I. Eskin
Boundary problems for elliptic pseudo–differential equations.
"Nauka", Moskva 1973 (Transl. of Math. Monographs $\underline{52}$,
Amer. Math. Soc. Providence, Rhode Island 1980)

[H1] T. Hirschmann
Functional analysis in cone and edge Sobolev spaces.
Ann. Global Anal. and Geom. $\underline{8}$, 2 (1990), 167–192

[J1] P. Jeanquartier
Transformation de Mellin et développements asymptotiques.
L'Enseignement mathématique $\underline{25}$, 3–4 (1979), 285–308

[K1] V.A. Kondrat'ev
Boundary value problems for elliptic equations in domains with conical points.
Trudy Mosk. Mat. Obšč. $\underline{16}$ (1967), 209–292

[M1] M.F. Morozov
Mathematical questions of the crack theory.
Moscow, "Nauka" 1984

[R1] S. Rempel, B.-W. Schulze
Index theory of elliptic boundary problems.
Akademie–Verlag Berlin 1982; North Oxford Acad. Publ. Comp. Oxford 1985
(Russ. transl. "Mir" Moscow 1986)

[R2] S. Rempel, B.-W. Schulze
Asymptotics for elliptic mixed boundary problems (Pseudo–differential and Mellin
operators in spaces whith conormal singularity).
Math. Research Series 50, Berlin 1989

[S1] B.-W. Schulze
Regularity with continuous and branching asymptotics for elliptic operators on
manifolds with edges.
Integral Equ. and Operator Theory 11 (1988), 557–602

[S2] B.-W. Schulze
Corner Mellin operators and reductions of order with parameters.
Ann. Sc. Norm. Sup. Pisa 16, 1 (1989),1–81

[S3] B.-W. Schulze
The Mellin pseudo–differential calculus on manifolds with corners. Proc. Sympos.
"Analysis in Domains and on Manifolds with Singularities"
Breitenbrunn 1990, Teubner Texte zur Mathematik 131, Leipzig 1992

[S4] B.-W. Schulze
Pseudo–differential operators and asymptotics on manifolds with corners.
Reports of the Karl–Weierstrass–Institute for Mathematics Berlin I, 07 (1989);
II, 02 (1990); III, 04 (1990), IV, 06 (1990); VI, 08 (1990); VII, 01 (1991); VIII,
02 (1991); IX, 03 (1991); XII, XIII, SFB 256–Preprints 214,220, Bonn 1992.

[S5] B.-W. Schulze
Elliptic regularity with continuous and branching edge asymptotics.
Int. Series of Numer. Math. Vol. 94, Birkhäuser Verlag, Basel 1990, 275–284

[S6] B.-W. Schulze
Pseudo–differential operators on manifolds with singularities.
North Holland, Amsterdam 1991

[S7] B.-W. Schulze
Crack problems in the edge pseudo–differential calculus.
Applicable Analysis (to appear 1992)

[V1] M.I. Višik, G.I. Eskin
Convolution equations in bounded domains in spaces with weighted norms.
Math. Sb. 69, 1 (1966), 65–110

Operator Theory:
Advances and Applications, Vol. 57
© 1992 Birkhäuser Verlag Basel

Schrödinger Operators with Arbitrary Nonnegative Potentials

Karl-Theodor Sturm

Abstract. We give a brief survey on the symmetric form approach to Schrödinger operators $-\frac{1}{2}\Delta + V$ with arbitrary potentials $V \geq 0$ and on the canonical generalizations to more general "nonnegative perturbations" of the free Hamiltonian. This leads to the investigation of generalized Schrödinger operators $H^\mu = -\frac{1}{2}\Delta + \mu$ with measures μ charging no polar sets. One of the main features will be that the associated symmetric forms will, in general, be not densely defined on $L^2(\mathbb{R}^d)$.

Introduction

In order to investigate Schrödinger operators $-\frac{1}{2}\Delta + V$ with singular potentials $V \geq 0$, the most reasonable (analytic) quantity to start with is the symmetric form

$$Q^V(f,g) \;:=\; \frac{1}{2}\int \nabla f \nabla g\, dm + \int fg\, V\, dm.$$

In the definition of this form, one can replace the measures $V \cdot m$ by general measures μ on $\mathbb{R}^d$. The appropriate condition on the measures μ is that they do not charge polar sets. Starting with the symmetric forms Q^μ associated with such measures μ, we obtain generalized Schrödinger operators $H^\mu = -\frac{1}{2}\Delta + \mu$ and generalized Schrödinger semigroups $(e^{-tH^\mu})_{t>0}$. The reason to consider not only usual Schrödinger operators $-\frac{1}{2}\Delta + V$ with potentials $V \geq 0$ but generalized Schrödinger operators $-\frac{1}{2}\Delta + \mu$ with measures μ charging no polar sets is that

- the class of these generalized Schrödinger operators turns out to be closed with respect to strong resolvent convergence

- this class contains all the Dirichlet Laplacians on open subsets of $\mathbb{R}^d$

- in this class, the set of usual Schrödinger operators $-\frac{1}{2}\Delta + V$ with smooth potentials $V \in C_0^\infty(\mathbb{R}^d)$ is dense with respect to strong resolvent convergence.

Note that, in general, the forms $(Q^\mu, \mathcal{D}(Q^\mu))$ are not densely defined on $L^2(\mathbb{R}^d, m)$. One of the main goals will be to derive a characterization of the form domains $\mathcal{D}(Q^\mu)$, in particular, criteria for $(Q^\mu, \mathcal{D}(Q^\mu))$ being densely defined. Let us mention some results in terms of the set $E^\mu \subset \mathbb{R}^d$ of permanent points for μ which plays a crucial rôle in various places:

- $\mathcal{D}(Q^\mu)$ is dense in $W^{1,2}(\mathbb{R}^d)$ if and only if $\text{cap}(\mathbb{R}^d \setminus E^\mu) = 0$

- $\mathcal{D}(\mathcal{Q}^\mu)$ is dense in $L^2(I\!\!R^d, m)$ if and only if $m(I\!\!R^d \setminus E^\mu) = 0$.

For the convenience of the reader, we add, at the end of the paper, a brief appendix on potential theoretic notions (like "regular", "fine", "quasi-", "polar", "capacity") which we use in the text without explicit definition. These notions are always understood with respect to the Laplace operator (or, equivalently, the Brownian motion) on $I\!\!R^d$.

Throughout this paper, m denotes the Lebesgue measure on $I\!\!R^d$. If not specified otherwise, *measurable* means measurable with respect to m. The *Borel* σ-field in $I\!\!R^d$ is denoted by $\mathcal{B}(I\!\!R^d)$. We recall that a measure μ on $(I\!\!R^d, \mathcal{B}(I\!\!R^d))$ is a *Radon measure* iff $\mu(K) < \infty$ for all compact sets $K \subset I\!\!R^d$.

1 Symmetric Forms with Measures Charging No Polar Sets.

If one wants to study "positive perturbations" of the free energy $\mathcal{Q}^0$ which are more general than the above forms $\mathcal{Q}^V$ with functions $V \geq 0$ one is lead in a natural way to consider forms of the type

$$\mathcal{Q}^\mu(f,g) \; := \; \frac{1}{2} \int \nabla f \nabla g \, dm + \int fg \, d\mu$$

where μ is some measure on $I\!\!R^d$. If μ charges no sets of Lebesgue measure 0 (that is, if $\mu = V \cdot m$ with a measurable function $V \geq 0$), this leads to the previous case. In some sense the minimal assumption on μ should be that it charges no polar sets.

(1.1) Definition. We say that a *measure μ on $(I\!\!R^d, \mathcal{B}(I\!\!R^d))$ charges no polar sets* if and only if
$$\mu(F) = 0 \qquad \text{for every polar set } F \in \mathcal{B}(I\!\!R^d).$$

The set of all measures on $(I\!\!R^d, \mathcal{B}(I\!\!R^d))$ which do not charge polar sets will be denoted by $\mathcal{M}_0$.

For instance, the d-dimensional Lebesgue measure m and the δ-dimensional Hausdorff measures for $\delta > d - 2$ are in $\mathcal{M}_0$ (cf. [3], [27]). For any measure $\mu \in \mathcal{M}_0$ and any Borel function $f \geq 0$ on $I\!\!R^d$ the measure $f \cdot \mu$ (having density f with respect to μ) is in $\mathcal{M}_0$, too. Special attention should be given to the fact that measures in $\mathcal{M}_0$ have neither to be regular nor to be σ-finite, in particular, they are not assumed to be Radon measures. A typical example is the measure

$$\overline{\infty}(.) := \infty \cdot \mathrm{cap}(.) : \quad F \mapsto \left\{ \begin{array}{ll} 0, & \text{if } F \text{ is polar,} \\ \infty, & \text{else.} \end{array} \right.$$

By this, $\mathcal{M}_0$ can be characterized as the set of all measures on $(I\!\!R^d, \mathcal{B}(I\!\!R^d))$ which are absolutely continuous with respect to $\overline{\infty}$.

We are now going to show that the condition $\mu \in \mathcal{M}_0$ *suffices* to define the form Q^μ on a reasonable domain $\mathcal{D}(Q^\mu)$. In order to see this, we make the following

(1.2) Remarks. a) Let $\overline{\mathcal{B}}(\mathbb{R}^d)$ denote the completion of $\mathcal{B}(\mathbb{R}^d)$ with respect to the measure $\overline{\infty}$, the so-called σ-field of *nearly Borel* sets. Then every measure $\mu \in \mathcal{M}_0$ can be extended in a trivial way to a measure on $(\mathbb{R}^d, \overline{\mathcal{B}}(\mathbb{R}^d))$. Note that every quasi-open or quasi-closed (in particular, every polar) set belongs to $\overline{\mathcal{B}}(\mathbb{R}^d)$. Similarly, all *quasi-continuous* functions on $\mathbb{R}^d$ are $\overline{\mathcal{B}}(\mathbb{R}^d)$-measurable, hence, they are μ-measurable for every $\mu \in \mathcal{M}_0$.

b) Every element f in the Sobolev space $W^{1,2}(\mathbb{R}^d)$ has a quasi-continuous version $\tilde{f}$. That is, there exists a quasi-continuous function $\tilde{f}$ which coincides m-a.e. with f. Such a function $\tilde{f}$ is q.e. uniquely determined (and can be defined arbitrarily on a polar set). Actually, one can choose $\tilde{f}$ to be the Lebesgue mean of f, i.e.

$$\tilde{f}(x) \;=\; \lim_{\epsilon \to 0} \frac{1}{m(B_\epsilon(x))} \int_{B_\epsilon(x)} f(y)\, dy$$

which converges for q.e. $x \in \mathbb{R}^d$ and coincides with $f(x)$ for m-a.e. $x \in \mathbb{R}^d$.

(1.3) Definition. For any measure $\mu \in \mathcal{M}_0$ we define the nonnegative symmetric form

$$Q^\mu(f,g) \;:=\; \frac{1}{2} \int \nabla f \nabla g \, dm + \int fg \, d\mu$$

$$\mathcal{D}(Q^\mu) \;:=\; \left\{ f \in W^{1,2}(\mathbb{R}^d) : \; \tilde{f} \in L^2(\mathbb{R}^d, \mu) \right\}.$$

According to the previous remarks, this form is always well-defined. The main observation ([31], Theorem 1.2) is

(1.4) Theorem.   *For every $\mu \in \mathcal{M}_0$ the form $(Q^\mu, \mathcal{D}(Q^\mu))$ is a closed form on $L^2(\mathbb{R}^d, m)$.*

(1.5) Remarks. a) In general, the form $(Q^\mu, \mathcal{D}(Q^\mu))$ with $\mu \in \mathcal{M}_0$ is not *densely defined* on $L^2(\mathbb{R}^d)$, in particular, it is not *regular*. Concerning the question of being densely defined, we note that, of course, the form is always densely defined on the closure of $\mathcal{D}(Q^\mu)$ in $L^2(\mathbb{R}^d, m)$, cf. section 2. Actually we will prove that this closure coincides with the set

$$L_0^2(E^\mu, m) \;:=\; \left\{ f \in L^2(\mathbb{R}^d, m) : \; f = 0 \; m\text{-a.e. on } \mathcal{C}E^\mu \right\}$$

where E^μ denotes the set of permanent points for μ, cf. section 6.

b) For a measure $\mu \in \mathcal{M}_0$ the form $(Q^\mu, \mathcal{D}(Q^\mu))$ is regular if and only if μ is a *Radon measure* ([4], Theorem 2.2.2). In this case, the set

$$C_0^\infty(\mathbb{R}^d) \text{ is a core for } (Q^\mu, \mathcal{D}(Q^\mu)).$$

In other words, the symmetric form $(Q^\mu, C_0^\infty(\mathbb{R}^d))$ is closable on $L^2(\mathbb{R}^d, m)$ with closure $(Q^\mu, \mathcal{D}(Q^\mu))$.

c)　Now let conversely μ be an arbitrary Radon measure on $\mathbb{R}^d$. In order that

$$(\mathcal{Q}^\mu, \mathcal{C}_0^\infty(\mathbb{R}^d)) \text{ is closable on } L^2(\mathbb{R}^d, m)$$

it is *necessary* (and, as already stated, also sufficient) that μ *does not charge polar sets*, i.e. that $\mu \in \mathcal{M}_0$ ([3]; [12]; [27] Theorem 12.4/1).

We close this section with an important

(1.6) Example. If $\mu = 1_{\mathcal{C}G} \cdot \overline{\infty}$ with a nearly Borel set $G \subset \mathbb{R}^d$, then

$$\mathcal{D}(\mathcal{Q}^\mu) \;=\; W_0^{1,2}(G) \;:=\; \left\{ f \in W^{1,2}(\mathbb{R}^d) : \; \tilde{f} = 0 \text{ q.e. on } \mathcal{C}G \right\}.$$

For more details concerning the Sobolev spaces $W_0^{1,2}(G)$ for not necessarily open sets $G \subset \mathbb{R}^d$ we refer to [19]. We restrict ourselves to the following

(1.7) Remarks. a)　If G is an open set, then this definition of the Sobolev space $W_0^{1,2}(G)$ coincides with the usual one, namely to be the closure of $\mathcal{C}_0^\infty(G)$ in $W^{1,2}(\mathbb{R}^d)$.

b)　For arbitrary nearly Borel sets $G \subset \mathbb{R}^d$, the Sobolev space $W_0^{1,2}(G)$ coincides with $W_0^{1,2}(G^{f-int})$ and with $W_0^{1,2}(\mathrm{reg}(G))$ where G^{f-int} is the fine interior of G and where $\mathrm{reg}(G)$ (the *regularization* of G) is the largest finely open set $G_0 \subset \mathbb{R}^d$ such that $G_0 \setminus G$ is polar, cf. Appendix. Note that in general $G \setminus \mathrm{reg}(G)$ need not to be polar. For instance, if $G = \overline{B_r(x)}$ then $\mathrm{reg}(G) = B_r(x)$.

c)　For quasi-open sets $G \subset \mathbb{R}^d$, the measure $\mu = 1_{\mathcal{C}G} \cdot \overline{\infty}$ can be used to produce complete absorption on the complement of G. In particular, it can be used to simulate homogeneous Dirichlet "boundary" conditions on $\mathcal{C}G$.

d)　One might be tempted to simulate Dirichlet "boundary" conditions on $\mathcal{C}G$ also by a potential V which is $\equiv \infty$ on $\mathcal{C}G$ and $\equiv 0$ in G, i.e. by $\mu = V \cdot m$ with $V = 1_{\mathcal{C}G} \cdot \infty$. This, however, leads to

$$\mathcal{D}(\mathcal{Q}^\mu) \;=\; \left\{ f \in W^{1,2}(\mathbb{R}^d) : \; f = 0 \text{ } m\text{-a.e. on } \mathcal{C}G \right\}$$

which in general is a proper superset of $W_0^{1,2}(G)$, even if G is assumed to be open. It coincides with $W_0^{1,2}(G)$ (and produces the right boundary condition) if and only if the measures $1_{\mathcal{C}G} \cdot \overline{\infty}$ and $1_{\mathcal{C}G} \cdot \infty \cdot m$ are *equivalent* (cf. section 3 and [24]).

2　Schrödinger Operators and Schrödinger Semi-groups

We recall that for arbitrary $\mu \in \mathcal{M}_0$ the closed form $(\mathcal{Q}^\mu, \mathcal{D}(\mathcal{Q}^\mu))$ is in general not densely defined on $L^2(\mathbb{R}^d)$. One goal will be to characterize the closure $\overline{\mathcal{D}}(\mathcal{Q}^\mu)$ of $\mathcal{D}(\mathcal{Q}^\mu)$ in $L^2(\mathbb{R}^d, m)$ and to state necessary and sufficient criteria for $\overline{\mathcal{D}}(\mathcal{Q}^\mu) = L^2(\mathbb{R}^d, m)$, that is, for $(\mathcal{Q}^\mu, \mathcal{D}(\mathcal{Q}^\mu))$ being densely defined on $L^2(\mathbb{R}^d, m)$. For instance, we will prove that this is the case if μ is a Radon measure on $\mathbb{R}^d$ (or more generally a Radon measure on an open set $G \subset \mathbb{R}^d$ with $m(\mathbb{R}^d \setminus G) = 0$). In general, $\overline{\mathcal{D}}(\mathcal{Q}^\mu)$ will turn out to coincide with the set

$$L_0^2(E^\mu, m) := \left\{ f \in L^2(\mathbb{R}^d, m) : \ f = 0 \ m\text{-a.e. on } \mathcal{C}E^\mu \right\}$$

where E^μ denotes the set of permanent points for μ.

At the moment we restrict ourselves with the fact that $(\mathcal{Q}^\mu, \mathcal{D}(\mathcal{Q}^\mu))$ is always a densely defined, closed form on the Hilbert space $\overline{\mathcal{D}}(\mathcal{Q}^\mu)$ (equipped with the inner product of $L^2(\mathbb{R}^d, m)$).

Since there is a one-to-one correspondence between closed symmetric forms and self-adjoint operators (cf. [26] and [28]) we obtain

(2.1) Theorem. *For every $\mu \in \mathcal{M}_0$ there exists a unique nonnegative selfadjoint operator $(H^\mu, \mathcal{D}(H^\mu))$ on $\overline{\mathcal{D}}(\mathcal{Q}^\mu)$ which corresponds to $(\mathcal{Q}^\mu, \mathcal{D}(\mathcal{Q}^\mu))$ in the sense that $\mathcal{D}(H^\mu) \subset \mathcal{D}(\mathcal{Q}^\mu)$ and*

$$(*) \qquad \int H^\mu f \cdot g \, d\mu \ = \ \mathcal{Q}^\mu(f, g) \qquad \forall \, f \in \mathcal{D}(H^\mu), \ g \in \mathcal{D}(\mathcal{Q}^\mu).$$

This operator H^μ is the *form sum* of the free Hamiltonian $H^0 = -\frac{1}{2}\Delta$ and of the operator of integration with respect to μ. It will also be denoted by $-\frac{1}{2}\Delta + \mu$ and is called the (generalized) *Schrödinger operator* associated with μ. In the case $\mu = V \cdot m$, it is simply denoted by H^V or $-\frac{1}{2}\Delta + V$. Note that for $f \in \mathcal{D}(H^V) \cap C^2(\mathbb{R}^d)$ one actually obtains from $(*)$ by means of Green's formula

$$\int H^V f \cdot g \, dm \ = \ \int \left(-\frac{1}{2}\Delta f + Vf \right) \cdot g \, dm \qquad \text{for all } g \in \mathcal{D}(\mathcal{Q}^V).$$

It turns out that for *every* measurable potential $V \geq 0$ the Schrödinger semigroup $(e^{-tH^V})_{t>0}$ on $L^2(\mathbb{R}^d, m)$ (defined analytically by means of the form sum $H^V = -\frac{1}{2}\Delta + V$) coincides with the extension to $L^2(\mathbb{R}^d, m)$ of the probabilistically defined *Feynman-Kac semigroup* $(P_t^V)_{t>0}$, cf. [2] and [34].

The alternative way to define $-\frac{1}{2}\Delta + V$ analytically is to define it as the *operator sum*. If the potential V is locally integrable, this leads to the same selfadjoint operator as defined above. In the general case, however, this alternative approach is not satisfactory: one obtains neither the existence of a selfadjoint (extension of this) operator nor the existence of a reasonable semigroup associated to it.

(2.2) Examples. a) If $\mu = V \cdot m$ with $V \in L^1_{loc}(\mathbb{R}^d, m)$ then H^μ is the usual Schrödinger operator $-\frac{1}{2}\Delta + V$. In other words, $(H^\mu, \mathcal{D}(H^\mu))$ is the Friedrichs extension of the operator $(-\frac{1}{2}\Delta + V, \ C_0^\infty(\mathbb{R}^d))$.

b) If $\mu = 1_{\mathcal{C}G} \cdot \infty$ with an open set $G \subset \mathbb{R}^d$ then H^μ is $(-\frac{1}{2}$ times$)$ the Dirichlet Laplacian on G. In other words, $(H^\mu, \mathcal{D}(H^\mu))$ is the Friedrichs extension of the operator $(-\frac{1}{2}\Delta, \ C_0^\infty(G))$.

There is also a one-to-one correspondence between self-adjoint operators and strongly continuous semigroups (resp. strongly continuous resolvents). That yields

(2.3) Corollary.   *a) For every $\mu \in \mathcal{M}_0$ there exists a unique strongly continuous contraction semigroup $(e^{-tH^\mu})_{t>0}$ on $\overline{\mathcal{D}}(\mathcal{Q}^\mu)$ with generator $-(H^\mu, \mathcal{D}(H^\mu))$. Defining e^{-tH^μ} to be 0 on the orthogonal complement of $\overline{\mathcal{D}}(\mathcal{Q}^\mu)$ in $L^2(I\!\!R^d, m)$ this semigroup trivially extends to a contraction semigroup on $L^2(I\!\!R^d, m)$, called* Schrödinger semigroup *and also denoted by $(e^{-tH^V})_{t>0}$.*

b) Similarly, one obtains the existence of a unique strongly continuous resolvent $(H^\mu + \alpha)^{-1}$, $\alpha > 0$, on $\overline{\mathcal{D}}(\mathcal{Q}^\mu)$ which in an analogous way will be extended to a resolvent on $L^2(I\!\!R^d, m)$.

(2.4) Remark.   According to [7] (Prop. 2.1), for every $f \in L^2(I\!\!R^d, m)$ the function $(H^\mu + \alpha)^{-1} f$ is given as the unique minimal point of the functional

$$g \mapsto \mathcal{Q}^\mu(g, g) - 2 \int f g \, dm$$

on $W^{1,2}(I\!\!R^d, m)$.

An essential observation now is that the symmetric form $(\mathcal{Q}^\mu, \mathcal{D}(\mathcal{Q}^\mu))$ as well as the Schrödinger operator $(H^\mu, \mathcal{D}(H^\mu))$ associated with a measure $\mu \in \mathcal{M}_0$ only depends on the equivalence class $(\mu, \sim)$ of μ under a certain equivalence relation $\sim$, — and not on the particular choice of the representant μ in $(\mu, \sim)$. Of course, the same is then also true for the associated semigroup and resolvent.

(2.5) Definition.   Two measures μ, ν in $\mathcal{M}_0$ are called *equivalent* $(\mu \sim \nu)$ iff

$$(\mathcal{Q}^\mu, \mathcal{D}(\mathcal{Q}^\mu)) \;=\; (\mathcal{Q}^\nu, \mathcal{D}(\mathcal{Q}^\nu)).$$

The set of equivalence classes in $\mathcal{M}_0$ is denoted by $(\mathcal{M}_0, \sim)$.

This equivalence relation $\sim$ and the induced set of equivalence classes $(\mathcal{M}_0, \sim)$ will be investigated in more details in the next section. At the moment we turn our attention to a quite natural notion of convergence in $(\mathcal{M}_0, \sim)$ which makes this set to a nice topological space.

(2.6) Definition.   Let $(\mu_n)_{n \in N}$ be a sequence in $\mathcal{M}_0$ and let $\mu \in \mathcal{M}_0$. We say that the sequence of Schrödinger operators $(H^{\mu_n})_{n \in N}$ *converges in the strong resolvent sense* to H^μ iff for some (hence all) $\alpha > 0$ the sequence of resolvent operators $([H^{\mu_n} + \alpha]^{-1})_{n \in N}$ on $L^2(I\!\!R^d, m)$ converges strongly to $[H^\mu + \alpha]^{-1}$.

In this case, we also say that the sequence of measures $(\mu_n)_{n \in N}$ is *γ-convergent* to μ (or, more precisely, that the sequence of equivalence classes $((\mu_n, \sim))_{n \in N}$ is γ-convergent to $(\mu, \sim)$).

For equivalent characterizations and many interesting properties of the γ-convergence we refer to [7] and literature cited there. We restrict ourselves to the follwing important result from [14].

(2.7) Theorem.
  a) Under the γ-topology the set $(\mathcal{M}_0, \sim)$ is compact and metrizable.

b) The set $\{V \cdot m : V \in C_0^\infty(\mathbb{R}^d)\}$ as well as the set $\{1_K \cdot \overline{\infty} : K \subset \mathbb{R}^d \text{ compact}\}$ is γ-dense in $(\mathcal{M}_0, \sim)$.

One of the reasons to consider generalized Schrödinger operators $-\frac{1}{2}\Delta + \mu$ with measures $\mu \in \mathcal{M}_0$ is that this class of operators is closed with respect to strong resolvent convergence.

(2.8) Corollary. *The strong resolvent limit of any sequence of Schrödinger operators $(-\frac{1}{2}\Delta + \mu_n)_{n \in N}$ with measures $\mu_n \in \mathcal{M}_0$, $n \in \mathbb{N}$, is (if it exists) again a Schrödinger operator $-\frac{1}{2}\Delta + \mu$ with a measure $\mu \in \mathcal{M}_0$.*

Actually, the above mentioned set of generalized Schrödinger operators is the smallest set which is closed with respect to strong resolvent convergence and which contains the usual Schrödinger operators $-\frac{1}{2}\Delta + V$ with smooth potentials $V \in C_0^\infty(\mathbb{R}^d)$.

(2.9) Corollary.

a) Every generalized Schrödinger operator $-\frac{1}{2}\Delta + \mu$ with a measure $\mu \in \mathcal{M}_0$ is the limit in the strong resolvent sense of a sequence of usual Schrödinger operators $(-\frac{1}{2}\Delta + V_n)_{n \in N}$ with smooth potentials $V_n \in C_0^\infty(\mathbb{R}^d)$, $n \in \mathbb{N}$.

b) On the other hand, every generalized Schrödinger operator $-\frac{1}{2}\Delta + \mu$ with a measure $\mu \in \mathcal{M}_0$ is also the limit in the strong resolvent sense of a sequence of Dirichlet Laplacians (times $-\frac{1}{2}$) on open sets $G_n \subset \mathbb{R}^d$ (with $\mathbb{R}^d \setminus G_n$ being compact), $n \in \mathbb{N}$.

3 Equivalence of Measures in $\mathcal{M}_0$

According to [7], the equivalence $\sim$ can also be expressed in terms of the fine topology.

(3.1) Theorem. *Two measures μ, ν in $\mathcal{M}_0$ are equivalent if and only if*

$$\mu(F) = \nu(F) \qquad \text{for all finely open sets } F \in \mathcal{B}(\mathbb{R}^d).$$

(3.2) Remarks. a) Obviously, always the following implications hold:

- if the measures μ, ν are identical (i.e. $\mu(F) = \nu(F)$ for all Borel sets $F \subset \mathbb{R}^d$), then they are equivalent

- if they are equivalent, then they satisfy $\mu(F) = \nu(F)$ for all open sets $F \subset \mathbb{R}^d$.

In general, however, none of the converse implications holds: An example of measures, which are equivalent but not identical, is given by the measures $\overline{\infty}$ and $\infty \cdot m$. On the other hand, let G be an open set which is dense but not finely dense in $\mathbb{R}^d$ (cf. [33], chap. 9). Then the measures $\overline{\infty}$ and $1_G \cdot \overline{\infty}$ obviously coincide for all open sets, but not for all finely open sets.

b) We emphasize that the equivalence of the measures μ and ν, in general, *does not* imply that for a given Borel set $G \subset \mathbb{R}^d$ the measures $1_{cG} \cdot \mu$ and $1_{cG} \cdot \nu$ are equivalent. For instance, consider the equivalent measures $\infty \cdot m$ and $\overline{\infty}$ and the open set $G = \mathbb{R}^d \setminus \partial B_1(0)$. Then $1_{cG} \cdot \infty \cdot m = 0$ (zero measure) but $1_{cG} \cdot \overline{\infty} \not\sim 0$.

(3.3) Proposition. *Let $\mu, \nu \in \mathcal{M}_0$ and let $F, G \in \overline{\mathcal{B}}(\mathbb{R}^d)$.*

    a) If μ is a Radon measure, then: $\mu \sim \nu \iff \mu = \nu$.

    b) If $G \subset \mathbb{R}^d$ is quasi-closed, then: $\mu \sim \nu \implies 1_{cG} \cdot \mu \sim 1_{cG} \cdot \nu$.

    c) $\mathrm{reg}(G) = \mathrm{reg}(F) \iff 1_{cG} \cdot \overline{\infty} \sim 1_{cF} \cdot \overline{\infty}$.

    d) $\mathrm{reg}(G) = \mathrm{reg}(\overline{G}) \implies 1_{cG} \cdot \overline{\infty} \sim 1_{cG} \cdot \infty \cdot m$.

Proof. a) - c) follows from [33]. In order to see d), note that, according to c), the assumption implies $1_{cG} \cdot \overline{\infty} \sim 1_{c\overline{G}} \cdot \overline{\infty}$. But the measures $\overline{\infty}$ and $\infty \cdot m$ are equivalent. Applying b) to these measures and to the set $\overline{G}$ yields $1_{c\overline{G}} \cdot \overline{\infty} \sim 1_{c\overline{G}} \cdot \infty \cdot m$, hence, $1_{cG} \cdot \overline{\infty} \sim 1_{c\overline{G}} \cdot \infty \cdot m$. This proves the claim since always $1_{c\overline{G}} \cdot \infty \cdot m \leq 1_{cG} \cdot \infty \cdot m \leq 1_{cG} \cdot \overline{\infty}$.
        $\square$

(3.4) Remarks. a) Let us call a measure $\mu \in \mathcal{M}_0$ *maximal* (with respect to $\sim$) iff all measures $\nu \in \mathcal{M}_0$ equivalent to μ satisfy $\nu \leq \mu$. According to (3.3.a), every Radon measure is maximal. In Lemma (4.3) below, we shall see that for every measure μ there exists a *unique maximal* measure $\overline{\mu}$ equivalent to μ. For instance, the measure $\overline{\infty}$ is the unique maximal measure equivalent to the measure $\infty \cdot m$. Thus there is a one-to-one correspondence between the set $\overline{\mathcal{M}}_0 \subset \mathcal{M}_0$ of maximal measures and the set $(\mathcal{M}_0, \sim)$ of equivalence classes in $\mathcal{M}_0$.

    b) Another way to obtain a unique, canonical representative in each equivalence class of $(\mathcal{M}_0, \sim)$ is to look at quasi-regular measures. Here a measure $\mu \in \mathcal{M}_0$ is called *quasi-regular* (from outside) iff $\mu(F) = \inf\{\mu(G) : G \supset F, G \text{ quasi-open}\}$ for all $F \in \mathcal{B}(\mathbb{R}^d)$. It is easy to see that for any measure $\mu \in \mathcal{M}_0$ there exists a *unique quasi-regular* measure $\mu^* \in \mathcal{M}_0$ equivalent to μ, namely

$$\mu^*(F) := \inf\{\mu(G) : G \supset F, G \text{ quasi-open}\} \qquad \text{for } F \in \mathcal{B}(\mathbb{R}^d).$$

Since any maximal measure is immediately proved to be quasi-regular, one actually obtains

$$\mu^* = \overline{\mu}.$$

We add parenthetically that any quasi-regular measure $\mu \in \mathcal{M}_0$ is not only quasi-regular from outside (as our definition states) but by itself also regular from inside, i.e. $\mu(F) = \sup\{\mu(K) : K \subset F, K \text{ compact}\}$ for every $F \in \mathcal{B}(\mathbb{R}^d)$ ([13], Theorem 4.4). This justifies our usage of "quasi-regular" instead of "quasi-regular from outside".

    For $r > 0$ let us define the kernel

$$k(r) := \begin{cases} 1 & \text{if } d = 1 \\ \sup\{-\ln r, 0\} & \text{if } d = 2 \\ r^{2-d} & \text{if } d \geq 3. \end{cases}$$

(3.5) Definition (cf. [1] and [18]). We say that a measure μ on $(\mathbb{R}^d, \mathcal{B}(\mathbb{R}^d))$ belongs to the *Dynkin class* ($\mu \in \mathcal{K}_\infty$) iff for some (hence all) $r > 0$

$$\sup_{x \in \mathbb{R}^d} \int_{B_r(x)} k(\|x - y\|)\, \mu(dy) < \infty.$$

In the case $d \geq 2$, μ is said to belong to the *Kato class* ($\mu \in \mathcal{K}_0$) iff

$$\lim_{r \to 0} \sup_{x \in \mathbb{R}^d} \int_{B_r(x)} k(\|x - y\|)\, \mu(dy) = 0.$$

In the case $d = 1$, the Kato class coincides by definition with the Dynkin class.

Of course, we always have $\mathcal{K}_0 \subset \mathcal{K}_\infty \subset \mathcal{M}_0$. Dynkin measures (in particular, Kato measures) $\mu \in \mathcal{M}_0$ play an important rôle since the associated symmetric forms $\mathcal{Q}^\mu$ have the same domain $\mathcal{D}(\mathcal{Q}^\mu)$ as the unperturbed form $\mathcal{Q}^0$ (cf. [5], [26], [36]). For an entirely different situation, recall Example (1.6).

(3.6) Proposition. *For every Dynkin measure $\mu \in \mathcal{K}_\infty$ we have*

$$\mathcal{D}(\mathcal{Q}^\mu) = W^{1,2}(\mathbb{R}^d).$$

Proof. We regard $\mathcal{Q}^\mu$ as a perturbation of $\mathcal{Q}^0$ (and vice versa!) by means of the form

$$q^\mu(f, g) := \frac{1}{2} \int fg\, d\mu$$

with $\mathcal{D}(q^\mu) := L^2(\mathbb{R}^d, \mu)$ which defines a nonnegative symmetric form on $L^2(\mathbb{R}^d, m)$ (cf. [2], (4.4)). It is easy to see that for $\mu \in \mathcal{K}_\infty$ the form q^μ is $\mathcal{Q}^0$-form bounded with relative form bound $a < \infty$ (cf. [1], Th. 4.9, for the particular case of absolutely continuous $\mu \in \mathcal{K}_0$). Since q^μ is nonnegative, we conclude from the preceding that it is $\mathcal{Q}^\mu$-form bounded with relative form bound $\frac{a}{1+a} < 1$. By KLMN-theorem it follows that $\mathcal{D}(\mathcal{Q}^0) = \mathcal{D}(\mathcal{Q}^\mu - q^\mu) = \mathcal{D}(\mathcal{Q}^\mu)$. $\qquad\square$

4 The Set of Permanent Points

(4.1) Definition ([7]). We define the set E^μ of *permanent points for μ* to be the set of all points $x \in \mathbb{R}^d$ which have a finely open neighbourhood $G \subset \mathbb{R}^d$ satisfying

$$(*) \qquad\qquad \int_G k(\|x - y\|)\, \mu(dy) < \infty.$$

E^μ is always a finely open Borel set (even a F_σ-set, [33]).

(4.2) Remarks. a) The conditon $(*)$ can equivalently be replaced by the condition

$$\int_G k(\|z - y\|)\, \mu(dy) < \infty \qquad \text{for all } z \in \mathbb{R}^d$$

or even by

$$\sup_{z \in \mathbb{R}^d} \int_G k(\|z - y\|)\, \mu(dy) < \infty,$$

which is to say $1_G \cdot \mu \in \mathcal{K}_\infty$ ([33], Theorem (5.1)).

b) The set E^μ is closely related with the set F^μ, the *set of finiteness* for μ, which is by definiton the union F^μ of all finely open sets G with finite μ-measure. In other words, F^μ is the set of all points $x \in \mathbb{R}^d$ which have a finely open neighbourhood $G \subset \mathbb{R}^d$ satisfying $\mu(G) < \infty$. Obviously, the set F^μ is finely open. It is also easy to see that $E^\mu \subset F^\mu$ and that $F^\mu \setminus E^\mu$ is always polar. Hence, the sets

$$E^\mu \text{ and } F^\mu \text{ differ at most by a polar set.}$$

In many of the following analytic statements one is therefore allowed to replace E^μ by F^μ. However, the probabilistic approach shows that the right quantity to look at is indeed E^μ.

The set of permanent points (as well as the set of finiteness) can also be used to characterize the equivalence relation $\sim$.

(4.3) Lemma ([33]). *Two measures μ and ν in $\mathcal{M}_0$ are equivalent if and only if*

$$E^\mu = E^\nu \qquad and \qquad 1_{E^\mu} \cdot \mu = 1_{E^\nu} \cdot \nu.$$

In particular, every measure $\mu \in \mathcal{M}_0$ is equivalent to the measure

$$\overline{\mu} := 1_{CE^\mu} \cdot \overline{\infty} + 1_{E^\mu} \cdot \mu$$

which is the unique maximal (resp. quasi-regular) measure equivalent to μ.

Be careful: the equivalence of μ and $\overline{\mu}$ *does not imply* that $1_{CE^\mu} \cdot \mu$ is equivalent to $1_{CE^\mu} \cdot \overline{\infty}$.

5 Decomposition Theorem

The measures $\mu \in \mathcal{M}_0$ which we have considered up to now have

- either been extremely singular, like $\mu_1 = 1_{CG} \cdot \overline{\infty}$ with $G \in \mathcal{B}(\mathbb{R}^d)$ (which implies $\mathcal{D}(Q^{\mu_1}) = W_0^{1,2}(G)$)

- or rather smooth, like $\mu_2 \in \mathcal{K}_\infty$ (which implies $\mathcal{D}(Q^{\mu_2}) = W^{1,2}(\mathbb{R}^d)$).

Of course, we can compose such measures in order to obtain a measure $\mu = \mu_1 + \mu_2$ which is "singular" on CG and "smooth" on G.

However, much more interesting is the converse question, namely whether one can decompose an *arbitrary measure* $\mu \in \mathcal{M}_0$ into a "singular" part μ_1 and a "smooth" part μ_2. The main result in this section is that such a decomposition is always possible.

(5.1) Theorem ([33]).

   a) On CE^μ the measure μ is "singular" in the sense that

$$\mu \sim 1_{CE^\mu} \cdot \overline{\infty} + 1_{E^\mu} \cdot \mu.$$

   b) On E^μ the measure μ is "smooth" in the sense that there exists an increasing sequence $(F_n)_{n\in N}$ of closed sets with the property that the fine interiors F_n^{f-int} of the sets F_n increase to E^μ and for $n \in \mathbb{N}$

$$1_{F_n} \cdot \mu \in \mathcal{K}_\infty.$$

This "smoothness" property of μ on E^μ is indeed closely related to the notion of smoothness in the sense of [21], cf. also [33], chap. 8.

(5.2) Definition. A measure μ is called *smooth* on a finely open set $G \subset \mathbb{R}^d$ iff $\mu \in \mathcal{M}_0$ and there exists an increasing sequence of compact sets $F_n \subset G$ satisfying

- $\operatorname{cap}(K \setminus F_n) \xrightarrow{n\to\infty} 0$ for any compact set $K \subset G$ and

- $1_{F_n} \cdot \mu$ is a Radon measure for any $n \in \mathbb{N}$.

(5.3) Remarks. a) Due to Ph. Blanchard and Zh. Ma ([9], Theorem 2.1), the condition "$1_{F_n} \cdot \mu$ is a Radon measure" may be replaced by the condition "$1_{F_n} \cdot \mu$ is a Kato measure (i.e. $1_{F_n} \cdot \mu \in \mathcal{K}_0)$".

   b) According to [33], the following conditions are equivalent:

- $\operatorname{cap}(K \setminus F_n) \xrightarrow{n\to\infty} 0$ for compact sets $K \subset G$;

- $G \setminus \bigcup_n F_n^{f-int}$ is polar;

- $\tau(F_n) \xrightarrow{n\to\infty} \tau(G)$ $\mathbf{P}^x$-a.s. for q.e. $x \in \mathbb{R}^d$.

(5.4) Corollary. *A measure μ is smooth on a finely open set $G \subset \mathbb{R}^d$ iff*

$$G \setminus E^\mu \text{ is polar.}$$

In particular, μ is smooth on $\mathbb{R}^d$ if and only if

$$\mathbb{R}^d \setminus E^\mu \text{ is polar.}$$

The condition that $\mathbb{R}^d \setminus E^\mu$ is polar plays also a crucial rôle in the proof of a multidimensional analogue to the 0-1-law of Engelbert-Schmidt ([25]). We refer to [25] for various analytic and probabilistic conditions equivalent to that condition. For further conditions equivalent to it, we refer to (6.3.a) and (7.5) below.

6   Characterization of $\mathcal{D}(Q^\mu)$

We are now in a position to characterize the form domain $\mathcal{D}(Q^\mu)$ for an arbitrary measure $\mu \in \mathcal{M}_0$. For that purpose, let G_n be the fine interiors of the sets F_n, $n \in I\!N$, from (5.1) and recall that $E^\mu = \bigcup_{n \in N} G_n$.

(6.1) Theorem.   *For every $n \in I\!N$ the following holds*

$$W_0^{1,2}(G_n) \subset \mathcal{D}(Q^\mu) \subset W_0^{1,2}(E^\mu).$$

Proof. The first inclusion follows immediately from Theorem (5.1.b) and Proposition (3.6), the second one from Theorem (5.1.a) and Example (1.6). $\square$

(6.2) Corollary.
a)   *The closure of $\mathcal{D}(Q^\mu)$ in $W^{1,2}(I\!R^d)$ is*

$$W_0^{1,2}(E^\mu) = \{f \in W^{1,2}(I\!R^d) : \ \tilde{f} = 0 \ q.e. \ on \ \mathcal{C}E^\mu\}.$$

b)   *The closure of $\mathcal{D}(Q^\mu)$ in $L^2(I\!R^d, m)$ is*

$$L_0^2(E^\mu, m) = \{f \in L^2(I\!R^d, m) : \ f = 0 \ m\text{-}a.e. \ on \ \mathcal{C}E^\mu\}.$$

(6.3) Corollary.
  a)   *$\mathcal{D}(Q^\mu)$ is dense in $W^{1,2}(I\!R^d)$* $\Longleftrightarrow$ *$cap(I\!R^d \setminus E^\mu) = 0$.*
  b)   *$\mathcal{D}(Q^\mu)$ is dense in $L^2(I\!R^d, m)$* $\Longleftrightarrow$ *$m(I\!R^d \setminus E^\mu) = 0$.*

In other words, the last assertion says that $(Q^\mu, \mathcal{D}(Q^\mu))$ is a densely defined form on $L^2(I\!R^d, m)$ if and only if the set $I\!R^d \setminus E^\mu$ is polar. A quite different criterion for $(Q^\mu, \mathcal{D}(Q^\mu))$ being densely defined was given by P. Stollmann [31].

7   Limits of Schrödinger Operators

The characterization of the form domain $\mathcal{D}(Q^\mu)$ allows to investigate the limits of the sequences $(\frac{1}{n} \cdot \mu)_{n \in N}$ as well as $(n \cdot \mu)_{n \in N}$ with respect to γ-convergence. In the case of the sequence $(n \cdot \mu)_{n \in N}$ we obviously have $n \cdot \mu \longrightarrow \infty \cdot \mu$ (for $n \to \infty$) in the sense of γ-convergence (and in the sense of monotone convergence). It is also easy to see that $\infty \cdot \mu \sim 1_{\mathcal{C}E^{\infty \cdot \mu}} \cdot \overline{\infty} = 1_{\mathcal{C}F^{\infty \cdot \mu}} \cdot \overline{\infty}$.

(7.1) Proposition.   *In the sense of γ-convergence, for $n \to \infty$*

$$n \cdot \mu \longrightarrow 1_{\mathcal{C}F^{\infty \cdot \mu}} \cdot \overline{\infty}$$

where $F^{\infty \cdot \mu}$ (the set of finiteness for the measure $\infty \cdot \mu$) coincides with the fine support $f\text{-}supp(\mu)$ of μ.

(7.2) Corollary. *The Schrödinger operators $H^{n\cdot\mu} = -\frac{1}{2}\Delta + n\cdot\mu$ converge for $n \to \infty$ in the strong resolvent sense to $(-\frac{1}{2}$ times) the Dirichlet Laplacian on the finely open set $f\text{-}supp(\mu)$.*

If $\mu = 1_{cG} \cdot m$ with a Borel set $G \subset I\!\!R^d$ which is either quasi-closed or satisfies $\mathrm{reg}(G) = \mathrm{reg}(\overline{G})$ (cf. (3.3)), then the operators $H^{n\cdot\mu}$ converge for $n \to \infty$ in the strong resolvent sense to $(-\frac{1}{2}$ times) the Dirichlet Laplacian on G (or, which comes to the same, on $\mathrm{reg}(G)$). For similar results we refer to [17] and [24].

On the other hand, we have in the case of the sequence $(\frac{1}{n} \cdot \mu)_{n \in N}$

(7.3) Proposition. *In the sense of γ-convergence, for $n \to \infty$*

$$\frac{1}{n} \cdot \mu \longrightarrow 1_{cE^\mu} \cdot \overline{\infty}.$$

(7.4) Corollary. *The Schrödinger operators $H^{\frac{1}{n}\cdot\mu} = -\frac{1}{2}\Delta + \frac{1}{n}\cdot\mu$ converge for $n \to \infty$ in the strong resolvent sense to $(-\frac{1}{2}$ times) the Dirichlet Laplacian on the finely open set E^μ.*

(7.5) Corollary. *The following statements are equivalent:*

* $I\!\!R^d \setminus E^\mu$ *is polar*

* $\frac{1}{n} \cdot \mu \longrightarrow 0$ *(for $n \to \infty$) in the sense of γ-convergence*

* $H^{\frac{1}{n}\cdot\mu} \longrightarrow H^0 = -\frac{1}{2}\Delta$ *(for $n \to \infty$) in the strong resolvent sense.*

For related results, cf. [31] and [36].

Appendix: Potential Theoretic Notions

All potential theoretic notions like *fine, polar,* etc. are those of classical potential theory. They can also be descripted by means of Brownian motion $(X_t, \mathbf{P}^x)_{t>0, x \in R^d}$ on $I\!\!R^d$ (with $\Omega = C(I\!\!R_+, I\!\!R^d)$).

By *cap* we denote the capacity (Newtonian resp. logarithmical). A statement is said to hold *q.e.* *(=quasi everywhere)* (on $I\!\!R^d$) if it holds except on a polar subset of $I\!\!R^d$. We recall that Borel sets $F \subset I\!\!R^d$ with $\mathrm{cap}(F) = 0$ are called *polar*. These sets can also be characterized by the property

$$T(F) = \infty \qquad \text{a.s.}$$

where the phrase *a.s.* is used to say that a statement holds $\mathbf{P}^x$-a.s. (on Ω) for every $x \in I\!\!R^d$ and where $T(F) := \inf\{t > 0 : X_t \in F\}$ is the hitting time of F.

It is well-known (and can also be used as a definition) that a Borel function u on $I\!\!R^d$ is *finely continuous* if and only if a.s. the map

$$t \mapsto u(X_t) \quad \text{is continuous on } [0, \infty[$$

and that a Borel set $F \subset I\!\!R^d$ is *finely open* if and only if a.s. the set

$$\{t \in I\!\!R_+ : X_t \in F\} \quad \text{is open in } I\!\!R_+.$$

The fine topology is the coarsest topology rendering all superharmonic (resp. all α-excessive) functions continuous.

There is a close connection between fine continuity and quasi-continuity. We recall that by definition a numerical function u on $I\!\!R^d$ is *quasi-continuous* iff for any $\epsilon > 0$ there exists an open set $D = D_\epsilon$ such that $\text{cap}(D) \leq \epsilon$ and

$$u|_{\mathcal{C}D} : \mathcal{C}D \to \overline{I\!\!R} \quad \text{is continuous.}$$

Indeed, a numerical function u on $I\!\!R^d$ is quasi-continuous if and only if it is finely continuous q.e. It should be clear that similar results also hold for quasi-open (resp. quasi-closed) sets and finely open (resp. finely closed) sets. For instance, a set $F \subset I\!\!R^d$ is quasi-open if and only if it is the union of a finely open set and a polar set. In particular, every finely open set G is quasi-open, that is, for any $\epsilon > 0$ there exists an open set D_ϵ such that $\text{cap}(D_\epsilon) \leq \epsilon$ and such that the set $G_\epsilon := G \cup D_\epsilon$ is open.

We define the *regularization* $\text{reg}(G)$ of a (nearly) Borel set $G \subset I\!\!R^d$ to be the finely open Borel set

$$\left\{ x \in I\!\!R^d : \mathbf{P}^x \left\{ T(I\!\!R^d \setminus G) > 0 \right\} = 1 \right\}.$$

Do not confuse $\text{reg}(G)$ with the set of regular points for the stopping time $T(I\!\!R^d \setminus G)$ (which in the literature is often denoted by $(I\!\!R^d \setminus G)^r$ and which coincides with $I\!\!R^d \setminus \text{reg}(G)$).

$\text{reg}(G)$ is the largest finely open set $G_0 \subset I\!\!R^d$ such that $G_0 \setminus G$ is polar. The initial set G is finely open if and only if $G \subset \text{reg}(G)$ and it is quasi-open if and only if $G \setminus \text{reg}(G)$ is polar. G is called regular iff $G = \text{reg}(G)$.

References

[1] M. Aizenman, B. Simon: *Brownian motion and Harnack inequality for Schrödinger operators*. Comm. Pure Appl. Math. **35** (1982), 209-273

[2] S. Albeverio, Ph. Blanchard, Zh. Ma: *Feynman-Kac semigroups in terms of signed smooth measures*. BiBoS-Preprint Nr. 424, Bielefeld 1990

[3] S. Albeverio, J. Brasche, M. Röckner: *Dirichlet forms and generalized Schrödinger operators*. In Lect. Notes in Physics **345**, 1-42, Springer 1989

[4] S. Albeverio, Zh. Ma: *Nowhere Radon smooth measures, perturbations of Dirichlet forms and singular quadratic forms*. In Lect. Notes Control and Inf. Sciences, Springer 1989

[5] S. Albeverio, Zh. Ma: *Perturbation of Dirichlet forms*. To appear in J. Funct. Anal.

[6] S. Albeverio, Zh. Ma: *Additive functionals, nowhere Radon and Kato class smooth measures associated with Dirichlet forms.* SFB 237 - Preprint Nr.66, Bochum 1989

[7] J. Baxter, G. Dal Maso, U. Mosco: *Stopping times and Γ-convergence.* Trans. Amer. Math. Soc. **303** (1987), 1-38

[8] Ph. Blanchard, Zh. Ma: *Semigroups of Schrödinger operators with potentials given by Radon measures.* To appear in "Stochastic Processes – Physics and Geometry" (ed. by S. Albeverio et al.). World Scient. Singapore

[9] Ph. Blanchard, Zh. Ma: *Smooth measures and Schrödinger semigroups.* BiBoS - Preprint Nr.295, Bielefeld 1987

[10] Ph. Blanchard, Zh. Ma: *New results on the Schrödinger semigroups with potentials given by signed smooth measures.* In Lect. Notes Math. **1444**, 213-243, Springer 1990

[11] A. Boukricha, W. Hansen, H. Hueber: *Continuous solutions of the generalized Schrödinger equation and perturbation of harmonic spaces.* Expo. Math. **5** (1987), 97-135

[12] J. Brasche: *Perturbations of self-adjoint operators supported by null sets.* Thesis, Bochum 1988

[13] G. Dal Maso: *Γ-convergence and μ-capacities.* Ann. Sc. Norm. Sup. Pisa Cl. Sci. (4) **14** (1987), 423-464

[14] G. Dal Maso, U. Mosco: *Wiener's criterion and Γ-convergence.* Appl. Math. Optim. **15** (1987), 15-63

[15] M. Demuth: *On topics in spectral and stochastic analysis for Schrödinger operators.* Preprint, Berlin 1990

[16] M. Demuth, J. A. van Casteren: *On spectral theory of self-adjoint Feller-generators.* Reviews Math. Phys. **1** (1989), 325-414

[17] M. Demuth, J. A. van Casteren: *On trace class properties of singularly perturbed Feynman-Kac semi-groups.* Preprint, Berlin/Antwerp 1992

[18] E. B. Dynkin: *Markov processes I, II.* Grundl. d. math. Wiss. **121, 122.** Springer 1965

[19] D. Feyel, A. de La Pradelle: *Le rôle des espaces de Sobolev en topolgie fine.* In: Lect. Notes Math. **563**, 43-61. Springer 1976

[20] D. Feyel, A. de La Pradelle: *Étude de l'équation $1/2\Delta u - u\mu = 0$ où μ est une mesure positive.* Ann. Inst. Fourier **38** (1988), 199-218

[21] M. Fukushima: *Dirichlet forms and Markov processes*. North-Holland/Kodansha 1980

[22] W. Hansen, H. Hueber: *Eigenvalues in potential theory*. To appear in J. Diff. Equ.

[23] W. Hansen, Zh. Ma: *Perturbation by differences of unbounded potentials*. Math. Ann. **287** (1990), 553-569

[24] I. W. Herbst, Z. Zhao: *Sobolev spaces, Kac regularity and the Feynman-Kac formula*. In: Seminar on Stochastic Processes 1987 (ed. by E. Cinlar et al.). Birkhäuser 1988

[25] R. Höhnle, K.-Th. Sturm: *A multidimensional analogue to the 0-1-law of Engelbert and Schmidt*. Preprint, Erlangen 1992

[26] T. Kato: *Perturbation theory for linear operators*. Springer 1966

[27] V. G. Maz'ja: *Sobolev spaces*. Springer 1985

[28] B. Simon: *A canonical decomposition for quadratic forms with applications to monotone convergence theorems*. J. Funct. Anal. **28** (1978), 377-385

[29] B. Simon: *Schrödinger semigroups*. Bull. Amer. Math. Soc. **7** (1982), 447-526

[30] P. Stollmann: *Admissible and regular potentials for Schrödinger forms*. J. Operator Theory **18** (1987), 139-151

[31] P. Stollmann: *Smooth and Kato class perturbations of regular Dirichlet forms*. Preprint, Oldenburg 1991

[32] K.-Th. Sturm: *Störung von Hunt-Prozessen durch signierte additive Funktionale*. Thesis, Erlangen 1989

[33] K.-Th. Sturm: *Measures Charging No Polar Sets and Additive Functionals of Brownian Motion*. Forum Math. **4** (1992)

[34] K.-Th. Sturm: *Schrödinger Semigroups on Manifolds*. Preprint, Erlangen 1991

[35] J. A. van Casteren: *Generators of strongly continuous semigroups*. Pitman 1985

[36] J. Voigt: *Absorption semigroups, their generators, and Schrödinger semigroups*. J. Funct. Anal. **67** (1986), 167-205

Karl-Theodor Sturm
Mathematisches Institut, Universität Erlangen-Nürnberg
Bismarckstraße 1 1/2, D – W – 8520 Erlangen, Germany

Operator Theory:
Advances and Applications, Vol. 57
© 1992 Birkhäuser Verlag Basel

ABEL SUMMABILITY OF THE SERIES OF EIGEN- AND ASSOCIATED FUNCTIONS OF THE INTEGRAL AND DIFFERENTIAL OPERATORS

By Dr. Igor Trooshin

Let us consider integral operator

$$(Af)(x) = (Mf)(x) + \sum_{k=1}^{m} (\int_0^1 f(t)v_k(t)dt) \, g_k(x),$$

where $(Mf)(x) = \int_0^x M(x,t)f(t)dt$, $\{v_k(x)\}_{k=1}^m$, $\{g_k(x)\}_{k=1}^m$ – lineary independent systems of continuous functions, $x \in [0,1]$.

For the first time such operators were introduced by A.P.Hromov. He investigated spectral properties of such operators (see A.P.Hromov 1973, L.B.Macnev and A.P.Hromov 1977): i.e. completeness questions of eigen– and associated function systems (e.a.f.), questions of convergence of expansion by e.a.f. of such operators etc.

Based on above mentioned result and using method, which developed A.G.Kostyuchenko and A.A.Shkalikov (1978) method, the following theorem is obtained in the paper of I.Yu.Trooshin (1988).

THEOREM: 1 *Let M be: $M = (E+N)I^n(E+N)^{-1}$, where E is identical operator, N is Volterra operator with continuous kernel,*

$$(J^n f)(x) = \int_0^x \frac{(x-t)^{n-1}}{(n-1)!} f(t)dt,$$

Then for any function $f = A^l f_1 (f_1 \in L(0,1)$, l is certain known natutal number depending on functions $v_k(x)$ and $g_k(x)$ behavior at end points of interval $[0,1])$ Fourier series in e.a.f. of operator A is summarized uniformly with respect to x on interval

$[0,1]$ by $\frac{1}{n} < \alpha < \frac{1}{n-2m}$ order Abel method in case

$$\frac{n-4}{3} \leq m < \frac{n}{2} \qquad \text{(if n is even)}$$

$$\frac{n-3}{3} \leq m < \frac{n}{2} \qquad \text{(if n is odd)} \tag{1}$$

It is nessesery to note, that if the condition (1) is not fulfilled, then summability in interval $[0,1]$ is not take place. However, in this case the interval $[0,b]$, $b < 1$, of uniformly summability exists, where b is taken as a solution of some minimizing problem. An example showing exactness of summarizing interval is designed.

Let consider now operator L raised by ordinary differential expression:

$$l(y) = y^{(n)} + \sum_{k=1}^{n-2} p_k(x)y^{(k)} \quad (x \in [0,1])$$

with disintegrating normalized boundary conditions:

$$U_j(y) = y^{(s_j)}(0) + \sum_{k=1}^{s_j-1} b_{j,k}y^{(k)}(0) = 0, \quad j = 1,...,n-p,$$

$$U_j(y) = y^{(s_j)}(1) + \sum_{k=1}^{s_j-1} b_{j,k}y^{(k)}(1) = 0, \quad j = n-p+1,...,n.$$

Let us define by D –a quadrangle in complex plane with apexes

$$[0, 1, (1 - e^{\frac{2\pi i}{n}})^{-1}, (1 - e^{\frac{-2\pi i}{n}})^{-1}].$$

Let us suppose, that $p_k(x)$ – are analytic in D and continuous right up to the boundary of D.

Let us note, that operator L^{-1} belongs to operators, which are considered in Theorem 1. So, the next theorem is followed from Theorem 1:

THEOREM: **2** *Let us suppose, that*

$$|\, n - 2p \,| \leq \ 4 + \min(p, n-p) \quad \text{(if n is even)}$$

$$|\, n - 2p \,| \leq \ 3 + \min(p, n-p) \quad \text{(if n is odd)}. \tag{2}$$

Then for any function from definition field of operator L^l (l is certain natural number) Fourier series in e.a.f. of operator L is summarized uniformly with respect to x on interval $[0,1]$ by $\frac{1}{n} < \alpha < \frac{1}{|n-2p|}$ order Abel method.

In Theorem 2, in comparison with A.G.Kostyuchenko and A.A.Shkalikov (1978) theorem, set of operators for which summability on all interval $[0,1]$ is take place is wider: limitations on boundary conditions (2) and on differential expression coefficients are weaked.

In case $|\, n - 2p \,| > 4 + \min(p, n-p)$ uniformly summability is take place only on smaller interval $[0,b]$, $b < 1$, where b is a solution of some minimizing problem.

References

[1] Hromov A.P. (1973) Finite–dimensional perturbations of Volterra operator. D.Sc.Thesis, Novosybirsk. (Russian)

[2] Macnev L.B., Hromov A.P. (1977) Assymptotic of Volterra operator resolvent kernel in indifferentiated case and its applications. Differential Equations and the Theory of Functions, Vyp.1, Izdat. Sarat. Univ., Saratov, pp. 36–65. (Russian)

[3] Kostyuchenko A.G., Shkalikov A.A. (1978) Summability of eigenfunction expansions of differential operators and convolution operators. Funct. Anal. 12 (1978), 262–276. Transl. from Funkc. Analiz i Ego Prilozh. 12 (1978) 4, 24–40.

[4] Trooshin I Yu. (1988) Summability of eigenfunction expansions of finite-dimensional perturbations of Volterra operator. Manuscript No.8826–B88, deposited at VINITI. (Russian)

Author's address:
Dr. Igor Trooshin
Saratov University
Radischeva 23A–9
Saratov, Russia

Operator Theory:
Advances and Applications, Vol. 57
© 1992 Birkhäuser Verlag Basel

The Relativistic Oscillator

André Unterberger

Abstract. The Mathieu oscillator L is a relativistic deformation of the harmonic oscillator. The Klein-Gordon symbolic calculus of operators permits to build families of operators that commute with L , from which one can derive information on the eigenfunctions of L.

One of the reasons that make the study of the harmonic oscillator $L_W = \pi \ \Sigma \ (x_j^2 + D_j^2)$

trivial is that one can make the Weyl symbol of the operator $\exp\text{-}t\,(L_W - \frac{n}{2})$ explicit as

$$(1) \qquad (\tfrac{1}{2}\,(1+e^{-t}))^{-n} \exp [\, -2\pi \ (\tanh \tfrac{t}{2})\,(\,|\,\vec{x}\,|^2 + |\,\vec{p}\,|^2\,)\,]$$

where we denote as $(\vec{x}, \vec{p})$ the generic point of the phase space $\mathbf{R}^{2n}$.

We proceed to do much of the same for the relativistic oscillator L defined as

$$(2) \qquad -4\pi L = \Sigma\ \frac{\partial^2}{\partial x_j^2} - 4\pi^2\,|\,\vec{x}\,|^2 + c^{-2}\Big[\,(\Sigma x_j \frac{\partial}{\partial x_j})^2 + (n\text{-}1)\ \Sigma x_j \frac{\partial}{\partial x_j}\,\Big]\ .$$

The operator L, initially defined on $\mathscr{S}(\mathbf{R}^n)$, is essentially self-adjoint both on the Sobolev space $H_c^{\frac12}(\,\mathbf{R}^n\,)$ defined as the space of distributions u such that

$$(3) \qquad \int |\,\hat{u}\,(\vec{p}\,)\,|^2\,(\,1 + c^{-2}\,|\,\vec{p}\,|^2\,)^{\frac12}\ d\vec{p} < \infty$$

and on the space $L^2(\,\mathbf{R}^n, (1 + c^{-2}\,|\,\vec{x}\,|^2\,)^{-\frac12}\ d\vec{x}\,)$. It has a discrete spectrum that can be arranged as a sequence $0 < \lambda_0 < \lambda_1 \ldots$ going to infinity, and each eigenspace E_k is finite-dimensional.

As $c \to \infty$, the space $H_c^{1/4}(\mathbb{R}^n)$ contracts to $L^2(\mathbb{R}^n)$ and L goes to L_W, i.e the non-relativistic limit of the relativistic oscillator is the harmonic oscillator.

A capital fact is that L commutes with the relativistic Fourier transformation $\mathcal{F}_c$ characterized as

$$(4) \qquad (\mathcal{F}_c\, u)(\vec{x}) = (1 + c^{-2}|\vec{x}|^2)^{1/4}\, \hat{u}(\vec{x}) \; .$$

It will be handy to use the notation $<\vec{x}>$ for $(1 + c^{-2}|\vec{x}|^2)^{1/4}$ and the related one $<D>$ for

$$\left[1 - (4\pi^2 c^2)^{-1} \sum \frac{\partial^2}{\partial x_j^2}\right]^{1/4} .$$

Denoting as P the parity operator such that $(P\, u)(\vec{x}) = u(-\vec{x})$,

one has $P\mathcal{F}_c^2 = <\vec{x}><D>$. Now, at least in the one-dimensional case, this raises the question of making the operator $\mathcal{F}_c$ or $<\vec{x}><D>$ an explicit function of the oscillator L. Also, can one generalize (1) ?

It turns out that one can deal with these two questions in a satisfactory way through the use of the Klein-Gordon symbolic calculus of operators, a relativistic substitute for the Weyl calculus developed in Unterberger (1991). Let $\mathfrak{M}$ be the mass hyperboloid (actually one sheet of it) in $\mathbb{R}^{n+1}$ of equation

$$(5) \qquad p_0 = (c^2|\vec{p}|^2 + c^4)^{1/2}, \qquad\qquad p = (p_0, \vec{p}) \; .$$

It is a Riemannian symmetric space of the non-compact type, which enables one to define the geodesic middle mid (p, p') of any two points on $\mathfrak{M}$; also, $p_0^{-1}\, d\vec{p}$ is a Lorentz-invariant measure on $\mathfrak{M}$. If $p, p' \in \mathfrak{M}$, let

$$(6) \qquad <Jp, p'> = c^{-2}\, p_0 p'_0 - <\vec{p}, \vec{p}'> :$$

then $<Jp, p'> = c^2 \cosh[c^{-1} d(p,p')]$ if d is the (Riemannian) distance on $\mathfrak{M}$.

Finally, given $u \in H_c^{1/4}(\mathbb{R}^n)$, let $\mathcal{U}\, u \in L^2(\mathfrak{M}; p_0^{-1}\, d\vec{p})$ be characterized as

$$(7) \qquad (\mathcal{U}\, u)(p) = (1 + c^{-2}|\vec{p}|^2)^{1/2}\, \hat{u}(\vec{p})$$

so that $\mathcal{F}_c$ would be just $\mathcal{U}$ if $\mathfrak{M}$ were identified to $\mathbb{R}^n$ by means of the coordinates $\vec{p}$.

Definition. *Given a function* $g = g(\vec{x}, p)$ *on* $\mathbb{R}^n \times \mathfrak{M}$, *the operator* A *with symbol* g *is defined as*

$$(\mathcal{G} Au)(p) = 2^{\frac{n-2}{2}} \int (\mathcal{F}_1 g)(\vec{p} - \vec{p}\,', \text{mid } (p, p'))(\mathcal{G} u)(p')(p_0 + p'_0)[\, 1+c^{-2} < Jp, p' >]^{-\frac{n}{2}}\, p'^{-1}_0\, d\vec{p}\,'$$

in which $\mathcal{F}_1 g$ denotes the Fourier transform of g with respect to the first n variables .

We have tried here to keep the conceptual developments at a minimum, which may make the definition look artificial : it is not, as explained in [*loc. cit.*] , where one can also find estimates on symbols that make the Klein-Gordon calculus useful as a pseudodifferential analysis. Let us just recall that, for a true understanding of this calculus, it is necessary to view $H_c^{\frac{1}{2}}(\mathbf{R}^n)$ as the

set of time-zero restrictions of some space of solutions, with positive energy, of the Klein-Gordon equation. Also (for the benefit of readers acquainted with the calculus), the species of symbol used here is the *active* symbol.

The relativistic oscillator L is a kind of Casimir operator for the Bargmann-Wigner representation of the orthochronous Poincaré group in $H_c^{\frac{1}{2}}(\mathbf{R}^n)$, i.e. it can be defined by

$$\pi^{-1} L = \Sigma\, (B_j^2 + D_j^2) - c^{-2} \sum_{j<k} R_{jk}^2$$

where the operators B_j and R_{jk} are the infinitesimal operators of the representation that correspond to relativistic boosts and to rotations respectively : the coefficient of ΣD_j^2 can be rescaled in an elementary way .

1 . The one - dimensional case .

In that case, L reduces to

(8)
$$L = - (4\pi)^{-1} [\, \frac{d^2}{dx^2} - 4\pi^2 x^2 + c^{-2} (x \frac{d}{dx})^2 \,] .$$

Letting $x = c \sinh t$, L becomes the operator

(9)
$$M = (- 4\pi c^2)^{-1} [\, \frac{d^2}{dt^2} - 4\pi^2 c^4 \sinh^2 t \,] ,$$

self-adjoint on $L^2(\mathbf{R})$, whose eigenfunctions are the (modified) Mathieu functions introduced by that author in 1868 to solve the Dirichlet problem in an ellipse. Several books, devoted in full or partly to this subject, have appeared, among which Whittaker-Watson (1902), Mac Lachlan (1947), Meixner-Schäfke (1954) and Campbell (1955) : nevertheless , the structure of the Mathieu equation is still , in our opinion, poorly understood ; this was also the reiterated opinion of R. Campbell, and the subject does not seem to have changed much since then.

Let us see which contribution the Klein-Gordon analysis can bring to it. Consider the function

$$(10) \qquad r = r(x, p_1) = p_1^2 + x^2 + c^{-2} p_1^2 x^2 \ .$$

Then the symbol ℓ of L is

$$(11) \qquad \ell = \pi \, r - (16\pi \, c^2)^{-1} \ .$$

It is proved in [*loc. cit.*] that the operators A whose symbols are functions of r , say h (r), are exactly the functions of L , i.e. are those that admit a decomposition along the eigenfunctions ψ_k of L :

$$(12) \qquad A\,u = \Sigma \ \alpha_k (u, \ \psi_k) \ \psi_k \ \ .$$

Moreover, the (Klein-Gordon) composition of such a symbol by that of $- 4\pi$ L is given by

$$(13) \qquad - 4\pi\,\ell \quad h = r\,(\frac{r}{c^2} + 1)\,h'' + (\frac{2r}{c^2} + 1)\,h' - 4\pi^2\,rh + \frac{1}{4c^2}\,h \ .$$

As an example, the symbol of the operator $(<x> <D>)^{-1} = (1 + \frac{D^2}{c^2})^{-\frac{1}{2}} (1 + \frac{x^2}{c^2})^{-\frac{1}{2}}$

alluded to above is just the function $(1 + \frac{r}{c^2})^{-\frac{1}{2}} = (1 + \frac{p_1^2}{c^2})^{-\frac{1}{2}} (1 + \frac{x^2}{c^2})^{-\frac{1}{2}}$.

Let us then consider, for any real number s , the operator G_s with symbol $g_s = \exp (-2\pi \, e^s \, r)$. To get rid of constants, let us introduce

$$(14) \qquad F_s = 2^{1/2} \, e^{s/2} \ e^{-2\pi c^2 \cosh s} \, G_s$$

as well. Then a computation shows that

$$(15) \ (\quad F_s \, u)\,(p) = c^2 \int \exp - 2\pi \ (c^{-2}\,p_0 p'_0 \cosh s + p_1 p'_1 \sinh s)\,(\quad u)\,(p')\,p'^{-1}_0\,dp'_1$$

for every $u \in \mathscr{S}(\mathbb{R})$. As we have found a one - (continuous) parameter family of operators F_s that commute with L , and since the eigenspaces of L have dimension 1 , we may expect a product $F_s \, F_t$ to be expressible as a linear superposition of the operators F_u : indeed, one has

$$(16) \quad F_s F_t = c \int_{-\infty}^{\infty} \exp - 2\pi c^2 [\, \cosh s \ \cosh t \ \cosh u + \sinh s \ \sinh t \ \sinh u \,] \ F_u \, du \ .$$

Also, $F_{-\frac{i\pi}{2}}$ is just $\mathscr{F}_c^{-1}$.

In some sense, we have generalized formula (1) : only , it remains to make F_s explicit as a function of L , i.e. to characterize the functions f_k ($k \in \mathbb{R}$) that make

$$(17) \qquad\qquad F_s \ \psi_k = f_k(s) \ \psi_k$$

valid. Now, denoting as $\mathscr{L}$ the differential operator (in r) that occurs on the right-hand side of (13) , one easily checks that

$$(18) \qquad\qquad \mathscr{L} \ g_s = c^{-2} \left[\frac{\partial^2}{\partial s^2} - 4\pi^2 \ c^4 \ \sinh^2 s \right] g_s \ ;$$

from (9) and (16) it follows that

$$(19) \qquad\qquad f_k(s) = \ \psi_k \ (c \sinh s) \ ,$$

provided that the eigenfunction ψ_k has been normalized so that

$$(20) \qquad\qquad \psi_k(-x) \sim \left(\frac{x}{c} \right)^{-\frac{1}{2}} \exp - 2\pi c^2 \left(1 + \frac{x^2}{c^2} \right)^{\frac{1}{2}}$$

as $x \to +\infty$.

Observe that $\psi_k(x)$ appears as an *eigenfunction* of L but that $\psi_k(c \sinh s)$ appears as an *eigenvalue* of F_s . A by - product of the analysis is the formula (useful for k even)

$$\int_{-\infty}^{\infty} \ \psi_k(x) < x >^{-1} dx = \psi_k(0) \ \psi_k(-ic)$$

which may be considered as a relativistic generalization of the formula $\int e^{-\pi x^2} dx = 1$. It is a special case of the identity (a consequence of (16))

$$(21) \qquad \psi_k(x) \ \psi_k(y) = \int e^{-2\pi(c^2 <x><y><z> + c^{-1}xyz)} \psi_k(z) < z >^{-1} dz \ .$$

2. The n - dimensional case .

In accordance with (14) and (15) , let us now set

(22) $(\mathcal{G} F_s u)(p) = c^2 \displaystyle\int_{\mathfrak{M}} \exp -2\pi (c^{-2} p_0 p'_0 \cosh s + <\vec{p}, \vec{p}'> \sinh s)(\mathcal{G} u)(p') \, p'^{-1}_0 \, d\vec{p}$

and

(23) $$G_s = 2^{-\frac{n}{2}} e^{-\frac{n s}{2}} e^{2\pi c^2 \cosh s} \, F_s \, .$$

As s describes $\mathbb{R}$, the operators F_s commute with L and with one another ; they cannot any more, however, be considered as " functions " of L since they do not act as scalars on the eigenspace E_k with $k \neq 0$. Two kinds of asymptotics are of interest.

In the first one , we let $c \to \infty$: then

$$c^{-2} p_0 p'_0 - c^2 = c^{-2}(c^2 | \vec{p} |^2 + c^4)^{1/2} (c^2 | \vec{p}' |^2 + c^4)^{1/2} - c^2$$

goes to $\frac{1}{2} (| \vec{p} |^2 + | \vec{p}' |^2)$ and the kernel $k (\vec{p}, \vec{p}')$ of $\mathcal{G} G_s \mathcal{G}^{-1}$ with respect to $d\vec{p}'$ contracts to

(24) $k_\infty (\vec{p}, \vec{p}') = 2^{-n/2} e^{-\frac{n s}{2}} \exp - \pi [(| \vec{p} |^2 + | \vec{p}' |^2) \cosh s + 2 <\vec{p}, \vec{p}'> \sinh s] :$

now, with $e^s = \tanh \frac{t}{2}$, this is just the kernel of the operator $(\frac{1}{2} (1 + e^{-t}))^n \exp - t (L_W - \frac{n}{2})$ (one may use (1) or recognize in (24) Mehler's formula) ; it is a reassuring fact that, as $c \to \infty$, our family F_s contracts to the semi-group of the harmonic oscillator (note that conjugation under the Fourier transformation is harmless in this respect).

We now fix c and let s go to $-\infty$, setting $\varepsilon = 2e^s$. Computing the (Klein-Gordon) symbol of F_s , one finds out that

(25) $$G_{\ell n \frac{\varepsilon}{2}} = (I + \varepsilon^2 h (\varepsilon)) \exp - \varepsilon \, (L - \frac{n(n-2)}{16 \pi c^2})$$

where $h (\varepsilon)$ is bounded on each eigenspace E_k as an operator-valued function as $\varepsilon \to 0$. This makes an expression of $\exp - \delta (L - \frac{n(n-2)}{16 \pi c^2})$ as a Feynman-like integral possible, writing it as the limit of a product

$$\prod_{j=0}^{m-1} G_{\ell n \frac{1}{2}(\delta_{j+1} - \delta_j)}$$

with $0 = \delta_0 < \delta_1 < ... < \delta_m = \delta$, and $\max (\delta_{j+1} - \delta_j) \to 0.$

Thus, in the one-dimensional case, we get the formula

$$(\mathcal{F}_c(\exp\text{-}\delta[\, L+(16\pi c^2)^{-1}]\,)\, \mathcal{F}_c^{-1}\, u)\,(c \sinh \xi_0)$$

$$= \lim c^m (\prod_{j=0}^{m-1} (\delta_{j+1} - \delta_j)^{-\frac{1}{2}}) \int \ldots \int \exp\text{-}\pi c^2 \sum(\delta_{j+1} - \delta_j) \sinh^2 (\frac{\xi_j + \xi_{j+1}}{2})$$

$$\exp\text{-}4\pi c^2 \sum(\delta_{j+1} - \delta_j)^{-1} \sinh^2 (\frac{\xi_j - \xi_{j+1}}{2}) . \, u(c \sinh \xi_m)\, d\xi_1 \ldots d\xi_m$$

in which one may also dispense with $\mathcal{F}_c^{\pm 1}$ if preferred .

Let us turn to the n-dimensional case again. There is no need to write such a product in full : instead, let us make the following observations. One may rewrite the kernel (with respect to $p_0^{-1} \, d\vec{p}\,'$) of $\mathcal{G}\, G_{\ell n \frac{\varepsilon}{2}}\, \mathcal{G}^{-1}$ as

(26) $c^2\, \varepsilon^{-n/2} \exp(\, -\pi[\, <Jp, p'> - c^2\,]\,(\frac{\varepsilon}{2} + \frac{2}{\varepsilon}\,))\exp\text{-}\pi\varepsilon\, <\vec{p}, \vec{p}\,'> .$

Also, one may note that

(27) $\mathcal{G} L \mathcal{G}^{-1} = -(4\pi)^{-1}\, \Delta_{\mathfrak{M}} + \pi\, |\,\vec{p}\,|^2$

where $\Delta_{\mathfrak{M}}$ is the Laplace-Beltrami operator of $\mathfrak{M}$. If one considers, by virtue of (25) , the operator $G_{\ell n \frac{\varepsilon}{2}}$ as an approximation of $\exp\text{-}\varepsilon\,(\,L - \frac{n(n-2)}{16\pi c^2}\,)$, the second exponential on the right-hand side of (26) just accounts for the term $\pi\,|\,\vec{p}\,|^2$ in (27) . On the other hand, the function

$$c^2\, \varepsilon^{-n/2}\exp\,(-\pi[\,<Jp, p'> - c^2\,]\,(\frac{\varepsilon}{2} + \frac{2}{\varepsilon}\,)) \; = \; c^2 \varepsilon^{-n/2} \exp\text{-}2\pi c^2 (\frac{\varepsilon}{2} + \frac{2}{\varepsilon}\,) \sinh^2 \frac{d(p,p')}{2c}$$

is the kernel of the operator

(28) $2c\varepsilon^{-n/2} (\frac{1}{2}\,(\frac{\varepsilon}{2} + \frac{2}{\varepsilon}\,))^{\frac{1-n}{2}} \exp\,(\pi c^2\,(\frac{\varepsilon}{2} + \frac{2}{\varepsilon}\,))\, K_{i\nu}\,(\pi c^2\,(\frac{\varepsilon}{2} + \frac{2}{\varepsilon}\,))$

with

(29) $\nu = [\,-c^2\, \Delta_{\mathfrak{M}} - (\frac{n-1}{2})^2\,]^{\frac{1}{2}}$

Asymptotics for $K_{i\nu}\,(x)$, $x \to \infty$, confirm that (29) is $(\, I + O\,(\varepsilon^2))$ times the operator

(30) $\exp \varepsilon\,[\,\frac{1}{4\pi}\, \Delta_{\mathfrak{M}} + \frac{n(n-2)}{16\pi c^2}\,]\, .$

As shown by (28), the "Feynman integral" which we have found is rather different from anything we might have expected as coming from Trotter's formula applied to a decomposition such as (27) : it also has much nicer properties since we are dealing at each time with a family of commuting operators $G_{\ell n \frac{1}{2}(\delta_{j+1}-\delta_j)}$ even before we go to the limit ; finally, formulas are

meaningful for any complex δ with Re $\delta \geq 0$.

Let us note that (22) was derived from the one-dimensional case by a generalization of the *kernel* of the operator $\mathcal{G} \, F_s \, \mathcal{G}^{-1}$. Instead, one may generalize its (Klein-Gordon) *symbol* .

The symbol ℓ of L is, this time,

$$\ell = \pi r - \frac{n}{16 \pi c^2} \, ,$$

with $r = |\vec{x}|^2 + |\vec{p}|^2 + c^{-2} < \vec{x}, \vec{p} >^2$, and one may consider the operator $\widetilde{G}_s$ with symbol $\exp(-2\pi e^s r)$. It is not too difficult to see that the kernel of the operator $\mathcal{G} \, \widetilde{G}_{\ell n \frac{s}{2}} \, \mathcal{G}^{-1}$ (with respect to $p_0'^{-1} \, d\vec{p}\,'$) is

$$c^2 \left[\cosh \frac{d(p,p')}{2c} \right]^{1-n} \varepsilon^{-n/2} \, \exp{-\pi} \left[\frac{4c^2}{\varepsilon} \sinh^2 \frac{d(p,p')}{2c} + \varepsilon \left| \overrightarrow{\mathrm{mid}(p,p')} \right|^2 \right] .$$

Then, a (quite formidable) computation shows that $\widetilde{G}_s$ commutes with L : thus, again, operators whose symbols are functions of ℓ commute with L .

3. The 2 - dimensional case .

This case has some specially nice features (is 2-dimensional space a particularly commendable place from where to observe Klein-Gordon particles ? of course, for us 3-dimensional-space dwellers, spin $\frac{1}{2}$ - particles are more fun to watch) .

This time, instead of a one-parameter family (F_s), let (F_μ) be parametrized by $\mathfrak{M}$ and defined as

$$(31) \qquad (\mathcal{G} \, F_\mu u) \, (p) = c^2 \int e^{-2\pi \{\mu, p, p'\}} \, \mathcal{G} u \, (p') \, p_0'^{-1} \, d\vec{p}\,'$$

with

$$(32) \qquad \{ \mu, p, p' \} = c^{-4} \, \mu_0 p_0 p_0' + c^{-1} \mu_1 \, (p_1 p_1' + p_2 p_2') + c^{-1} \mu_2 \, (p_1 p_2' - p_2 p_1')$$

The notation is consistent with (22) if $s \in \mathbb{R}$ is identified with $(c^2 \cosh s, c \sinh s, 0) \in \mathfrak{M}$.

Again, the operators F_μ commute with L. Indeed, setting $\Lambda_p = \mathcal{C}_y(-4\mu L)\mathcal{C}_y^{-1}$ (this is the operator made explicit in (27)) to recall that p is the variable, and $\Theta = \{\mu, p, p'\}$, one can see after some calculations that

$$(33) \qquad \Lambda_p\,(e^{-2\pi\Theta}) = 4\pi^2 c^2\, e^{-2\pi\Theta}\,[\Theta(\Theta-\pi^{-1}) + 2c^4 - \mu_0^2 - p_0^2 - q_0^2]\,.$$

Now this is the same as $\Lambda_{p'}\,(e^{-2\pi\Theta})$, which proves that F_μ commutes with L.

Much more, it is the same as $\Lambda_\mu\,(e^{-2\pi\Theta})$. As a consequence, as functions of μ, the operators F_μ are also eigenvectors of the relativistic oscillator : only, they are not scalar, but so to speak matrix-valued functions on μ when restricted to any eigenspace E_k.

It can be proved that the operators F_μ ($\mu \in \mathfrak{M}$) commute with one another and that

$$(34) \qquad F_\mu\, F_\nu = c^2 \int_{\mathfrak{M}} e^{-2\pi\{\mu,\nu,\lambda\}}\, F_\lambda\, \lambda_0^{-1}\, d\vec{\lambda}\,,$$

a formula whose proof has some intriguing combinatorial features .

References

[1]   Campbell, R. (1955), Théorie générale de l'équation de Mathieu et de quelques autres équations différentielles de la mécanique, Masson, Paris.

[2]   Mac Lachlan, N.W. (1947), Theory and applications of Mathieu functions, Clarendon Press, Oxford.

[3]   Meixner, J., Schäfke, F.W. (1954), Mathieusche Funktionen und Sphäroidfunktionen, Springer-Verlag, Berlin.

[4]   Unterberger, A. (1991), Quantification relativiste, Mémoires (nouvelle série, 44-45) de la Soc. Math. de France, Paris.

[5]   Whittaker, E.T., Watson, G.N. (1902), A course of modern analysis, Cambridge Univ. Press, Cambridge.

André Unterberger
Mathématiques , Université de Reims
Moulin de la Housse, BP 347
F 51062 Reims Cedex

Operator Theory:
Advances and Applications, Vol. 57
© 1992 Birkhäuser Verlag Basel

ON THE RATIO OF ODD AND EVEN SPECTRAL COUNTING FUNCTIONS

Michiel van den Berg

1.　Introduction

Let D be an open set in euclidean space $\mathbb{R}^m$ ($m = 1, 2, \ldots$). Suppose that D is symmetric with respect to a $(m - 1)$ - dimensional hyperplane P ($x \in D$ if and only if $Px \in D$, where Px denotes the reflection of x into P). Then

$$L^2(D) = L^2_o(D) \oplus L^2_e(D), \tag{1}$$

where

$$L^2_o(D) = \left\{ f \in L^2(D) \; : \; f(x) = - f(Px), \quad x \in D \right\}, \tag{2}$$

$$L^2_e(D) = \left\{ f \in L^2(D) \; : \; f(x) = f(Px), \quad x \in D \right\}. \tag{3}$$

Let Δ_D denote the Dirichlet laplacian for D, and suppose that $- \Delta_D$ has discrete spectrum $\lambda_1 \leq \lambda_2 \leq \ldots$ with a corresponding complete orthonormal set of eigenfunctions $\{\varphi_1, \varphi_2, \ldots\}$ in $L^2(D)$. Then for each $j \in \mathbb{Z}^+$ either $\varphi_j \in L^2_o(D)$ or $\varphi_j \in L^2_e(D)$. We define the odd spectral counting function N^o_D, the even spectral counting function N^e_D, and N_D by

$$N^o_D(\lambda) = \#\left\{ j \; : \; \lambda_j < \lambda, \quad \varphi_j \in L^2_o(D) \right\}, \tag{4}$$

$$N^e_D(\lambda) = \#\left\{ j \; : \; \lambda_j < \lambda, \quad \varphi_j \in L^2_e(D) \right\}, \tag{5}$$

$$N_D(\lambda) = \#\left\{ j \; : \; \lambda_j < \lambda \right\}. \tag{6}$$

Since $\varphi_1(x) \geq 0$ for $x \in D$, we have $\varphi_1 \in L^2_e(D)$. For $\lambda > \lambda_1$ we define the ratio

$$r_D(\lambda) = N^o_D(\lambda)/N^e_D(\lambda). \tag{7}$$

Proposition 1. Let D symmetric with respect to a $(m - 1)$ - dimensional hyperplane P, and suppose that $-\Delta_D$ has discrete spectrum. Then $r_D(\lambda) \leq 1$ for $\lambda > \lambda_1$.

Proof: The hyperplane P partitions $\mathbb{R}^m$ into two open half spaces H^+ and H^- respectively. The odd eigenfunctions of $-\Delta_D$ satisfy Dirichlet conditions on $P \cap D$, while the even eigenfunctions of $-\Delta_D$ satisfy Neumann conditions on $P \cap D$. Let $D^+ = H^+ \cap D$, $D^- = H^- \cap D$. It follows that

$$N^o_D(\lambda) = N_{D^+}(\lambda) = N_{D^-}(\lambda). \tag{8}$$

Moreover N^e_D is the spectral counting function associated to a Laplace operator with Dirichlet conditions on $\partial D^+ \backslash P$, and Neumann conditions on $\partial D^+ \cap P$. By Neumann bracketing (Proposition XI.2.4 in [1]) $N^e_D(\lambda) \geq N_{D^+}(\lambda)$, which completes the proof.

Proposition 2. Let D be symmetric with respect to a hyperplane P. Suppose that the volume $|D|_m$ of D is finite. Then $-\Delta_D$ has discrete spectrum and

$$\lim_{\lambda \to \infty} r_D(\lambda) = 1. \tag{9}$$

Proof: By Weyl's theorem (e.g. Theorem 10.6 in [2]) we have for $\lambda \to \infty$

$$N_D(\lambda) = (4\pi)^{-m/2}(\Gamma((m + 2)/2))^{-1}|D|_m \lambda^{m/2}(1 + o(1)), \tag{10}$$

$$N_{D^+}(\lambda) = (4\pi)^{-m/2}(\Gamma((m + 2)/2))^{-1}|D^+|_m \lambda^{m/2}(1 + o(1)), \tag{11}$$

and (9) follows by (4-8, 10, 11) and $|D^+|_m = 2^{-1}|D|_m$.

In this note we give examples of symmetric open sets D with infinite volume, for which $-\Delta_D$ has discrete spectrum, and for which (9) is, or is not valid.

2. Symmetric horn-shaped regions in $\mathbb{R}^2$

Theorem 3. Let $f : [0,\infty) \to \mathbb{R}^+$ be right continuous and decreasing to 0. Let

$$D = \left\{(x_1, x_2) \in \mathbb{R}^2 : x_1 > 0, \quad -f(x_1) < x_2 < f(x_1)\right\}, \tag{12}$$

and suppose there exists a constant $\rho > 0$ and a function $L : \mathbb{R}^+ \to \mathbb{R}^+$, slowly varying at infinity, such that for $t \to 0^+$

$$\int_{[0,\infty)} dx \sum_{k=1}^{\infty} e^{-t\pi^2 k^2/(4f^2(x))} = t^{-1/(2\rho)} L(1/t). \tag{13}$$

Then $-\Delta_D$ has discrete spectrum and

$$\lim_{\lambda \to \infty} r_D(\lambda) = (2^{1/\rho} - 1)^{-1}, \tag{14}$$

with respect to the plane $P = \{(x_1, x_2) : x_2 = 0\}$.

Example 4. Let D be given by (12), where

$$f(x) = (1 + x)^{-\alpha}(\log(e + x))^{-\beta}, \tag{15}$$

with constants $\alpha > 0$ and $\beta \geq 0$. Then f satisfies (13) for some function L, slowly varying at infinity, and a constant ρ given by

$$\rho = \min\{\alpha, 1\}. \tag{16}$$

Note that $\lim_{\lambda \to \infty} r_D(\lambda) = 1$ for $\alpha = 1$, $0 \leq \beta \leq 1$, even though $|D|_2 = \infty$ in this case.

Example 5. Let D be given by (12), where $f \in L^1[0,\infty)$, and satisfies the conditions of Theorem 3. Then $|D|_2 = 2\|f\|_1 < \infty$, and we recover the conclusion of Proposition 2 as follows. The inequality

$$(\pi t)^{-1/2}f(x) - 1 \leq \sum_{k=1}^{\infty} e^{-t\pi^2 k^2/(4f^2(x))} \leq (\pi t)^{-1/2}f(x), \tag{17}$$

implies that

$$(\pi t)^{-1/2} \|f\|_1 - (\pi t)^{-1/2} \int\limits_{\{x:f(x)<t^{1/2}\}} dx \, f(x) - \text{meas}\left\{x \in \mathbb{R}^+ : f(x) \geq t^{1/2}\right\}$$

$$\leq \int\limits_{[0,\infty)} dx \sum_{k=1}^{\infty} e^{-t\pi^2 k^2/(4f^2(x))} \leq (\pi t)^{-1/2} \|f\|_1. \tag{18}$$

By Lebesgue's dominated convergence theorem we conclude that f satisfies (13) with $\rho = 1$ and a function L satisfying

$$\lim_{t \to 0} L(1/t) = \pi^{-1/2} \|f\|_1. \tag{19}$$

Hence L is slowly varying at infinity and (14) agrees with the conclusion of Proposition 2.

Proof of Theorem 3: Define for $t > 0$

$$Z_D(t) = \text{trace } (e^{t\Delta_D}). \tag{20}$$

Then, by Example 1 in [3], or by an application of Theorem 3 in [4],

$$Z_D(t) = (4\pi t)^{-1/2} \int\limits_{[0,\infty)} dx \sum_{k=1}^{\infty} e^{-t\pi^2 k^2/(4f^2(x))} (1 + o(1)), \tag{21}$$

for $t \to 0^+$. By (13) we have for some L, slowly varying at infinity,

$$Z_D(t) = (4\pi t)^{-1/2} t^{-1/2\rho)} L(1/t). \tag{22}$$

Similarly we have for $t \to 0^+$

$$Z_{D^+}(t) = (4\pi t)^{-1/2} \int\limits_{[0,\infty)} dx \sum_{k=1}^{\infty} e^{-t\pi^2 k^2/f^2(x)} (1 + o(1))$$

$$= (4\pi t)^{-1/2} (4t)^{-1/(2\rho)} L(1/(4t))(1 + o(1)). \tag{23}$$

Since

$$Z_D(t) = \int\limits_{[0,\infty)} e^{-t\lambda} dN_D(\lambda), \tag{24}$$

$$Z_{D^+}(t) = \int\limits_{[0,\infty)} e^{-t\lambda} dN_{D^+}(\lambda), \tag{25}$$

we have by the tauberian theorems 1 and 2 on pp. 443–445 in [5]

$$N_D(\lambda) = (4\pi)^{-1/2}(\Gamma((1 + \rho)/(2\rho)))^{-1}\lambda^{(1+\rho)/(2\rho)}L(\lambda)(1 + o(1)), \tag{26}$$

$$N_{D^+}(\lambda) = (4\pi)^{-1/2}2^{-1/\rho}(\Gamma((1 + \rho)/(2\rho)))^{-1}\lambda^{(1+\rho)/(2\rho)}L(\lambda/4)(1 + o(1)) \tag{27}$$

as $\lambda \to \infty$. Hence by (4-8, 26,27)

$$r_D(\lambda) = \frac{2^{-1/\rho}L(\lambda/4)}{L(\lambda) - 2^{-1/\rho}L(\lambda/4)} \ (1 + o(1)), \tag{28}$$

and Theorem 3 follows since L is slowly varying at infinity.

It is possible to have a symmetric open set D for which $r_D(\lambda) \to 0$ for $\lambda \to \infty$.

Example 6. Let D be as in Example 4 with $\alpha = 0$ and $\beta > 0$. Then by Theorem 2 and Corollary 4 in [6] one can show that

$$\lim_{\lambda\to\infty} \exp\left\{(2^{1/\beta} - 1)\pi^{-1/\beta}\lambda^{1/(2\beta)}\right\}r_D(\lambda) = 2^{1/(2\beta)}, \tag{29}$$

so that an exponentially small fraction of eigenfunctions are odd with respect to the plane $x_2 = 0$.

We already noted that $N_D^e(\lambda)$ is the spectral counting function for the Laplace operator with Neumann conditions on $P \cap \partial D^+$, and Dirichlet conditions on $\partial D^+\backslash P$. We use this observation in the proof of the following.

Theorem 7. Let $f : [0,\infty) \to \mathbb{R}^+$ be right continuous and decreasing to 0. Let $D^+ = D \cap \{x_2 > 0\}$, where D is given by (12), and let $\Delta_{D^+}^e$ denote the Laplace operator with Neumann conditions on the positive x_1 - axis and with Dirichlet conditions on the remaining boundary of ∂D^+. Then $- \Delta_{D^+}^e$ has discrete spectrum, and the corresponding spectral counting function satisfies

$$\left| N_{D^+}^e(\lambda) - \pi^{-1}\int_{[0,\infty)} dx \sum_{k=0}^{\infty}\left\{\left\{\lambda - \pi^2(k + \tfrac{1}{2})^2 f^{-2}(x)\right\}^+\right\}^{1/2}\right|$$

$$\leq f(0)\lambda^{1/2} + (f(0))^{1/2}\lambda^{3/4}\left\{\int_{\{x:f(x)\geq\pi/(2\lambda^{1/2})\}} dx\ f(x)\right\}^{1/2}, \tag{30}$$

where $\left\{\ldots\right\}^+$ denotes the positive part of $\left\{\ldots\right\}$.

Proof: By Theorem 2 in [6]

$$\left| N_D(\lambda) - \pi^{-1} \int_{[0,\infty)} dx \sum_{k=1}^{\infty} \left\{ \left\{ \lambda - 4^{-1}\pi^2 k^2 f^{-2}(x) \right\}^+ \right\}^{1/2} \right|$$

$$\leq 2\pi^{-1} f(0)\lambda^{1/2} + (2/\pi)^{3/2}(f(0))^{1/2}\lambda^{3/4} \left\{ \int_{\{x:f(x)\geq\pi\lambda^{-1/2}\}} dx\ f(x) \right\}^{1/2}, \qquad (31)$$

$$\left| N_{D^+}(\lambda) - \pi^{-1} \int_{[0,\infty)} dx \sum_{k=1}^{\infty} \left\{ \left\{ \lambda - 4^{-1}\pi^2 k^2 f^{-2}(x) \right\}^+ \right\}^{1/2} \right|$$

$$\leq \pi^{-1} f(0)\lambda^{1/2} + 2\pi^{-3/2}(f(0))^{1/2}\lambda^{3/4} \left\{ \int_{\{x:f(x)\geq\pi\lambda^{-1/2}\}} dx\ f(x) \right\}^{1/2}. \qquad (32)$$

Since

$$N_{D^+}^e(\lambda) = N_D(\lambda) - N_{D^+}(\lambda), \qquad (33)$$

and

$$\sum_{k=0}^{\infty} \left\{ \left\{ \lambda - \pi^2(k + \tfrac{1}{2})^2 f^{-2}(x) \right\}^+ \right\}^{1/2}$$

$$= \sum_{k=1}^{\infty} \left\{ \left\{ \left\{ \lambda - 4^{-1}\pi^2 k^2 f^{-2}(x) \right\}^+ \right\}^{1/2} - \left\{ \left\{ \lambda - \pi^2 k^2 f^{-2}(x) \right\}^+ \right\}^{1/2} \right\}, \qquad (34)$$

Theorem 7 follows from (31-(34) and the inequalities $3 < \pi$ and $2 + 8^{1/2} < \pi^{3/2}$.

Remark 8.   Note that Theorem 7 can be proved directly by Dirichlet-Neumann bracketing. The set $\{\pi^2(k + \tfrac{1}{2})^2 f^{-2}(x) : k \in \mathbb{Z}^+ \cup \{0\}$ is the spectrum of Laplace operator on an interval of length $f(x)$ with Neumann conditions and Dirichlet conditions at the respective end points.

Finally we give an example of a horn-shaped region in $\mathbb{R}^2$, where Theorem 2 in [6] (or e.g. (31) in this note) does not give the leading asymptotic behaviour of $N_D(\lambda)$, $\lambda \to \infty$.

Example 9. Let D be given by (12), and

$$f(x) = 1/(2a_n), \qquad a_n \leq x < a_{n+1}, \qquad n = 1, 2, \ldots \qquad (35)$$

$$f(x) = 1/(2a_n), \qquad 0 \leq x < a_1, \qquad (36)$$

where

$$a_n = 2^{(p^n)}, \qquad n = 1, 2, \ldots, \tag{37}$$

and $p > 4$ is a constant. Let (λ_n) be given by

$$\lambda_n = (\pi a_n)^2. \tag{38}$$

The right hand side of (31) is bounded from below by

$$(2/\pi)^{3/2}(2a_1)^{-1/2}\lambda_n^{3/4}\left\{\int_{\{x:f(x)\geq\pi\lambda_n^{-1/2}\}} dx\, f(x)\right\}^{1/2} \geq 2a_1^{-1/2}a_n(a_{n+1})^{1/2}, \tag{39}$$

while

$$\int_{[0,\infty)} dx \sum_{k=1}^{\infty} \pi^{-1}\left\{\left\{\lambda_n - 4^{-1}\pi^2 k^2 f^{-2}(x)\right\}^+\right\}^{1/2}$$

$$= \int_{[0,a_n)} dx \sum_{k=1}^{\infty}\left\{\left\{a_n^2 - k^2 f^{-2}(x)\right\}^+\right\}^{1/2}$$

$$\leq \int_{[0,a_n)} dx\, a_n \cdot \#\left\{k \in \mathbb{Z}^+ : k \leq a_n f(x)\right\} \leq a_n^3/(2a_1). \tag{40}$$

But $a_{n+1}^{1/2}/a_n^2 \to \infty$ as $n \to \infty$ since $p > 4$, so that Theorem 2 in [6] does not determine the leading asymptotic behaviour of the spectral counting function for this region.

References

[1] Edmunds D.E. and Evans W.D., Spectral theory and differential operators, Clarendon, Oxford 1987.

[2] Simon B., Functional integration and quantum physics, Academic Press, New York 1979.

[3] van den Berg M., On the spectrum of the Dirichlet laplacian for horn-shaped regions in $\mathbb{R}^n$ with infinite volume, J. Funct. Anal. **58** (1984), 150-156.

[4] van den Berg M., On the asymptotics of the heat equation and bounds on traces associated with the Dirichlet laplacian. J. Funct. Anal. **71** (1987), 279-293.

[5] Feller W., An introduction to probability theory and its applications II, Wiley, New York 1971.

[6] van den Berg M., Dirichlet-Neumann bracketing for horn-shaped
     regions, J. Funct. Anal. **104** (1992), 110-120.

Author's address
Dr M van den Berg
Department of Mathematics
Heriot-Watt University
Riccarton
Edinburgh EH14 4AS
United Kingdom

Operator Theory:
Advances and Applications, Vol. 57
© 1992 Birkhäuser Verlag Basel

A Trace Class Property of Singularly Perturbed Generalized Schrödinger Semigroups

Jan A. van Casteren

Abstract. Let E be a locally compact second countable Hausdorff space and let Σ be an open subset of E. In addition let m be a Radon measure on E. Let K_0 be the self-adjoint generator in $L^2(E, m)$ of a symmetric strongly continuous $L^1(E, m)$-$L^\infty(E, m)$-smoothing Feller semigroup $\{\exp(-tK_0) : t \geq 0\}$ with continuous density and let V be a Kato-Feller potential with the property that its negative part belongs to $L^1(E, m)$. Suppose that $E \setminus \Sigma$ has finite capacity. Also suppose that the carré du champ operator exists in a suitable sense. Then the operators $\exp\left(-t(K_0 \dotplus V)\right) - J^* \exp\left(-t(K_0 \dotplus V)_\Sigma\right) J$, $t \geq 0$, are trace class. Here Jf restricts f to Σ and $J^* f$ extends f with 0 on the complement of Σ.

1. Introduction.

The introduction is almost the same as the one in van Casteren (1992a). In fact the whole paper is a shortened version of (1992a) in the sense that outlines of proofs are given, but by lack of space complete proofs will appear elsewhere: also see (1992b). In particular this is true for the identity in formula (3.6) of Proposition 3.10. The proof of this identity is quite technical. The identity will only be used for estimates on surface integrals. Let E be a locally compact second countable Hausdorff space with a reference Radon measure m, let K_0 be a self-adjoint generator in $L^2(E, m)$ of a strongly continuous symmetric $L^1(E, m)$-$L^\infty(E, m)$-smoothing Feller semi-group with a continuous transition density $p_0(t, x, y)$, let V be a so-called Kato-Feller potential and let Γ be a closed subset of E. Here L^1-L^∞-smoothing means that $\|\exp(-tK_0)\|_{1,\infty} = \sup_{x,y \in E} p_0(t, x, y) < \infty$ for all $t > 0$. If the negative part of V belongs to $L^1(E, m)$ and if Γ has finite capacity, then the semigroup difference

$$\left\{\exp\left(-t\left(K_0 \dotplus V\right)\right) - J^* \exp\left(-t\left(K_0 \dotplus V\right)_\Sigma\right) J : t \geq 0\right\}, \tag{1.1}$$

where $\Sigma = E \setminus \Gamma$, consists trace class operators. It is noticed that this result will be proved under the additional assumption, that the carré du champ operator exists in a sense as explained in section 3. Here $K_0 \dotplus V$ generates the ordinary Feynman-Kac or generalized Schrödinger semigroup in $L^2(E, m)$ and $\left(K_0 \dotplus V\right)_\Sigma$ the corresponding Dirichlet semigroup in $L^2(\Sigma, m)$. From the point of view of scattering theory, this kind of result is interesting: if the operators in (1.1) are trace class, then the wave operators exist and are complete. For more details we refer the reader to Baumgärtel and Wollenberg (1983) and Reed and Simon (1979). For examples of self-adjoint Feller generators and Kato-Feller potentials we refer the reader to forthcoming paper(s) by Demuth and van Casteren (1989, 1991, 1992). If we are dealing with ν-dimensional Brownian motion (i.e. if $K_0 = -\frac{1}{2}\Delta$), then a subset Γ of $\mathbf{R}^\nu$ has finite capacity if its Lebesgue measure is finite (if $\nu = 1$), if it can be covered by a sequence of discs $B(x_j, r_j)$ with $\sum_{j=1}^\infty 1/\log r_j^{-1} < \infty$ and $\lim_{j\to\infty} r_j = 0$ (if $\nu = 2$) and if it can be covered be a sequence of balls $B(x_j, r_j)$ with $\sum_{j=1}^\infty r_j^{\nu-2} < \infty$ (if $\nu \geq 3$). In this introduction we give some references to papers that treat examples of generators of (self-adjoint) Feller semigroups in spaces of the form $L^2(E, m)$. Relevant and interesting generators of examples of essentially self-adjoint generators of Markov processes can be found in Kochubeĭ (1986a) Theorem 2. and in (1986b). Examples of relativistic Hamiltonians were introduced by Ichinose (see e.g. 1987, 1988, 1989) and also Ichinose and Tamura (1986a, 1986b). For systems without electromagnetic fields we refer to Carmona, Masters and Simon (1990). The previous result in (1990) can be generalized for negative definite functions F defined on a locally compact, second countable, abelian group G. In that case the variable p varies over the dual group. It is also noticed that these results fit in the theory of Lévy processes. Taira (1992), Jacob (1990a, 1990b) and Sturm (1990, 1991) also describe relevant examples. Theorem 10.3. in Ikeda and Watanabe (1989) states that under appropriate conditions (boundedness of certain vector fields V_j, $0 \leq j \leq \nu$, and Hörmander's hypo-ellipticity condition) the $\mathbf{P}_x$-distribution of the solution $(X(t) : t \geq 0)$ of the stochastic differential equation $dX(t) = \sigma(X(t))dB(t) + b(X(t))dt$, $X(0) = x$, defines a Markov process $\{(\Omega, \mathcal{F}, \mathbf{P}_x), (X(t) : t \geq 0), (\vartheta_t : t \geq 0), (E, \mathcal{E})\}$. The latter example has its counterpart for Riemannian manifolds. In fact instead of the Laplace operator on $\mathbf{R}^\nu$ we can also consider the Laplace-Beltrami operator on a Riemannian manifold. For details we refer the reader to Elworthy (1982, 1988), Azencott et al (1981), Bismut (1984) and several others. The authors also establish existence results for and bounds on the corresponding heat kernels. Recent and very interesting papers are (1991, 1988) written by Davies. It provides the reader with much insight into the behavior of heat kernels. Of course his book (1989) should be consulted also. We also mention the generator $K_0 := -\frac{1}{2}\Delta + x.\nabla$ of the so-called Ornstein-Uhlenbeck process in $L^2\left(\mathbf{R}^\nu, \exp\left(-|y|^2\right)\pi^{-\nu/2}dy\right)$ and the generator $K_0 := -\frac{1}{2}\Delta + \frac{1}{2}|x|^2$ of the oscillator process. For more details the reader is referred to e.g. Simon (1979).

2. Basic Assumptions of Stochastic Spectral Analysis (BASSA)

As mentioned above the state space (or configuration space) will be a second countable locally compact Hausdorff space E with Borel field $\mathcal{E}$. A non-negative Radon measure m (reference measure) on $\mathcal{E}$ is given. Instead of $dm(x)$ or $m(dx)$ we usually write dx. In what follows the function $p_0(t, x, y)$ defined on $(0, \infty) \times E \times E$ will be a continuous density function with the following properties:

A1. It is non-negative and it satifies the Chapman-Kolmogorov identity, i.e.

$$\int p_0(s, x, z) p_0(t, z, y) dz = p_0(s + t, x, y), \quad s,\ t > 0,\ x,\ y \in E,$$

and its total mass is less than or equal to 1, i.e. $\int p_0(t, x, y) dy \leq 1$, $t > 0$, $x \in E$;

A2. (Feller property) For every $f \in C_\infty(E)$ the function $x \mapsto \int f(y) p_0(t, x, y) dm(y)$ belongs to $C_\infty(E)$;

A3. (continuity) For every $f \in C_\infty(E)$ and for every $x \in E$ the following identity is true: $\lim_{t \downarrow 0} \int f(y) p_0(t, x, y) dm(y) = f(x)$;

A4. The function $p_0(t, x, y)$ is symmetric: $p_0(t, x, y) = p_0(t, y, x)$ for all $t > 0$ and for all x and y in E.

Remark. It is well-known that there exists a strong Markov process

$$\{(\Omega, \mathcal{F}, \mathbf{P}_x), (X(t) : t \geq 0), (\vartheta_t : t \geq 0), (E, \mathcal{E})\}$$

(see e.g. Blumenthal and Getoor 1968) with the following properties. The one-dimensional distributions are given by $\mathbf{P}_x(X(t) \in B) = \int_B p_0(t, x, y) dy$, $t > 0$, B Borel subset of E. Its sample paths are $\mathbf{P}_x$-almost surely right continuous and possess $\mathbf{P}_x$-almost sure left limits in E on its life time. In other words the process $\{X(t), \mathbf{P}_x\}$ is cadlag on its life time. Moreover we may assume that the closure of the (random) set $\{X(s) : 0 \leq s < t\}$ is a compact subset of E, whenever $X(t-)$ belongs to E. In other the process does not re-enter E once it has hit δ, the point at infinity.

This is perhaps the right place to fix some notation. Let K_0 be the L^2-generator of the Markov process

$$\{(\Omega, \mathcal{F}, \mathbf{P}_x), (X(t) : t \geq 0), (\vartheta_t : t \geq 0), (E, \mathcal{E})\}$$

and let a be a strictly positive real number. For any Borel function g, defined on E, whenever it makes sense we write

$$[\exp(-sK_0)g](x) = \mathbf{E}_x(g(X(s)) = \int p_0(s, x, y) g(y) dy \quad \text{and}$$

$$[(aI + K_0)^{-1}g](x) = \int_0^\infty e^{-as}[\exp(-sK_0)g](x) ds = \int_0^\infty e^{-as} p_0(s, x, y) g(y) dy.$$

For a concise formulation of our results we introduce the following definitions.

2.1. Definition. Let $V : E \to [0, \infty]$ be a Borel measurable function on E.

(a) The function V is said to belong to $K(E)$ if

$$\limsup_{t \downarrow 0} \left\| \int_0^t P_0(s) V ds \right\|_{\infty, \infty} = \limsup_{t \downarrow 0} \sup_{x \in E} \int_0^t \left(\int_E p_0(s, x, y) V(y) dm(y) \right) ds = 0.$$

(b) The Borel measurable function $V : E \to [0, \infty]$ belongs to $K_{\mathrm{loc}}(E) = K_{\mathrm{loc}}(E, A_0)$ if $1_K V$ belongs to $K(E)$ for all compact subsets K of E.

(c) The Borel measurable function $V = V_+ - V_-$ is said to be a Kato-Feller potential if its positive part $V_+ = \max(V, 0)$ belongs to $K_{\mathrm{loc}}(E)$ and if its negative part $V_- = \max(-V, 0)$ belongs to $K(E)$.

The following general result can be proved. For details in the symmetric case see (1985) and (1989). For the Gaussian semigroup the reader may consult Simon (1982) and (1979).

2.2. Theorem. Suppose that $V = V_+ - V_-$ is a Borel measurable function defined on E such that V_- belongs to $K(E)$ and such that V_+ belongs to $K_{\mathrm{loc}}(E)$.

(a) There exists a closed, densely defined linear operator $K_0 \dotplus V$ in $C_\infty(E)$, extending $K_0 + V$, which generates a strongly continuous positivity preserving semigroup $\{\exp(t(K_0 \dotplus V)) : t \geq 0\}$ in $C_\infty(E)$. Every operator $\exp(-t(K_0 \dotplus V))$, $t > 0$, is of the form

$$\left[\exp(-t(K_0 \dotplus V))f\right](x) = \int_E \exp(-t(K_0 + V))(x, y)f(y)dm(y), \quad f \in C_\infty(E),$$

where $\exp(-t(K_0 + V))(x, y)$ is a continuous function which satisfies the identity of Chapman-Kolmogorov ($t > 0$, $x, y \in E$):

$$\exp(-t(K_0 + V))(x, y) = \int_E \exp(-s(K_0 + V))(x, z) \exp(-t(K_0 + V))(z, y)dz.$$

(b) The semigroup $\{\exp(-t(K_0 \dotplus V)) : t \geq 0\}$ also acts as a strongly continuous semigroup in $L^p(E, m)$, $1 \leq p < \infty$.

(c) If $\exp(-tK_0)$ maps $L^1(E, m)$ into $L^\infty(E, m)$ for all $t > 0$ (i.e. if $\sup\{p_0(t, x, y) : x, y \in E\} < \infty$ for all $t > 0$), then $\exp(-t(K_0 \dotplus V))$, $t > 0$, maps $L^p(E, m)$ into $L^q(E, m)$, for $1 \leq p \leq q \leq \infty$. If $t > 0$ and if $1 \leq p \leq q < \infty$, then $\exp(-t(K_0 + V))$ maps $L^p(E, m)$ into $L^q(E, m) \cap C_\infty(E)$.

(d) In $L^2(E, m)$ the family $\{\exp(-t(K_0 \dotplus V)) : t \geq 0\}$ is a self-adjoint positivity preserving strongly continuous semigroup with a self-adjoint generator.

(e) The Feynman-Kac semigroup in $L^2(E, m)$ coincides with the semigroup corresponding to the quadratic form Q with $D(Q) = D\left(K_0^{1/2}\right) \cap D\left(V_+^{1/2}\right)$ and defined by $Q(f, g) = \left\langle K_0^{1/2} f, K_0^{1/2} g \right\rangle - \left\langle V_-^{1/2} f, V_-^{1/2} g \right\rangle + \left\langle V_+^{1/2} f, V_+^{1/2} g \right\rangle$, where f and g belong to $D(Q)$.

Next let Γ be a Borel subset of the second countable locally compact Hausdorff space E. In relation to the set Γ we employ the following stopping times:

$$S = \inf \left\{ s > 0 : \int_0^s 1_\Gamma(X(\sigma))d\sigma > 0 \right\}, \quad T = \inf \{ s > 0 : X(s) \in \Gamma \}.$$

It readily follows that $S \geq T$, $\mathbf{P}_x$-almost surely, for all $x \in E$. A point $x \in E$ belongs to Γ^r if $\mathbf{P}_x(T = 0) = 1$. Some authors call the time S the *penetration time*: see e.g. Herbst and Zhongxin Zhao (1988). Suppose $\Gamma^r = (\mathrm{int}(\Gamma))^r$. Then $S = T$, $\mathbf{P}_x$-almost surely for all $x \in E$. For a proof we refer the reader to van Casteren (1992a).

2.3. Definition. Let the time S be the penetration time of Γ. The integral kernel $\exp\left(-t\left(K + V\right)_\Sigma\right)(x, y)$ is defined by

$$\exp(-t(K_0 + V)_\Sigma)(x, y) = \lim_{t' \uparrow t} \mathbf{E}_x \left(\exp\left(-\int_0^{t'} V(X(\sigma))d\sigma \right) p_0(t - t', X(t'), y) : S > t' \right).$$

In the results below we let $\Sigma = E^\Delta \setminus \Gamma$ be an open subset of E and $\left(K_0 \dot{+} V\right)_\Sigma$ denotes the Feynman-Kac generator of the semigroup killed in the complement of Σ, i.e. the semigroup $\left\{ \exp\left(-t(K_0 \dot{+} V)_\Sigma\right) : t \geq 0 \right\}$, defined by

$$\left[\exp(-t(K_0 \dot{+} V)_\Sigma)f\right](x) = \mathbf{E}_x \left(\exp\left(-\int_0^t V(X(s))ds \right) f(X(t)) : S > t \right)$$
$$= \int_E \exp\left(-t(K_0 + V)_\Sigma\right)(x, y)f(y)dy.$$

If $\Gamma = (\mathrm{int}(\Gamma))^r$, then the penetration time S may be replaced with the exit time T.

2.4. Definition. The *carré du champ operator* $\widetilde{\gamma}$ is defined as follows. The (bilinear) operator $\widetilde{\gamma}$ is given by the following identity:

$$\left[\widetilde{\gamma}(v, w)\right](x) = \lim_{s \downarrow 0} \frac{1}{s} \mathbf{E}_x \left((v(X(s)) - v(x))(w(X(s)) - w(x)) \right),$$

for whatever functions v and w these limits make sense. If $v \equiv w$ is a real-valued function, then $\widetilde{\gamma}(v, v) \geq 0$. We also observe that $\int_E \widetilde{\gamma}(f, \overline{f})dx = \int_E \left| \left[K_0^{1/2}f\right](x) \right|^2 dx$ for f belonging to $D\left(K_0^{1/2}\right)$. If $-K_0 = \frac{1}{2}\Delta$, then $\widetilde{\gamma}(v, w) = \nabla v . \nabla w$.

Let f, g be functions in the domain of K_0 such that the pointwise product fg also belongs to the domain $D(K_0)$. The following statements can be verified (also see Bouleau and Lamberton 1989):

(a) The identity $\widetilde{\gamma}(f, g) = -K_0(fg) + f(K_0 g) + (K_0 f)g$ holds.

(b) The process $(f(X(t)) - f(X(0)))(g(X(t)) - g(X(0))) - \int_0^t \widetilde{\gamma}(f, g)(X(u))du$ is a $\mathbf{P}_x$-martingale for every $x \in E$.

(c) For almost all $x \in E$ the following equality is true:

$$\left[\widetilde{\gamma}(f, g)\right](x) = \lim_{s \downarrow 0} \frac{1}{s} \mathbf{E}_x \left((f(X(s)) - f(X(0)))(g(X(s)) - g(X(0))) \right).$$

(d) The inequality $\left|\widetilde{\gamma}(f, g)\right|^2 \leq \widetilde{\gamma}(f, f)\widetilde{\gamma}(g, g)$ is valid.

For more details see Dellacherie and Meyer [11, pp. 244-260] and also a recent book (1991) by Bouleau and Hirsch.

3. Singular Perturbations: a Trace Class Property

In what follows we shall assume that there exists a function $v \geq 0$, $v \in D(K_0^{1/2})$, such that $\Sigma = \{v < 1\}$. We also take it for granted that the function v satisfies equality (3.1), i.e.

$$\lim_{s \downarrow 0} \frac{\mathbf{P}_x\left(v(X(s)) > b\right)}{s} = 0, \tag{3.1}$$

for all $x \in E$ with $v(x) < b$. Intuitively, this assumption means that the underlying process $\{X(t) : t \geq 0\}$ behaves like a diffusion in the direction of increasing v. It will also be a blanket assumption that for every F in the domain of $K_0^{1/2}$ the pointwise limits

$$\tilde{\gamma}(v, F)(x) = \lim_{s \downarrow 0} \frac{1}{s} \int_E dy p_0(s, x, y)(v(y) - v(x))(F(y) - F(x))$$

exists for m-almost all $x \in E$.

3.1. Definition. Let v be as above. The stopping time T_x^v is given by $T_x^v = \inf\{s \geq 0 : v(X(s)) > v(x)\}$ and also $(\xi = v(x))$:

$$\exp\left(-t(K_0 + V)_\xi\right)(x, y)$$

$$= \lim_{s \downarrow 0} \mathbf{E}_x\left(\exp\left(-\int_0^{t-s} V(X(u))du\right) p_0\left(s, X(t-s), y\right) : \sup_{0 \leq u < t-s} v(X(u)) \leq \xi\right)$$

$$= \lim_{s \downarrow 0} \mathbf{E}_x\left(\exp\left(-\int_0^{t-s} V(X(u))du\right) p_0\left(s, X(t-s), y\right) : T_x^v > t - s\right).$$

Here, as usually, V is a Kato-Feller potential. In the remainder of this paper we shall give an indication of the proof of the following main result. It is an improvement and a correction of results in section 5 (Theorem 5.10 and 5.12) of Demuth and van Casteren [12].

3.2. Theorem. Suppose that the negative part V_- of the potential belongs to $L^1(\Sigma)$. Also suppose that V is a Kato-Feller potential and that the integral $\int_E dx \mathbf{P}_x(S < 1)$ is finite. Then the semigroup difference $\exp\left(-t\left(K_0 + V\right)\right) - J^* \exp\left(-t\left(K_0 + V\right)_\Sigma\right) J$ is a trace class operator for all $t > 0$, provided that $\sup_{x \in E} p_0(t, x, x) < \infty$, for all $t > 0$.

Next we introduce, what could be called *surface integrals* and *normal derivatives* for Markov processes.

3.3. Definition. Let Σ be an open subset of E and let F be a function with the property that $\tilde{\gamma}(v, F)(x)$ exists for m-almost all $x \in \Sigma$. Let G be another appropriate function. The surface integral $\int_{\partial \Sigma} D_n F(x) G(x) dS(x)$ is defined by

$$\int_{\partial \Sigma} D_n F(x) G(x) dS(x) = \lim_{\epsilon \downarrow 0} \frac{-1}{\epsilon} \int_{\{1-\epsilon \leq v < 1\}} \tilde{\gamma}(v, F)(x) G(x) dx,$$

whenever this limit exists. In fact the quantity $D_n F(x) := -\widetilde{\gamma}(v, F)(x)$ can be considered as the *normal derivative* at x on the surface $\{v = v(x)\}$. If F depends on two state variables, then the integrals $\int_{\partial \Sigma} D_n F(\cdot, x)(x) dS(x)$ and $\int_{\partial \Sigma} D_n F(\cdot, x)(x) dS(x)$ are defined similarly.

It turns out that, for appropriate functions F, these integrals do not depend on the particular choice of v. The notation $D_n F$ suggests *normal* derivatives. This is the case, whenever we are dealing with Brownian motion. A consequence of Proposition 3.4. is that the surface integrals do not depend on the particularly chosen function v. A proof of Proposition 3.4. is to be found in (1992a).

3.4. Proposition. Let F and G be functions belonging to $D(K_0^{1/2})$ and let Σ be as in the definition. Then Green's theorem is true in the sense that the following identities are valid:

$$
\int_{\partial \Sigma} D_n F(x) G(x) dS(x) = - \int_{\Sigma} dx \, \widetilde{\gamma}(F, G)(x) + 2 \int_{\Sigma} dx \, K_0 F(x) G(x);
$$

$$
\int_{\partial \Sigma} D_n F(x) G(x) dS(x) - \int_{\partial \Sigma} F(x) D_n G(x) dS(x)
$$
$$
= 2 \int_{\Sigma} dx \, K_0 F(x) G(x) - 2 \int_{\Sigma} dx \, F(x) K_0 G(x). \tag{3.2}
$$

Here $\partial \Sigma$ is the topological boundary of Σ. Moreover the following inequality is valid:

$$
\left| \int_{\partial \Sigma} D_n F(x) G(x) dS(x) \right| \leq \left\| K_0^{1/2} F \right\|_2 \left\| K_0^{1/2} G \right\|_2^{1/2} + 2 \left\| K_0 F \right\|_2 \left\| G \right\|_2.
$$

Remark. The inequality for the surface integrals in terms of the norms of the functions $K_0^{1/2} F$, $K_0^{1/2} G$, $K_0 F$ and G shows that the surface integrals make sense as soon as the latter functions belong to $L^2(E, m)$. The inequality is a consequence of the identity $\int_E \widetilde{\gamma}(F, \overline{F}) dx = \int_E \left| \left[K_0^{1/2} F \right](x) \right|^2 dx$ for F belonging to $D\left(K_0^{1/2} \right)$. See also Definition 2.4.

Proof. Formula (3.2) can be proved as follows. We suppose that for all $0 < b \leq 1$ and for all $x \in E$ with $v(x) < b$ the identity in (3.1) is valid. Then, by virtue of (3.1), the symmetry of $p_0(\tau, x, y)$ and after some calculation we get for $0 < b \leq 1$:

$$
\lim_{\epsilon \downarrow 0} \frac{-1}{\epsilon} \int_\epsilon^{b-\epsilon} d\xi \int_{\{\xi - \epsilon \leq v < \xi\}} \widetilde{\gamma}(v, F)(x) G(x) dx
$$
$$
= - \int_0^b d\xi \int_{\{v < \xi\}} dx \, \widetilde{\gamma}(F, G)(x) + 2 \int_0^b d\xi \int_{\{v < \xi\}} dx \, K_0 F(x) G(x).
$$

Consequently, since the integrands in the latter formula are left continuous functions in ξ, Green's formula will follow. For details, again see (1992a).

3.5. Proposition. Let U be a linear operator of finite rank in $L^2(E,m)$. The following identities are valid:

$$\text{trace}\,(U 1_\Sigma D_\Sigma(t))$$

$$= \int_\Sigma \left\{ \overline{[U^* \exp\,(-t(K_0+V))\,(x,\cdot)]\,(x)} - [U \exp\,(-t(K_0+V)_\Sigma)\,(\cdot,x)]\,(x) \right\} dx$$

$$= \frac{1}{2} \int_0^t d\tau \int_{\partial\Sigma} dS(x) \int_\Sigma dy\,[D_n \exp\,(-\tau(K_0+V)_\Sigma)\,(\cdot,y)]\,(x)$$

$$\times \overline{[U^* \exp\,(-(t-\tau)(K_0+V))\,(\cdot,x)]\,(y)} \tag{3.3}$$

$$= \int_\Sigma \left\{ [U \exp\,(-t(K_0+V))\,(x,\cdot)]\,(x) - [U \exp\,(-t(K_0+V)_\Sigma)\,(\cdot,x)]\,(x) \right\} dx.$$

Proof. A proof may be based on Green's theorem: see (3.2). For more details we refer the reader to (1992a).

As a corollary we obtain the following result (in one dimension part of this result also appears in Simon and Spencer 1989).

3.6. Corollary. Suppose that m is a strictly positive measure in the sense that $m(U) > 0$ for U open and non-emty. The integral kernel of $D_\Sigma(t)$, $D_\Sigma(t)(x,y)$, x, $y \in \Sigma$, can be rewritten as follows:

$$D_\Sigma(t)(x,y) = \frac{1}{2} \int_0^t d\tau \int_{\partial\Sigma} dS(z)\,[D_n \exp\,(-\tau(K_0+V)_\Sigma)\,(\cdot,y)]\,(z)$$

$$\times \exp\,(-(t-\tau)(K_0+V))\,(x,z)$$

$$= \lim_{\epsilon \downarrow 0} \frac{-1}{\epsilon} \int_0^t d\tau \int_{\{1-\epsilon<v<1\}} dz\,[\widetilde{\gamma}\,(v(\cdot), \exp\,(-\tau(K_0+V)_{v(z)})\,(\cdot,y))]\,(z)$$

$$\times \exp\,(-(t-\tau)(K_0+V))\,(x,z).$$

Proof Let F and G be arbitrary continuous functions in $L^2(\Sigma,m)$ and let U be defined by $Uf = \int_E f(y)F(y)dyG$, $f \in L^2(E,m)$. For x fixed, the first formula in Corollary 3.6. follows for m-almost all $y \in \Sigma$ and similarly, for y fixed, the first formula follows for m-almost all $x \in \Sigma$. Employing symmetry and separate continuity yields the desired result for all x, $y \in \Sigma$. The second equality will follow from Proposition 3.10 and Theorem 3.7 in van Casteren (1992a).

We formulate some basic equalities in Theorem 3.7. The proof of the result is based on Corollary 3.6. and on Proposition 3.5.

3.7. Theorem. Let the notation and assumptions be as above. The following identities are valid:

$$\text{trace}\,(D_\Sigma(t)U) - \text{trace}\,(1_\Gamma \exp\,(-t(K_0 \dot{+} V))\,U) = \text{trace}\,(1_\Sigma D_\Sigma(t)U)$$

$$= \lim_{\epsilon \downarrow 0} \frac{1}{\epsilon} \int_0^t d\tau \int_{\{1-\epsilon<v<1\}} dx \int_E dy\,[-\widetilde{\gamma}\,(v(\cdot), \exp\,(-\tau(K_0+V)_{v(x)})\,(\cdot,y))]\,(x)$$

$$\times \overline{\left[U^* \exp\left(-(t-\tau)(K_0+V)\right)(\cdot,x)\right](y)} \qquad (3.4)$$

$$= \frac{1}{2}\int_0^t d\tau \int_{\partial\Sigma} dS(x) \int dy \left[D_n \exp\left(-\tau(K_0+V)_\Sigma\right)(\cdot,y)\right](x)$$

$$\times \overline{\left[U^* \exp\left(-(t-\tau)(K_0+V)\right)(\cdot,x)\right](y)}. \qquad (3.5)$$

Proof. The equality in (3.4) follows from Theorem 3.7 in van Casteren (1992a) together with the following observations:

$$\int_{\{1\le v\}} dx \overline{\left[U^* \exp\left(-t(K_0+V)\right)(\cdot,x)\right](x)}$$

$$= \mathrm{trace}\left(U 1_\Gamma \exp(-t(K_0\dot{+}V))\right) = \mathrm{trace}\left(1_\Gamma \exp\left(-t(K_0\dot{+}V)\right)U\right).$$

The equality in (3.5) follows from (3.3) in Proposition 3.5.

Next we want to introduce, what could be called *one-sided* surface integrals for normal derivatives. Let F and G be appropriate functions on E.

3.8. Definition. The one-sided surface integral $\int_{\partial\Sigma} dS(x) D_n^1 F(x)G(x)$ is defined by ($U(\epsilon) := \{1-\epsilon < v < 1\}$)

$$\int_{\partial\Sigma} dS(x)D_n^1 F(x)G(x) = \lim_{t\downarrow 0}\lim_{\epsilon\downarrow 0} \frac{-1}{t\epsilon} \int_0^t d\tau \int_{U(\epsilon)} dx \tilde{\gamma}\left(v, \mathsf{E}_{(\cdot)}\left(F(X(\tau)), T_x^v > \tau\right)\right)(x)G(x).$$

Here v is a "nice" function with $\Sigma = \{v < 1\}$ and T_x^v is the hitting time defined by $T_x^v = \inf\left\{s > 0 : v(X(s)) > v(x)\right\}$. An advantage of these one-sided surface integrals is the fact that for non-negative functions G we have Cauchy-Schwarz inequalities of the following kind. In fact, we will use a functional integration version of the Cauchy-Schwarz' inequality.

3.9. Proposition. Let F_1, F_2 and G, $G \ge 0$, be appropriate functions (this means that the expressions involved have to make sense). Then the following inequality is valid (F_1 and F_2 vanish on $\partial\Sigma$):

$$\left|\int_{\partial\Sigma} \left[D_n^1 F_1 F_2\right](x)G(x)dS(x)\right|^2 \le \int_{\partial\Sigma} D_n^1 |F_1|^2 (x)G(x)dS(x) \int_{\partial\Sigma} D_n^1 |F_2|^2 (x)G(x)dS(x).$$

A proof of this fact is left to the reader. On the other hand there also exists a precise relationship between *two-sided* and one-sided integrals.

3.10. Proposition. Again let F and G be appropriate real functions. The function F supposedly belongs to the domain of $K_0^{1/2}$ and vanishes on Γ and G belongs to $L^2(E,m)$. Then

$$\int_{\partial\Sigma} D_n F(x)G(x)dS(x) = 2 \int_{\partial\Sigma} D_n^1 F(x)G(x)dS(x). \qquad (3.6)$$

Proof. Let U be an appropriate operator in $L^2(E, m)$. From (3.4) and (3.5) we infer the following identities:

$$\text{trace}\left(1_\Sigma D_\Sigma(t)U\right) = \frac{1}{2}\int_\Sigma dy \int_0^t d\tau \int_{\partial\Sigma} dS(x)\left[D_n \exp\left(-\tau(K_0 + V)_\Sigma\right)(\cdot, y)\right](x)$$
$$\times \overline{\left[U^* \exp\left(-(t - \tau)(K_0 + V)\right)(\cdot, x)\right](y)}$$
$$= \int_\Sigma dy \int_0^t d\tau \int_{\partial\Sigma} dS(x)\left[D_n \exp\left(-\tau(K_0 + V)_v\right)(\cdot, y)\right](x)$$
$$\times \overline{\left[U^* \exp\left(-(t - \tau)(K_0 + V)\right)(\cdot, x)\right](y)}. \tag{3.7}$$

Let the operator U be defined as $Uf = \int_E f(x)F(x)dx\, G$. We also take $V \equiv 0$. The previous equalities in (3.7) may be rewritten as follows:

$$\text{trace}\left(1_\Sigma D_\Sigma(t)U\right)$$
$$= \frac{1}{2}\int_0^t d\tau \int_{\partial\Sigma} dS(x)\left[D_n \mathsf{E}_{(\cdot)}\left(F(X(\tau)), S > \tau\right)\right](x) \int_E \exp\left(-(t - \tau)K_0\right)(z, x)G(z)dz$$
$$= \int_0^t d\tau \int_{\partial\Sigma} dS(x)\left[D_n \mathsf{E}_{(\cdot)}\left(F(X(\tau)), T_x^v > \tau\right)\right](x) \int_E \exp\left(-(t - \tau)K_0\right)(z, x)G(z)dz.$$

The result will follow upon first dividing by t and then letting t tend to zero.

Remark. Green's formula gets the following form now (see Proposition 3.4.):

$$\int_{\partial\Sigma} D_n^1 F(x)G(x)dS(x) = -\frac{1}{2}\int_\Sigma dx\, \widetilde{\gamma}(F, G)(x) + \int_\Sigma dx\, K_0 F(x)G(x).$$

3.11.   Corollary. Let Σ etc. be as above. Suppose that the integral $\int_\Gamma \exp\left(-t(K_0 + V)\right)(x, x)dx$ is finite. Also suppose that the expression $C_V(t/2)$, determined by $C_V(t/2)^2 := \sup_x \exp\left(-\frac{1}{2}t(K_0 + V)\right)(x, x)$, is finite. Then, if the functions F and G, defined by respectively

$$y \mapsto \int_0^{t/2} d\tau \int_{\partial\Sigma} dS(x)\left[D_n \exp\left(-\tau(K_0 + V)_\Sigma\right)(\cdot, y)\right](x)$$

and

$$y \mapsto \int_0^{t/2} d\tau \int_{\partial\Sigma} dS(x)\exp\left(-\tau(K_0 + V)\right)(x, y)\left[D_n \exp\left(-\frac{t}{4}(K_0 + V)_\Sigma\right)1\right](x)$$

belong to $L^2(E, m)$, then the operator $D_\Sigma(t)$ is trace class. Moreover

$$\|D_\Sigma(t)\|_{\text{trace}} \leq C_V(t/2)\left\{\|F\|_2 + \|G\|_2\right\} + \int_\Gamma \exp\left(-t(K_0 + V)\right)(x, x)dx. \tag{3.8}$$

Proof. In order to obtain the announced result we split the integral between 0 and $t/2$ and on the interval $[t/2, t]$ we use the dual of the operator U^*. However first we introduce the functions $\psi_U^{(1)}$, $\psi_U^{(2)}$, $\chi_U^{(1)}$ and $\chi_U^{(2)}$ in the following manner:

$$\psi_U^{(1)}(\tau, x, y) = \overline{[U^* \exp(-\tau(K_0 + V))(\cdot, x)](y)};$$

$$\psi_U^{(2)}(\tau, w, y) = [U \exp(-\tau(K_0 + V)_\Sigma)(w, \cdot)](y);$$

$$\chi_U^{(1)}(\tau, x, y) = \frac{\psi_U^{(1)}(\tau, x, y)}{\sqrt{\int_E \left|\psi_U^{(1)}(\tau, x, y)\right|^2 dy}} \quad \text{and} \quad \chi_U^{(2)}(\tau, w, y) = \frac{\psi_U^{(2)}(\tau, w, y)}{\sqrt{\int_E \left|\psi_U^{(1)}(\tau, w, y)\right|^2 dy}}.$$

Moreover for $t/4 \le \tau \le t$ we have, for $i = 1, 2$,

$$\int_E \left|\psi_U^{(i)}(\tau, x, y)\right|^2 dy \le \|U\|^2 \exp(-2\tau(K_0 + V))(x, x) \le \|U\|^2 C_V(t/2)^2.$$

The following identities are then self-explanatory:

$$2\text{trace}\left(D_\Sigma(t)U\right) - 2\text{trace}\left(1_\Gamma \exp\left(-t(K_0 \dot+ V)\right) U\right)$$

$$= \int_\Sigma dy \int_{\partial\Sigma} dS(x) \int_0^{t/2} d\tau \left[D_n \exp\left(-\tau(K_0 + V)_\Sigma\right)(\cdot, y)\right](x)$$

$$\times \chi_U^{(1)}(t - \tau, x, y) \left(\int_E \left|\psi_U^{(1)}(t - \tau, x, y)\right|^2 dy\right)^{1/2}$$

$$+ \int_E dw \int_E dy \int_{\partial\Sigma} dS(x) \int_0^{t/2} d\tau \exp\left(-\tau(K_0 + V)\right)(x, y)$$

$$\times \left[D_n \exp\left(-\frac{t}{4}(K_0 + V)_\Sigma\right)(\cdot, w)\right](x)\chi_U^{(2)}\left(\frac{3}{4}t - \tau, w, y\right)$$

$$\times \left(\int_E \left|\psi_U^{(2)}\left(\frac{3}{4}t - \tau, w, y\right)\right|^2 dy\right)^{1/2}.$$

It follows that the absolute value of the trace of the operator

$$2\left(D_\Sigma(t)U\right) - 2\left(1_\Gamma \exp\left(-t(K_0 \dot+ V)\right) U\right)$$

can be estimated by (again use (3.6), which implies that the normal derivative $D_n F$ may be replaced by $2D_n^1 F$) the constant $\|U\| C_V(t/2)$ times

$$\sup_h \left|\int_\Sigma dy \int_{\partial\Sigma} dS(x) \int_0^{t/2} d\tau \left[D_n \exp\left(-\tau(K_0 + V)_\Sigma\right)(\cdot, y)\right](x) h(t - \tau, x, y)\right|$$

$$+ \sup_h \left|\int_E dw \int_E dy \int_{\partial\Sigma} dS(x) \int_0^{t/2} d\tau \exp\left(-\tau(K_0 + V)\right)(x, y)\right.$$

$$\left.\times \left[D_n \exp\left(-\frac{t}{4}(K_0 + V)_\Sigma\right)(\cdot, w)\right](x) h\left(\frac{3}{4}t - \tau, w, y\right)\right|,$$

where both suprema are taken over all functions h, defined on $[t/4, t] \times E \times E$ with the property that $\int_E |h(\tau, x, y)|^2 \, dy = 1$, $\tau \geq t/4$, $x \in E$. Since such a supremum coincides with a supremum over all functions h of the form $h(\tau, x, y) = h_1(\tau, x)h_2(y)$, where $|h_1(\tau, x)| \leq 1$ and where $\int_E |h_2(y)|^2 \, dy = 1$, we conclude that the absolute value of $2\,\mathrm{trace}\,(1_\Sigma D_\Sigma(t)U)$ is dominated by the constant $\|U\|\, C_V(t/2)$ times

$$
\sup_{\{h:\|h\|_2=1\}} \left| \int_\Sigma dy \int_{\partial\Sigma} dS(x) \int_0^{t/2} d\tau \, [D_n \exp\left(-\tau \left(K_0 + V\right)_\Sigma\right)(\cdot, y)]\,(x)h(y) \right|
$$

$$
+ \sup_{\{h:\|h\|_2=1\}} \left| \int_E dw \int_E dy \int_{\partial\Sigma} dS(x) \int_0^{t/2} d\tau \exp\left(-\tau \left(K_0 + V\right)\right)(x, y) \right.
$$

$$
\left. \times \left[D_n \exp\left(-\frac{t}{4}\left(K_0 + V\right)_\Sigma\right)(\cdot, w)\right](x)h\,(y) \right|
$$

$$
= \sup_{\{h:\|h\|_2=1\}} \left| \int_\Sigma dy \int_{\partial\Sigma} dS(x) \int_0^{t/2} d\tau \, [D_n \exp\left(-\tau \left(K_0 + V\right)_\Sigma\right)(\cdot, y)]\,(x)h(y) \right|
$$

$$
+ \sup_{\{h:\|h\|_2=1\}} \left| \int_E dy \int_{\partial\Sigma} dS(x) \int_0^{t/2} d\tau \exp\left(-\tau \left(K_0 + V\right)\right)(x, y) \right.
$$

$$
\left. \times \left[D_n \exp\left(-\frac{t}{4}\left(K_0 + V\right)_\Sigma\right)1\right](x)h\,(y) \right|
$$

$$
\leq \|F\|_2 + \|G\|_2 \,.
$$

From our hypotheses the result (3.8) follows.

We are going to finish with a proof of Theorem 3.2. We shall use the sufficient conditions given in Corollary 3.11. Moreover we shall need the following lemma. For more details the reader should consult (1992a).

3.12. Lemma. (a) Let S and Σ be as in the previous results. Suppose that the function $t \mapsto \int_E dx\mathbf{P}_x(S < t)$ is finite for some $t > 0$. Then it is finite for all $t >$ and moreover it is an increasing concave function. Then it follows that this function has a right and left derivative for all $t > 0$.

(b) For eached fixed $\tau > 0$ and for each fixed $x \in E$ the function $t \mapsto f_{\tau, x}(t) = \int_E \exp\left(-\tau K_0\right)(x, z)\mathbf{P}_z(S < t)dz = \mathbf{E}_x\left(\mathbf{P}_{X(\tau)}(S < t)\right)$ is an increasing concave function. Consequently, for $t_0 > 0$ fixed the function $t \mapsto \int_E f_{\tau, x}(t)d\tau$ is a concave function, that is uniformly bounded in $x \in E$ and in $0 \leq t \leq t_0$.

Proof. (a) Put $f(t) = \int_E dx\mathbf{P}_x(S < t)$. From the Markov property it follows that $f(2t) = f(t) + \int_E \mathbf{E}_x\left(\mathbf{P}_{X(t)}(S < t), S > t\right)dx$ and hence $f(2t) \leq 2f(t)$. Next we fix $0 < h < t$ and with the aid of Fubini's theorem, Markov property and symmetry we obtain:

$$
2f(t) - f(t - h) - f(t + h) = \int_E dz\mathbf{P}_z(t > S > t - h)\mathbf{P}_z(S < h) \geq 0.
$$

(b) As in (a) we have, for $0 < h < t$,

$$2f_{\tau,x}(t) - f_{\tau,x}(t-h) - f_{\tau,x}(t+h)$$
$$= \int_E dw \int_E dy P_y (S < h) E_y (\exp(-\tau K_0)(x, X(t-h)) : S > t - h) P_w(S < h) \geq 0.$$

Remark. Since $P_x(S < t) \leq e^{at} E_x \left(e^{-aS}, S < \infty\right)$, it follows that the function $f(t)$ in Lemma 3.12. is finite whenever Γ possesses finite a-capacity for some $a > 0$.

Proof of Theorem 3.2. By virtue of Corollary 3.11. and because of the relations, that exist between one-sided and two-sided surface integrals of normal derivatives, it suffices to prove that the functions F and G in Corollary 3.11 with "D_n" replaced by "D_n^1" belong to $L^2(E, m)$. For $g \in L^2(E, m)$ we shall estimate the integrals:

$$\left| \int_0^{t/2} d\tau \int_{\partial\Sigma} dS(x) \left[D_n^1 \exp\left(-\tau\left(K_0 + V\right)_\Sigma\right) g\right](x) \right| \quad \text{and}$$

$$\left| \int_0^{t/2} d\tau \int_{\partial\Sigma} dS(x) \left[\exp\left(-\tau\left(K_0 \dot{+} V\right)\right) g\right](x) \left[D_n^1 \exp\left(-\frac{t}{4}\left(K_0 + V\right)_\Sigma\right) 1\right](x) \right|.$$

Let g be a function in $L^2(E, m)$ such that $\|g\|_2 \leq 1$. From the Feynman-Kac formula together with the Cauchy-Schwarz inequality for surface integrals and Green's theorem we see that

$$\left| \int_0^{t/2} d\tau \int_{\partial\Sigma} dS(x) \left[D_n^1 \exp\left(-\tau\left(K_0 + V\right)_\Sigma\right) g\right](x) \right|^2$$
$$\leq \left(\int_\Sigma dx \left[(K_0 - 2V_-)_\Sigma \int_0^{t/2} d\tau \exp\left(-\tau\left(K_0 - 2V_-\right)_\Sigma\right) 1\right](x) \right.$$
$$\left. + 2 \int_\Sigma dx V_-(x) \int_0^{t/2} \left[\exp\left(-\tau\left(K_0 - 2V_-\right)_\Sigma\right) 1\right](x) \right)$$
$$\times \int_\Sigma dx \left[(K_0)_\Sigma \int_0^{t/2} d\tau \exp\left(-\tau(K_0)_\Sigma\right) |g|^2\right](x). \tag{3.9}$$

From (3.9) it follows that

$$\int_E \left| \int_0^{t/2} d\tau \int_{\partial\Sigma} dS(x) \left[D_n^1 \exp\left(-\tau\left(K_0 + V\right)_\Sigma\right)(\cdot, y)\right](x) \right|^2 dy \tag{3.10}$$

$$\leq \int_\Sigma dx P_x \left(S < \frac{t}{2}\right) + 2 \int_\Sigma dx V_-(x) \int_0^{t/2} d\tau \left[\exp\left(-\tau\left(K_0 - 2V_-\right)_\Sigma\right) 1\right](x).$$

Next we consider the second integral:

$$\left| \int_0^{t/2} d\tau \int_{\partial\Sigma} dS(x) \left[\exp\left(-\tau\left(K_0 \dotplus V\right)\right) g\right](x) \left[D_n^1 \exp\left(-\frac{t}{4}\left(K_0 + V\right)_\Sigma\right) 1\right](x)\right|^2$$

(Feynman-Kac)

$$= \left| \int_{\partial\Sigma} dS(x) \int_0^{t/2} d\tau \mathsf{E}_x\left(\exp\left(-\int_0^\tau V(X(u))du\right) g(X(\tau))\right)\right.$$

$$\left. \times \left[D_n^1 \exp\left(-\frac{t}{4}\left(K_0 + V\right)_\Sigma\right) 1\right](x)\right|^2$$

(Cauchy-Schwarz)

$$\leq \int_{\partial\Sigma} dS(x) \int_0^{t/2} d\tau \mathsf{E}_x\left(\exp\left(-2\int_0^\tau V(X(u))du\right)\right)$$

$$\times \left[D_n^1 \exp\left(-\frac{t}{4}\left(K_0 + V\right)_\Sigma\right) 1\right](x)$$

$$\times \int_{\partial\Sigma} dS(x) \int_0^{t/2} d\tau \mathsf{E}_x\left(|g(X(\tau))|^2\right) \left[D_n^1 \exp\left(-\frac{t}{4}\left(K_0 + V\right)_\Sigma\right) 1\right](x).$$

A combination of Green's formula, the self-adjointness of the operators $\exp(-\tau K_0)$, $\tau > 0$, and identities like

$$\frac{\partial}{\partial s} \exp\left(-s\left(K_0 \dotplus W\right)_\Sigma\right) g = -\left(K_0 \dotplus W\right)_\Sigma \exp\left(-s\left(K_0 \dotplus W\right)_\Sigma\right) g$$

gives rise to the equality:

$$\int_{\partial\Sigma} dS(x) \left[D_n^1 \exp\left(-\frac{t}{4}\left(K_0 - V_-\right)_\Sigma\right) 1\right](x)$$

$$\times \int_{\partial\Sigma} dS(x) \int_0^{t/2} d\tau \exp\left(-\tau K_0\right) |g|^2 (x) \left[D_n^1 \exp\left(-\frac{t}{4}\left(K_0 - V_-\right)_\Sigma\right) 1\right](x)$$

$$= 4 \int_\Sigma dx \left\{-\frac{\partial}{\partial t}\left[\exp\left(-\frac{t}{4}\left(K_0 - V_-\right)_\Sigma\right) 1\right](x) + V_-(x) \exp\left(-\frac{t}{4}\left(K_0 - V_-\right)_\Sigma\right) 1(x)\right\}$$

$$\times \left(4 \int_\Sigma dx\, |g(x)|^2 \left\{\int_0^{t/2} d\tau \left[\exp\left(-\tau K_0\right) - \frac{\partial}{\partial t}\exp\left(-\frac{t}{4}\left(K_0 - V_-\right)_\Sigma\right) 1\right](x)\right.\right.$$

$$\left. + \int_0^{t/2} d\tau \left[\exp\left(-\tau K_0\right) V_- \exp\left(-\frac{t}{4}\left(K_0 - V_-\right)_\Sigma\right) 1\right](x)\right\}$$

$$+ \int_\Sigma dx \left\{\exp\left(-\frac{t}{2}K_0\right) |g|^2 (x) - |g|^2 (x)\right\} \left[\exp\left(-\frac{t}{4}\left(K_0 - V_-\right)_\Sigma\right) 1\right](x)\right).$$

Consequently, it follows that

$$
\int_E dy \left| \int_0^{t/2} d\tau \int_{\partial \Sigma} dS(x) \exp\left(-\tau\left(K_0 + V\right)\right)(x,y) \left[D_n^1 \exp\left(-\frac{t}{4}\left(K_0 - V_-\right)_\Sigma\right)1\right](x) \right|^2
$$

$$
\leq 4 \sup_{x \in E} \int_0^{t/2} d\tau \mathbf{E}_x \left(\exp\left(-2\int_0^\tau V(X(u))du\right)\right)
$$

$$
\times \int_\Sigma dx \left\{ -\frac{\partial}{\partial t}\left[\exp\left(-\frac{t}{4}\left(K_0 - V_-\right)_\Sigma\right)1\right](x) + V_-(x)\exp\left(-\frac{t}{4}\left(K_0 - V_-\right)_\Sigma\right)1(x)\right\}
$$

$$
\times \left(4 \sup_{x \in E} \int_0^{t/2} d\tau \left[\exp\left(-\tau K_0\right)\left(-\frac{\partial}{\partial t}\right)\exp\left(-\frac{t}{4}\left(K_0 - V_-\right)_\Sigma\right)1\right](x) \right.
$$

$$
+ \sup_{x \in E} \int_0^{t/2} d\tau \left[\exp\left(-\tau K_0\right)V_- \exp\left(-\frac{t}{4}\left(K_0 - V_-\right)_\Sigma\right)1\right](x)
$$

$$
+ \sup_{x \in E} \left[\exp\left(-\frac{t}{4}\left(K_0 - V_-\right)1\right)\right](x)\Big). \tag{3.11}
$$

From (3.10) and (3.11) we conclude that the integrals in Corollary 3.11. are functions in $L^2(E, m)$ as soon as we can show that the following expressions are bounded from above:

$$
\int_\Sigma dx \mathbf{P}_x\left(S < \frac{t}{2}\right); \tag{3.12}
$$

$$
\int_E dx V_-(x) \int_0^{t/2} d\tau \left[\exp\left(-\tau\left(K_0 - 2V_-\right)_\Sigma\right)1\right](x); \tag{3.13}
$$

$$
\int_\Sigma dx \left[-\frac{\partial}{\partial t}\exp\left(-\frac{t}{4}\left(K_0 - V_-\right)_\Sigma\right)1\right](x); \tag{3.14}
$$

$$
\int_\Sigma dx V_-(x)\left[\exp\left(-\frac{t}{2}\left(K_0 - V_-\right)_\Sigma\right)1\right](x); \tag{3.15}
$$

$$
\sup_{x \in E} \int_0^{t/2} d\tau \left[\exp(-\tau K_0)\left(-\frac{\partial}{\partial t}\right)\exp\left(-\frac{t}{4}\left(K_0 - V_-\right)_\Sigma\right)1\right](x); \tag{3.16}
$$

$$
\sup_{x \in E} \int_0^{t/2} d\tau \left[\exp\left(-\tau K_0\right)V_- \exp\left(-\frac{t}{4}\left(K_0 - V_-\right)_\Sigma\right)1\right](x). \tag{3.17}
$$

The integral in (3.17) is dominated by

$$
\sup_{x \in E} \int_0^{t/2} d\tau \left[V_- \exp\left(-\frac{t}{4}\left(K_0 - V_-\right)_\Sigma\right)1\right](x)
$$

$$
\leq \sup_{x \in E} \exp\left(-\frac{t}{4}(K_0 - V_-)\right)1(x) \times \sup_{x \in E} \int_0^{t/2} d\tau \exp(-\tau K_0)V_-(x) < \infty,
$$

because V is a Kato-Feller potential. By hypothesis the integral in (3.12) is finite. Since V_- belongs to $L^1(E, m)$ and since V is a Kato-Feller potential the integrals in (3.13) and (3.15) are finite. The integral in (3.14) can be treated in the following fashion. Using a combination of the Markov property and Dyson expansions will show:

$$\int_\Sigma dx \left[-\frac{\partial}{\partial t} \exp\left(-\frac{t}{4}(K_0 - V_-)_\Sigma \right) 1 \right](x) - \frac{\partial}{\partial t} \int_E dx_0 \mathbf{P}_{x_0}\left(S < \frac{t}{4} \right)$$

$$= \int_E dx \left\{ -\frac{\partial}{\partial t} \right\} \mathbf{E}_x \left(\exp\left(\int_0^{t/2} V_-(X(u))du \right) : S > \frac{t}{4} \right)$$

$$= \left(-\frac{\partial}{\partial t} \right) \int_{0<t_1<\ldots<t_n<t/4} \int dt_1 \ldots dt_n \int \ldots \int dx_1 \ldots dx_n \prod_{j=1}^n V_-(x_j)$$

$$\times \prod_{j=1}^n \exp\left(-(t_j - t_{j-1})(K_0)_\Sigma \right)(x_{j-1}, x_j) \le 0.$$

By virtue of Lemma 3.12. it follows that the quantity in (3.14) is finite. In order to prove the finiteness of (3.16) we notice the following identity and inequality

$$-\frac{\partial}{\partial t} \exp\left(-\frac{t}{4}(K_0 - V_-)_\Sigma \right) 1(x)$$

$$= -\frac{1}{4}\mathbf{E}_x \left(\exp\left(\int_0^{t/4} V_-(X(u))du \right) V_-(X(t/4)) : S > \frac{t}{4} \right)$$

$$+ \frac{\partial}{\partial t'}\mathbf{E}_x \left(\exp\left(\int_0^S V_-(X(u))du \right) \mathbf{E}_{X(S)} \exp\left(\int_0^{t/4-S} V_-(X(u))du \right) : S < \frac{t'}{4} \right)\Big|_{t'=t}$$

$$\le \sup_{y\in E} \exp\left(-\frac{t}{4}(K_0 - V_-) \right) 1(y)\frac{\partial}{\partial t}\mathbf{E}_x \left(\exp\left(\int_0^S V_-(X(u))du \right) : S < \frac{t}{4} \right).$$

Consequently, with $C(t/4) = \sup_y \exp\left(-\frac{1}{4}t(K_0 - V_-) \right) 1(y)$ we obtain, upon invoking Dyson expansions and by using the Markov property, that the quantity

$$\int_0^{t/2} d\tau \left[\exp(-\tau K_0) \left(-\frac{\partial}{\partial t} \right) \exp\left(-\frac{t}{4}(K_0 - V_-)_\Sigma \right) 1 \right](x)$$

is dominated by $C(t/4)$ times

$$\int_0^{t/2} d\tau \frac{\partial}{\partial t}\mathbf{E}_x \left(\mathbf{P}_{X(\tau)}\left(S < \frac{t}{4} \right) \right)$$

$$+ \sum_{n=1}^\infty \mathbf{E}_x \left(\int_0^{t/2} d\tau \int_{\tau<t_1<t_2\ldots<t_n<t/4+\tau} \int dt_1 \ldots dt_n \right.$$

$$V_-(X(t_1))\ldots V_-(X(t_n))\frac{\partial}{\partial t}\mathsf{P}_{X(t_n)}\left(S < \frac{t}{4} + \tau - t_n\right), S > t_n - \tau\right)$$

$$\leq \int_0^{t/2} d\tau \frac{\partial}{\partial t}\mathsf{E}_x\left(\mathsf{P}_{X(\tau)}\left(S < t\right)\right) d\tau$$

$$+ \frac{1}{4}\int_0^t d\tau \mathsf{E}_x\left(\exp\left(\int_0^\tau V_-(X(u))du\right) V_-(X(\tau))\mathsf{P}_{X(\tau)}\left(\frac{t}{2} - \tau < S < \frac{3}{4}t - \tau\right)\right)$$

$$\leq \int_0^{t/2} d\tau \frac{\partial}{\partial t}\mathsf{E}_x\left(\mathsf{P}_{X(\tau)}\left(S < t\right)\right) d\tau + \frac{1}{4}\mathsf{E}_x\left(\exp\left(2\int_0^t V_-(X(u))du\right)\right).$$

Hence the boundedness of (3.16) follows from Lemma 3.12. (b) and the fact that V is Kato-Feller potential.

To conclude this paper we want to mention some problems.
- What kind of additive functionals may replace $\int_0^t V(X(u))du$?
- Can the non-symmetric case be treated as well?
- Replace, if possible, the Dirichlet semigroup $\{\exp\left(-t\left(K_0\dotplus V\right)_\Sigma\right) : t \geq 0\}$ by the Neumann semigroup.
- Is the equality $D_n F(x) = 2D_n^1 F(x)$ always true?
- Put $v(x) = \mathsf{E}_x\left(\exp(-aS) : S < \infty\right)$ and let F belong to $D(K_0^{1/2})$. Does the expression $\widetilde{\gamma}(v, F)$ make sense?

Acknowledgement. The author is indebted to the University of Antwerp (UIA), to the Belgian Fund for Scientific Research (NFWO) and to European Science project "Evolutionary systems: deterministic and stochastic evolution equations, control theory and mathematical biology" for their support. The author is grateful to W. Kirsch for the reference to Simon and Spencer (1989). He also thanks M. Demuth for the many stimulating discussions on the subject treated in this paper.

References.

Azencott R., P. Baldi, A. Bellaiche, C. Bellaiche, P. Bougerol, M. Chaleyat-Maurel, L. Elie, J. Granara (1981), Géodésiques et diffusions en temps petit, séminaire de probabilités, Université de Paris VII, Astérisque 84-85, Soc. Math. de France.

Baumgärtel H. and M. Wollenberg (1983), Mathematical scattering theory, Birkhäuser, Boston, Basel.

Bismut J.-M (1984), Large deviations and the Malliavin Calculus, Birkhäuser, Basel.

Blumenthal R.M. and R.K. Getoor (1968), Markov processes and potential theory, Academic Press, New York.

Bouleau N. and F. Hirsch (1991), Dirichlet forms and analysis on Wiener space, de Gruyter Studies in Mathematics 14, De Gruyter, Berlin.

Bouleau N. and D. Lamberton (1989), Residual risks and hedging strategies in Markovian markets, *Stochastic processes and their applications*.

Carmona R., W.C. Masters and B. Simon (1990), Relativistic Schrödinger operators, *J. Funct. Analysis* **51**, 117-142.

Davies E.B. (1988), Gaussian upper bounds for the heat kernels of some second-order operators on Riemannian manifolds, *J. of Functional Analysis* **80**, 16-32.

Davies E.B. (1989), Heat kernels and spectral theory, Cambridge University Press, Cambridge.

Davies E.B. (1991), Heat kernel bounds, conservation of probability and the Feller property, preprint King's College London.

Dellacherie C. et P.-A. Meyer (1987), Probabilités et Potentiel, Chapitres XII à XVI, édition entièrement refondue, Théorie du potentiel associée à une résolvente, Théorie des processus de Markov, Hermann, Paris.

Demuth M. and J.A. van Casteren (1989), On spectral theory for selfadjoint Feller generators, *Reviews in Mathematical Physics*, Vol. **1**, no. 4, 325-414.

Demuth M. and J.A. van Casteren (1991), Perturbations of generalized Schrödinger operators in stochastic spectral analysis, UIA-research report, Antwerp (submitted).

Demuth M. and J.A. van Casteren (1992), On spectral theory for selfadjoint Feller generators II, in preparation.

Elworthy K.D. (1982), Stochastic differential equations on manifolds, London Math. Soc. Lecture Notes in Mathematics **70**, Cambridge University Press.

Elworthy K.D. (1988), Geometric aspects of diffusions on manifolds, in École d'été de Probabilités de Saint-Flour XV-XVII, 1985-1987, editor P.L. Hennequin, Lecture Notes in Math. 1362, Springer-Verlag, Berlin.

Herbst I.W. and Zhongxin Zhao (1988), Sobolov spaces, Kac regularity and the Feynman-Kac formula, in Seminar on Stochastic Processes 1987, Progress Prob. Statist. **15** edited by E. Cinlar, K.L. Chung, R.K. Getoor and J. Glover, Birkhäuser Basel, 171-191.

Ichinose T. (1987), The nonrelativistic limit problem for a relativistic spinless particle in an electromagnetic field, *J. Functional Analysis* **73**, 233-257.

Ichinose T. (1988), Path integral for a Weyl quantized relativistic Hamiltonian and the nonrelativistic limit problem, in Differential Equations and Mathematical Physics, Lecture Notes in Mathematics **1285**, 205-210.

Ichinose T. (1989), Essential selfadjointness of the Weyl quantized relativistic Hamiltonian, *Ann. Inst. Henri Poincaré*, Vol. **51**, no. 3, 265-298.

Ichinose T. and H. Tamura (1986a), Imaginary-time path integral for a relativistic spinless particle in a electromagnetic field, *Comm. in Math. Physics Vol.* **105**, 239-257.

Ichinose T. and H. Tamura (1986b), Path integral for Weyl quantized relativistic Hamiltonian, *Proc. Japan Acad.* **62A**, 91-93.

Ikeda N. and S. Watanabe (1989), Stochastic differential equations and diffusion processes, second edition, North-Holland.

Jacob N. (1990a), Feller semigroups, Dirichlet forms, and pseudo differential operators, preprint Universität Erlangen, to appear in *Forum Math.*.

Jacob N. (1990b), A class of elliptic differential operators generating symmetric Dirichlet forms, preprint Universität Erlangen.

Kochubeĭ A.N. (1985), Singular parabolic equations and Markov processes, *Math. USSR Izvestiya* Vol. 24, no. 1, 73-97.

Kochubeĭ A.N. (1989), Parabolic pseudodifferential equations, hypersingular integrals and Markov processes, *Math. USSR Izvestiya* Vol. **33**, no. 2, 233-259.

Reed M. and B. Simon (1979), Methods of Modern mathematical physics, Vol. 3, Scattering theory; Academic Press, New York, London.

Simon B. (1979), Functional integration and quantum physics, Academic Press, New York.

Simon B. (1982), Schrödinger semigroups, *Bulletin (New Series) of the Amer. Math. Soc.*, Vol. 7, no. 3, p. 447-526.

Simon B. and T.C. Spencer (1989), Trace class perturbations and the absence of absolutely continuous spectra, *Comm. Math. Phys.* **125**, no. 1, 113-125.

Sturm K.-T (1990), Heat kernel bounds on manifolds, preprint Universität Erlangen.

Sturm K.-Th. (1991), Schrödinger semigroups on manifolds, Preprint, Univ. Erlangen-Nürnberg.

Taira K. (1992), On the existence of Feller semigroups with boundary conditions, Memoirs of the Am. Math. Soc. 267, Providence.

van Casteren J.A. (1985), Generators of strongly continuous semigroups, Research Notes in Math. **115**, Pitman, London.

van Casteren J.A. (1989), On generalized Schrödinger semigroups, in Markov Processes and Control Theory (H. Langer and V. Nollau, editors), Mathematical Research Vol. 54, Akademie der Wissenschaften der DDR, Berlin, 16-39.

van Casteren J.A. (1992a), On trace class properties of singularly perturbed Feynman-Kac semigroups, UIA-Research report 92-01.

van Casteren J.A. (1992b), Sur des différences de semi-groupes de Feynman-Kac: une propriété de trace, C. R. Acad. Sc. Paris, to appear.

Department of Math. and Comp. Science
University of Antwerp (UIA)
Universiteitsplein 1
2610 Wilrijk/Antwerp, Belgium

Operator Theory:
Advances and Applications, Vol. 57
© 1992 Birkhäuser Verlag Basel

RADIATION CONDITIONS AND SCATTERING THEORY
FOR N-PARTICLE HAMILTONIANS
(MAIN IDEAS OF THE APPROACH)

D.Yafaev *
Université de Nantes

Abstract

The correct form of radiation conditions is found in scattering problem for N-particle quantum systems. The estimates obtained allow us to give an elementary proof of asymptotic completeness for such systems in the framework of the theory of smooth perturbations. Here we outline main ideas of this proof.

1. One of the main problems of scattering theory is a description of asymptotic behaviour of N interacting quantum particles for large times. The complete classification of all possible asymptotics (channels of scattering) is called asymptotic completeness. The final result can easily be formulated in physics terms. Two particles can either form a bound state or are asymptotically free. In the case $N \geq 3$ a system of N particles can additionally be decomposed for large times into non-trivial subsystems (clusters). Particles of the same cluster form a bound state and different clusters do not interact with each other.

There are two essentially different approaches to a proof of asymptotic completeness for multiparticle ($N \geq 3$) quantum systems. The first of them, suggested by L. D. Faddeev [1], relies on the detailed study of a set of equations derived by him for the resolvent of the corresponding Hamiltonian. This approach was developped in [1] for the case of three particles and was further elaborated by J. Ginibre and M. Moulin [2] and L. Thomas [3]. The attempts [4, 5] towards a straightforward generalization of Faddeev's method to an arbitrary number of particles meet with numerous difficulties. However, the results of [6] for weak interactions are quite elementary.

Another approach relies on the commutator method [7] of T. Kato. In the theory of N-particle scattering it was introduced by R. Lavine [8, 9] for

---

*Permanent address: Math. Inst., Fontanka 27, St. Petersburg, 191011 Russia

repulsive potentials. The proof of asymptotic completeness in the general case is much more complicated and is due to I. Sigal and A. Soffer [10] (see also the article [11] by J. Derezinski for the proof of intermediary analytical results and the article [12] by H. Tamura for more careful presentation of the method of [10]). In the recent paper [13] G. M. Graf gave an accurate proof of asymptotic completeness in the time-dependent framework. The distinguishing feature of [13] is that all intermediary results are also purely time-dependent and most of them have a direct classical interpretation. Papers [10, 13] were to a large extent inspired by V. Enss (see e.g. [14]) who was the first to apply a time-dependent technique for the proof of asymptotic completeness.

The aim of the present paper is to give an elementary proof of asymptotic completeness for N-particle Hamiltonians with short-range potentials which fits into the theory of smooth perturbations [7, 15]. This proof is quite similar to the one [16] suggested by the author for three-particle Hamiltonians. One of the advantages of the theory of smooth perturbations is that it admits two equivalent formulations. The first of them, time-dependent, is given in terms of unitary groups of the Hamiltonians considered. Another, the stationary one, is based on their resolvents. In particular, the stationary version automatically gives (see e.g. [17]) formulas for basic objects of scattering theory: wave operators, scattering matrix etc. Properties of these representations, specific for N-particle systems, will hopefully be discussed elsewhere.

Our proof of asymptotic completeness relies on new estimates which establish some kind of radiation conditions for N-particle systems. Compared to the limiting absorption principle (see below) radiation conditions-estimates give us an additional information on the asymptotic behaviour of a quantum system for large distances and large times. The limiting absorption principle suffices for a proof of asymptotic completeness in the case of two-particle Hamiltonians with short-range potentials. However, radiation conditions-estimates are crucial in scattering for long-range potentials (see e.g. [18]), in scattering by unbounded obstacles [19, 20] and in scattering for anisotropically decreasing potentials [21]. In the latter paper the role of radiation conditions was also advocated for three-particle Hamiltonians.

2. Our interpretation of radiation conditions is, of course, different from the two-particle case. Before discussing their precise form let us introduce the generalized N-particle Hamiltonians. We consider the self-adjoint Schrödinger operator $H = -\Delta + V(x)$ in the Hilbert space $\mathcal{H} = L_2(\mathbf{R}^d)$. Suppose that some finite number α_0 of subspaces X^α of $X := \mathbf{R}^d$ is given and let x^α, x_α be the orthogonal projections of $x \in X$ on X^α and $X_\alpha = X \ominus X^\alpha$ respectively. We assume that

$$V(x) = \sum_{\alpha=1}^{\alpha_0} V^\alpha(x^\alpha), \tag{1}$$

where V^α are decreasing real functions of variables x^α. Without loss of generality, one can suppose that the linear sum of all subspaces X^α exhausts X. The two-particle Hamiltonian H is recovered if (1) consists of only one term with $X^\alpha = X$. The three-particle problem is distinguished from the general situation by the condition $X_\alpha \cap X_\beta = \{0\}$ for $\alpha \neq \beta$. We prove asymptotic completeness under the assumption that V^α are short-range functions of x^α but many intermediary results (in particular, radiation conditions-estimates) are as well true for long-range potentials. Clearly, $V^\alpha(x^\alpha)$ tends to zero as $|x| \to \infty$ outside of any conical neighbourhood of X_α and $V^\alpha(x^\alpha)$ is constant on planes parallel to X_α. Due to this property the structure of the spectrum of H is much more complicated than in the two-particle case. Operators H considered here were introduced in [22] and are natural generalizations of N-particle Hamiltonians. Consideration of a more general class of operators allows to unravel better the geometry of the problem.

The spectral theory of the operator H starts with the following geometrical construction. Let us introduce the set $\mathcal{X}$ of linear sums

$$X^a = X^{\alpha_1} + X^{\alpha_2} + \ldots + X^{\alpha_k}$$

of subspaces X^{α_j}. The zero subspace $X^0 = \{0\}$ is included in the set $\mathcal{X}$ and X itself is excluded. The index a (or b) labels all subspaces $X^a \in \mathcal{X}$ and can be interpreted as the collection of all those α_j for which $X^{\alpha_j} \subset X^a$. Let x^a and x_a be the orthogonal projections of $x \in X$ on the subspaces X^a and

$$X_a := X \ominus X^a = X_{\alpha_1} \cap X_{\alpha_2} \cap \ldots \cap X_{\alpha_k},$$

respectively. Since $X = X_a \oplus X^a$, $\mathcal{H}$ splits into a tensor product

$$L_2(X) = L_2(X_a) \otimes L_2(X^a). \tag{2}$$

In the multiparticle terminology, index a parametrizes decompositions of an N-particle system into noninteracting clusters; x^a is the set of "internal" coordinates of all clusters, x_a describes the relative motion of clusters.

Let us introduce for each a an auxiliary operator $H_a = T + V^a$, $T = -\Delta$, with a potential

$$V^a = \sum_{X^\alpha \subset X^a} V^\alpha, \tag{3}$$

which does not depend on x_a. In the representation (2)

$$H_a = T_a \otimes I + I \otimes H^a. \tag{4}$$

where $T_a = -\Delta_{x_a}$ acts in the space $\mathcal{H}_a = L_2(X_a)$ and

$$H^a = T^a + V^a, \quad T^a = -\Delta_{x^a}$$

are the operators in the space $\mathcal{H}^a = L_2(X^a)$. Set $\mathcal{H}^0 = \mathbb{C}, V^0 = 0, H^0 = 0$. The operator H^a corresponds to the Hamiltonian of clusters with their centers-of-mass fixed at the origin, T_a is the kinetic energy of the center-of-mass motion of these clusters and H_a describes an N-particle system with interactions between different clusters neglected. Eigenvalues of the operators H^a are called thresholds for the Hamiltonian H.

3. Let us introduce the following notation: $E(\Lambda) = E(\Lambda; H)$ is the spectral projection of the operator H corresponding to a Borel set $\Lambda \subset \mathbb{R}$; $\mathcal{H}^{(ac)}(H)$ is the absolutely continuous subspace of H; $P^{(ac)}(H)$ is the orthogonal projection on $\mathcal{H}^{(ac)}(H)$; $\mathcal{H}^{(p)}(H)$ is the subspace spanned by all eigenvectors of the operator H; Q is the operator of multiplication by $(x^2 + 1)^{1/2}$.

The limiting absorption principle asserts that the operator Q^{-r} is locally H-smooth (in the sense of T. Kato) for any $r > 1/2$. The term "locally" means that actually only the operator $Q^{-r}E(\Lambda)$ is H-smooth for an arbitrary bounded interval Λ which is separated from all thresholds and eigenvalues of H. A definition of H-smoothness of the operators $Q^{-r}E(\Lambda)$ can be given either in terms of the resolvent $(H - z)^{-1}$, $\operatorname{Im} z \neq 0$, of the operator H or of its unitary group $U(t) = \exp(-iHt)$. This is discussed e.g. in [17] or [23]. We recall here the definition in terms of $U(t)$: an H-bounded operator K is called H-smooth if for every $f \in \mathcal{D}(H)$

$$\int_{-\infty}^{\infty} \|K \exp(-iHt)f\|^2 \, dt \leq C\|f\|^2.$$

The limiting absorption principle ensures, in particular, that the singular continuous spectrum of H is empty,that is $\mathcal{H} = \mathcal{H}^{(p)}(H) \oplus \mathcal{H}^{(ac)}(H)$. Furthermore, in the case $N = 2$ (but not $N > 2$) it suffices for construction of scattering theory. There are many different proofs of the limiting absorption principle for $N = 2$ but the only one applicable for $N > 2$ relies on the Mourre estimate (see articles [24, 25, 26]).

The fundamental result of N-particle scattering theory called asymptotic completeness is the assertion that the evolution governed by the Hamiltonian H is decomposed as $t \to \pm\infty$ into a sum of simpler evolutions governed by the Hamiltonians H_a. This means that for every $f \in H^{(ac)}$ there exist $f_a^{\pm}$ such that

$$U(t)f \sim \sum_a U_a(t)f_a^{\pm}, \quad U_a(t) = \exp(-iH_a t), \quad t \to \pm\infty, \tag{5}$$

where " $\sim$ " denotes that the difference between left and right sides tends to zero. This relation is also called sometimes asymptotic clustering. Using separation of variables (4) and applying (5) to the Hamiltonians H^a (in place of H) one can desribe the asymptotics of $U(t)f$ in terms of the free operators T_a and of eigenvalues λ_n^a and eigenvectors ψ_n^a of the operators H^a. Actually, by inductive procedure, (5) yields

$$U(t)f \sim \sum_a \sum_n \exp(-i(T_a + \lambda_n^a)t)f_{a,n}^{\pm} \otimes \psi_n^a, \quad f_{a,n}^{\pm} \in \mathcal{H}_a. \tag{6}$$

In particular, in the two-particle case the right side of (6) consists of the single term $\exp(-iTt)f^{\pm}$ where $f^{\pm} \in \mathcal{H}$.

More detailed formulation of the scattering problem for N-particle Hamiltonians is given in terms of wave operators. Recall that for a couple of self-adjoint operators H_j, $j = 1, 2$, in a Hilbert space $\mathcal{H}$ and a bounded operator (identification) $J : \mathcal{H} \to \mathcal{H}$ the wave operator is defined by the relation

$$W^{\pm}(H_2, H_1; J) = s - \lim_{t \to \pm\infty} \exp(iH_2 t)J \exp(-iH_1 t)P^{(ac)}(H_1) \tag{7}$$

under the assumption that this limit exists. In this case the intertwining property

$$E(\Omega; H_2)W^{\pm}(H_2, H_1; J) = W^{\pm}(H_2, H_1; J)E(\Omega; H_1)$$

($\Omega \subset \mathbf{R}$ is any Borel set) holds. It follows that the range $R(W^{\pm}(H_2, H_1; J))$ of the operator (7) is contained in $\mathcal{H}^{(ac)}(H_2)$ and its closure is an unvariant subspace of H_2. Moreover, if the wave operator is isometric on some subspace $\mathcal{H}_1$, then the restrictions of H_1 and H_2 on the subspaces $\mathcal{H}_1$ and $\mathcal{H}_2 = W^{\pm}(H_2, H_1; J)\mathcal{H}_1$ respectively are unitarily equivalent. This equivalence is realized by the wave operator. Clearly, for every $f_2 = W^{\pm}(H_2, H_1; J)f_1$

$$\exp(-iH_2 t)f_2 \sim J \exp(-iH_1 t)f_1, \quad t \to \pm\infty.$$

In the case $J = I$ we omit dependence of wave operators on J. The operator $W^{\pm}(H_2, H_1)$ is obviously isometric on $\mathcal{H}^{(ac)}(H_1)$. The operator $W^{\pm}(H_2, H_1)$ is called complete if $R(W^{\pm}(H_2, H_1)) = \mathcal{H}^{(ac)}(H_2)$. This is equivalent to existence of the wave operator $W^{\pm}(H_1, H_2)$.

Let P^a be the orthogonal projection in $\mathcal{H}^a$ on the subspace spanned by all eigenvectors of H^a. Set $P_a = I \otimes P^a$ where the tensor product is defined by (2). According to (4) the orthogonal projection P_a commutes with the operator $H_a = T + V_a$ and its functions. Set also $V^0 = 0, H_0 = T, P_0 = I$. The basic result of the scattering theory for N-particle Schrödinger operators is the following

Theorem. *Suppose that operators $V^{\alpha}(T^{\alpha} + I)^{-1}$ are compact in the space $\mathcal{H}^{\alpha}$ and*

$$(|x^{\alpha}| + 1)^{\rho}V^{\alpha}(T^{\alpha} + I)^{-1}$$

are bounded in $\mathcal{H}^{\alpha}$ for some $\rho > 1$. Then the wave operators

$$W_a^{\pm} = W^{\pm}(H, H_a; P_a)$$

exist and are isometric on the ranges $R(P_a)$ of projections P_a. The ranges $R(W_a^\pm)$ of $W_a^\pm$ are mutually orthogonal and the asymptotic completeness holds:

$$\sum_a \oplus R(W_a^\pm) = \mathcal{H}^{(ac)}(H).$$

4. Our proof of this assertion relies on new estimates which we call radiation conditions-estimates. Actually, there is only one estimate which looks differently in different regions of the configuration space X. Denote by $\langle \cdot, \cdot \rangle$ the scalar product in the space $\mathbb{C}^d$. Let $\nabla_a = \nabla_{x_a}$ be the gradient in the variable x_a (i.e. $\nabla_a u$ is the orthogonal projection of ∇u on X_a) and let $\nabla_a^{(s)}$,

$$(\nabla_a^{(s)} u)(x) = (\nabla_a u)(x) - |x_a|^{-2}\langle (\nabla_a u)(x), x_a \rangle x_a,$$

be its orthogonal projection in X_a on the plane orthogonal to the vector x_a. Let Γ_a be a cone in $\mathbf{R}^d$ such that $\overline{\Gamma}_a \cap X_b = \{0\}$ if $X_a \not\subset X_b$ and let $\mathbf{Y}_a$ be an intersection of Γ_a with some conical neighbourhood of X_a. In other words, $\mathbf{Y}_a$ is a neighbourhood of X_a with some neighbourhoods of all X_b, $X_a \not\subset X_b$, removed from it. Denote by $\chi(\cdot)$ the characteristic function of a corresponding set. Our main analytical result is that for every a the operator

$$\mathcal{G}_a = \chi(\Gamma_a) Q^{-1/2} \nabla_a^{(s)}$$

is locally H-smooth. This result is formulated as a certain estimate (expressed either in terms of the resolvent of H or of its unitary group) which, by analogy with the two-particle problem, we call the radiation conditions-estimate. Actually, it suffices to verify local H-smoothness of operators

$$G_a = \chi(\mathbf{Y}_a) Q^{-1/2} \nabla_a^{(s)}.$$

Considering the collection of these operators for all a we obtain H-smoothness of the operators $\mathcal{G}_a E(\Lambda)$.

Let us compare the limiting absorption principle with the radiation conditions-estimates. Note that the operator $Q^{-1/2}$ is definitely not H-smooth even in the free case $H = -\Delta$. Thus the radiation conditions-estimates show that the differential operators $\nabla_a^{(s)}$ improve the fall-off of functions $(U(t)f)(x)$ for large t and $x \in \Gamma_a$. In particular, in the free region Γ_0, where all potentials V^α are vanishing, we have that the operator $Q^{-1/2}\nabla^{(s)}$ is H-smooth. This result is not very astonishing from the viewpoint of analogy with the classical mechanics. Indeed, for the free motion the vector $x(t)$ of the position of a particle is directed asymptotically as its momentum p (corresponding to the operator $-i\nabla$). So the projection of p on the plane orthogonal to $x(t)$ tends to zero. According to the conjecture (6) in the region Γ_a (for arbitrary a) the evolution in the variable x_a (corresponding to the relative motion of clusters of particles) is also asymptotically free. Therefore one can expect that the operator $\chi(\boldsymbol{\Gamma}_a)\nabla_a^{(s)}$ is "improving" for all a.

Our proof of H-smoothness of the operators G_a hinges on the commutator method rather than the integration-by-parts machinery which is used (see e.g. [18]) to derive the radiation conditions-estimates in the two-particle case. Actually, we construct such an H-bounded operator M that the commutator $[H, M] := HM - MH$ satisfies locally the estimate

$$i[H, M] \geq c(G_a^* G_a - Q^{-\rho}), \quad \rho > 1, \quad \forall\, a. \tag{8}$$

The arguments of [7] show that H-smoothness of the operator G_a is a direct consequence of this estimate and of the limiting absorption principle. We look for an operator M in a form of a first-order differential operator

$$M = \sum_{j=1}^{d}(m_j D_j + D_j m_j), \quad m_j = \partial m/\partial x_j, \tag{9}$$

with a suitably chosen real function m which we call generating for M. Note that m is a homogeneous function of degree 1 so that coefficients m_j of the operator M are bounded. The leading term $G_a^* G_a$ in the right side of (8) comes from $i[H_0, M]$. We emphasize that due to the operator $Q^{-\rho}$ values of m in a compact domain are inessential. The operator $Q^{-\rho}$ controls in (8) also the commutator

$$i[V^\alpha, M] = -2\langle \nabla m(x), \nabla V^\alpha(x^\alpha)\rangle. \tag{10}$$

To give an idea of the choice of m suppose for a moment that $m(x) = |x|$. Then there is the identity

$$i[H_0, M] = 4\nabla^{(s)}|x|^{-1}\nabla^{(s)}, \quad H_0 = T = -\Delta.$$

Furthermore, by (10),

$$i[V^\alpha, M] = -2|x|^{-1}\langle x^\alpha, \nabla V^\alpha(x^\alpha)\rangle. \tag{11}$$

Thus, under proper assumptions on V^α, we have that in the case $X^\alpha = X$

$$[V^\alpha, M] = O(|x|^{-\rho}), \quad |x| \to \infty, \tag{12}$$

for some $\rho > 1$. This yields the estimate (8) and hence smoothness of the operator $Q^{-1/2}\nabla^{(s)}$ with respect to the two-particle Schrödinger operator H.

However, if $X^\alpha \neq X$, then functions (11) decrease only as $|x|^{-1}$ at infinity. Actually, one can not expect that the operator $Q^{-1/2}\nabla^{(s)}$ is smooth with respect to the N-particle Hamiltonian H. To prove a weaker result about H-smoothness of the operators G_a the function $m(x)$ should be modified in such a way that the estimate (12) with $\rho > 1$ holds for all α. According to (10), this is true if $m(x)$ depends only on the variable x_α in some cone where $V^\alpha(x^\alpha)$ is concentrated. A similar idea was applied by G. M. Graf [13] in the time-dependent context. We emphasize that our requirement on the function $m(x)$ ensures that $m(x) = m(x_a)$ in some conical neighbourhood of every X_a. In other words, a level surface $m(x) = const$ (which is a sphere for $m(x) = |x|$) should be flattened in a neighbourhood of each X_a. Another restriction on $m(x)$ is that the commutator $i[H_0, M]$ should be positive (up to an error $O(|x|^{-\rho})$, $\rho > 1$). This demands that $m(x)$ be a convex function. In this case we can neglect the region $X \setminus \mathbf{Y}_a$ by the derivation of the estimate (8). It turns out that flattening and convexity are compatible. However, the commutator $i[H_0, M]$ gets smaller compared to the case $m(x) = |x|$ so that radiation conditions-estimates in the N-particle case are weaker for $N > 2$ than for $N = 2$. Note also that due to localization in energy in this estimate we can easily dispense with derivatives of V^α and prove H-smoothness of the operators G_a both in short-range and long-range cases.

5. Our approach to the N-particle scattering theory starts with consideration of the wave operators

$$W^\pm(H, H_a; M^{(a)} E_a(\Lambda)), \quad W^\pm(H_a, H; M^{(a)} E(\Lambda)) \tag{13}$$

with "identifications"

$$M^{(a)} = \sum_{j=1}^{d}(m_j^{(a)} D_j + D_j m_j^{(a)}), \quad m_j^{(a)} = \partial m^{(a)}/\partial x_j, \tag{14}$$

which are first-order differential operators with suitably chosen "generating" functions $m^{(a)}$. The "effective perturbation" equals

$$HM^{(a)} - M^{(a)} H_a = [T, M^{(a)}] + [V^a, M^{(a)}] + V_a M^{(a)}, \tag{15}$$

where V^a is defined by (3) and

$$V_a = V - V^a = \sum_{X^a \not\subset X^a} V^\alpha.$$

To prove existence of the wave operators (13) it suffices (see e.g. [17] or [23]) to verify that every term in the right side of (15) can be factorized into a product $K^* K_a$ where K is H-smooth and K_a is H_a-smooth (locally). Functions $m^{(a)}$ are chosen as homogeneous functions of order 1 (for $|x| \geq 1$). Therefore coefficients of the second-order differential operator $[T, M^{(a)}]$ decrease only as $|x|^{-1}$ at infinity. It turns out that this term can be considered with the help of the radiation conditions-estimates. Furthermore, similarly to $m(x)$, the function $m^{(a)}(x)$ depends only on x_α in some conical neighbourhood of X_α.

This ensures that $[V^{(a)}, M^{(a)}] = O(|x|^{-\rho})$ where $\rho > 1$. Finally, it is required that $m^{(a)}(x)$ equals zero in some conical neighbourhood of X_α such that $X_a \not\subset X_\alpha$. So coefficients of the operator $V_a M^{(a)}$ also vanish as $O(|x|^{-\rho})$, $\rho > 1$, at infinity. Thus the second and third terms in the right side of (15) can be taken into account by the limiting absorption principle. The obtained representation for the operator (15) ensures existence of both wave operators (13) (for all a).

Existence of the second wave operator (13) implies that for every vector $f^\pm \in E(\Lambda)\mathcal{H}$ and some $f_a^\pm$

$$M^{(a)}U(t)f^\pm \sim U_a(t)f_a^\pm, \quad t \to \pm\infty. \tag{16}$$

If the sum of $M^{(a)}$ over all a were equal the identity operator I, then summing up, as advised in [27], the relations (16) we would have obtained the asymptotic completeness (5). However, the equality $\sum_a M^{(a)} = I$ is incompatible with the definition (14). We choose functions $m^{(a)}$ in such a way that

$$\sum_a M^{(a)} = M,$$

where M is the same operator as in (9). Summing up the relations (16) we find only that

$$MU(t)f^\pm \sim \sum_a U_a(t)f_a^\pm, \quad t \to \pm\infty. \tag{17}$$

At the final step of the proof of the asymptotic completeness we get rid of the operator M in the left side of (17). To that end we introduce the observable

$$M^\pm(\Lambda) = W^\pm(H, H; ME(\Lambda))$$

and verify that the range of the operator $M^\pm(\Lambda)$ coincides with the subspace $E(\Lambda)\mathcal{H}$. Actually, we show that the operator $\pm M^\pm(\Lambda)$ is positively definite on $E(\Lambda)\mathcal{H}$. In virtue of the inequality $m(x) \geq |x|$ for $|x| \geq 1$, this can be derived from the Mourre estimate. Here we shall explain this result by analogy with classical mechanics. Let us consider a particle (of mass $1/2$) in an external field. In this case the observable $U^*(t)MU(t)$ corresponds, in the Heisenberg picture of motion, to the projection $\mathcal{M}(t) = |x(t)|^{-1}\langle \xi(t), x(t)\rangle$ of the momentum $\xi(t)$ of a particle on a vector $x(t)$ of its position. For positive energies λ and large t we have that $\xi(t) \sim \xi_\pm$, $\xi_\pm^2 = \lambda$, and $x(t) \sim 2\xi_\pm t + x_\pm$. Therefore $\mathcal{M}(t)$ tends to $\pm\lambda^{1/2}$ as $t \to \pm\infty$.

The asymptotic completeness in the form (5) can be easily deduced from these results. Actually, for every $f = M^\pm(\Lambda)f^\pm$ we have that

$$U(t)f \sim MU(t)f^\pm, \quad t \to \pm\infty. \tag{18}$$

Comparing (17) with (18) and taking into account that the range of $M^\pm(\Lambda)$ equals $E(\Lambda)\mathcal{H}$ we arrive at (5). It remains to establish existence of the wave operators $W^{\pm}(H, H_a)$. This is derived from existence of the first set of the wave operators (13). At this step we assume validity of the main Theorem for all operators H_a (in place of H). This additional assumption is, finally, removed by an inductive procedure.

REFERENCES

[1] L. D. Faddeev, *Mathematical Aspects of the Three Body Problem in Quantum Scattering Theory*, Trudy MIAN **69**, 1963. (Russian)

[2] J. Ginibre and M. Moulin, Hilbert space approach to the quantum mechanical three body problem, Ann. Inst. H.Poincaré, A **21**(1974), 97-145.

[3] L. E. Thomas, Asymptotic completeness in two- and three-particle quantum mechanical scattering, Ann. Phys. **90** (1975), 127-165.

[4] K. Hepp, On the quantum-mechanical N-body problem, Helv. Phys. Acta **42**(1969), 425-458.

[5] I. M. Sigal, *Scattering Theory for Many-Body Quantum Mechanical Systems*, Springer Lecture Notes in Math. **1011**, 1983.

[6] R. J. Iorio and M. O'Carrol, Asymptotic completeness for multi-particle Schrödinger Hamiltonians with weak potentials, Comm. Math. Phys. **27**(1972), 137-145.

[7] T. Kato, Smooth operators and commutators, Studia Math. **31**(1968), 535-546.

[8] R. Lavine, Commutators and scattering theory I: Repulsive interactions, Comm. Math. Phys. **20**(1971), 301-323.

[9] R. Lavine, Completeness of the wave operators in the repulsive N-body problem, J. Math. Phys. **14** (1973), 376-379.

[10] I. M. Sigal and A. Soffer, The N-particle scattering problem: Asymptotic completeness for short-range systems, Ann. Math. **126**(1987), 35-108.

[11] J. Derezinski, A new proof of the propagation theorem for N-body quantum systems, Comm. Math. Phys. **122** (1989), 203-231.

[12] H. Tamura, Asymptotic completeness for N-body Schrödinger operators with short-range interactions, Comm. Part. Diff. Eq. **16** (1991), 1129-1154.

[13] G. M. Graf, Asymptotic completeness for N-body short-range quantum systems: A new proof, Comm. Math. Phys. **132** (1990), 73-101.

[14] V. Enss, Completeness of three-body quantum scattering, in: *Dynamics and processes*, P. Blanchard and L. Streit, eds., Springer Lecture Notes in Math. **103** (1983), 62-88.

[15] T. Kato, Wave operators and similarity for some non-self-adjoint operators, Math. Ann. **162** (1966), 258-279.

[16] D. R. Yafaev, Radiation conditions and scattering theory for three-particle Hamiltonians, Preprint 91-01, Nantes University, 1991.

[17] D. R. Yafaev, *Mathematical Scattering Theory*, Amer. Math. Soc., 1992.

[18] Y. Saito, *Spectral Representation for Schrödinger Operators with Long-Range Potentials*, Springer Lecture Notes in Math. **727**, 1979.

[19] P. Constantin, Scattering for Schrödinger operators in a class of domains with noncompact boundaries, J. Funct. Anal. **44** (1981), 87-119.

[20] E. M. Il'in, Scattering by unbounded obstacles for elliptic operators of second order, Proc. of the Steklov Inst. of Math. **179** (1989), 85-107.

[21] D. R. Yafaev, Remarks on the spectral theory for the multiparticle type Schrödinger operator, J. Soviet Math. **31** (1985), 3445-3459 (translated from Zap. Nauchn. Sem. LOMI **133** (1984), 277-298).

[22] S. Agmon, *Lectures on Exponential Decay of Solutions of Second-Order Elliptic Equations*, Math. Notes, Princeton Univ. Press, 1982.

[23] M. Reed and B. Simon, *Methods of Modern Mathematical Physics* III, Academic Press, 1979.

[24] E. Mourre, Absence of singular spectrum for certain self-adjoint operators, Comm. Math. Phys. **78** (1981), 391-400.

[25] P. Perry, I. M. Sigal and B. Simon, Spectral analysis of N-body Schrödinger operators, Ann. Math. **144** (1981), 519-567.

[26] R. Froese, I. Herbst, A new proof of the Mourre estimate, Duke Math. J. **49** (1982), 1075-1085.

[27] P. Deift and B. Simon, A time-dependent approach to the completeness of multiparticle quantum systems, Comm. Pure Appl. Math. **30** (1977), 573-583.

Titles previously published in the series

OPERATOR THEORY: ADVANCES AND APPLICATIONS
BIRKHÄUSER VERLAG

40. **H. Dym, S. Goldberg, P. Lancaster, M.A. Kaashoek** (Eds.): The Gohberg Anniversary Collection, Volume I, 1989, (3-7643-2307-8)

41. **H. Dym, S. Goldberg, P. Lancaster, M.A. Kaashoek** (Eds.): The Gohberg Anniversary Collection, Volume II, 1989, (3-7643-2308-6)

42. **N.K. Nikolskii** (Ed.): Toeplitz Operators and Spectral Function Theory, 1989, (3-7643-2344-2)

43. **H. Helson, B. Sz.-Nagy, F.-H. Vasilescu, Gr. Arsene** (Eds.): Linear Operators in Function Spaces, 1990, (3-7643-2343-4)

44. **C. Foias, A. Frazho:** The Commutant Lifting Approach to Interpolation Problems, 1990, (3-7643-2461-9)

45. **J.A. Ball, I. Gohberg, L. Rodman:** Interpolation of Rational Matrix Functions, 1990, (3-7643-2476-7)

46. **P. Exner, H. Neidhardt** (Eds.): Order, Disorder and Chaos in Quantum Systems, 1990, (3-7643-2492-9)

47. **I. Gohberg** (Ed.): Extension and Interpolation of Linear Operators and Matrix Functions, 1990, (3-7643-2530-5)

48. **L. de Branges, I. Gohberg, J. Rovnyak** (Eds.): Topics in Operator Theory. Ernst D. Hellinger Memorial Volume, 1990, (3-7643-2532-1)

49. **I. Gohberg, S. Goldberg, M.A. Kaashoek:** Classes of Linear Operators, Volume I, 1990, (3-7643-2531-3)

50. **H. Bart, I. Gohberg, M.A. Kaashoek** (Eds.): Topics in Matrix and Operator Theory, 1991, (3-7643-2570-4)

51. **W. Greenberg, J. Polewczak** (Eds.): Modern Mathematical Methods in Transport Theory, 1991, (3-7643-2571-2)

52. **S. Prössdorf, B. Silbermann:** Numerical Analysis for Integral and Related Operator Equations, 1991, (3-7643-2620-4)

53. **I. Gohberg, N. Krupnik:** One-Dimensional Linear Singular Integral Equations, Volume I, Introduction, 1991, (3-7643-2584-4)

54. **I. Gohberg, N. Krupnik** (Eds.): One-Dimensional Linear Singular Integral Equations, 1992, (3-7643-2796-0)

55. **R.R. Akhmerov, M.I. Kamenskii, A.S. Potapov, A.E. Rodkina, B.N. Sadovskii:** Measures of Noncompactness and Condensing Operators, 1992, (3-7643-2716-2)

56. **I. Gohberg** (Ed.): Time-Variant Systems and Interpolation, 1992, (3-7643-2738-3)

If you have any concerns about our products,
you can contact us on
ProductSafety@springernature.com

In case Publisher is established outside the EU,
the EU authorized representative is:
Springer Nature Customer Service Center GmbH
Europaplatz 3, 69115 Heidelberg, Germany

Printed by Libri Plureos GmbH
in Hamburg, Germany